Total Quality Management

This book has been developed to provide significant information about the usage and application of the Total Quality Management (TQM) concept in a construction project environment. The content spans from the inception through to the closing of the project focusing on the TQM approach in each phase of the project.

Total Quality Management: Applications and Concepts for Construction Projects focuses on the application of the TQM concept in construction projects and contains many quick-reference figures and tables for easy comprehension. It offers a concise and complete implementation process for the application of TQM and helps achieve competitive advantages in the global marketplace resulting in the construction project being qualitatively competitive and economical. The book highlights the standards for TQM and gives a brief introduction to the quality management system along with providing an overview of the project, the quality, the types of project delivery systems, and the principles involved. Discussions of quality and the different steps it moves through within the project setting including inspection, statistical quality control, and quality assurance round out the book's offerings.

Construction and quality professionals, industrial engineers, project managers, students, academics, and trainers will find that this book satisfies their needs and meets their requirements for a book that specifically uses TQM in construction projects.

Quality Management and Risk Series

Series Editor: Abdul Razzak Rumane, Senior Engineering Consultant, Kuwait

This new series will include the latest and innovative books related to quality management and risk related topics. The definition of quality relating to manufacturing, processes, and the service industries, is to meet the customer's need, satisfaction, fitness for use, conforms to requirements, and a degree of excellence at an acceptable price. With globalization and competitive markets, the emphasis on quality management has increased. Quality has become the most important single factor for the survival and success of any company or organization. The demand for better products and services at the lowest costs have put tremendous pressure on organizations to improve the quality of products, services, and processes, in order to compete in the marketplace. Because of these changes, the ISO 9001 now lists that risk-based thinking must be incorporated into the management system by considering the context of the organization. Quality management and risk management now play an important role in the over all quality management system. This means that books which cover quality, need to also cover risk to update practices/processes, tools, and techniques, per ISO 9001. The goal of this new series is to include the books that will meet this need and demand.

Quality Management in Oil and Gas Projects
Abdul Razzak Rumane

Risk Management Applications to Sustain Quality in Projects: A Practical Guide
Abdul Razzak Rumane

Quality Management: How to Achieve Sustainability in Projects
Abdul Razzak Rumane

Total Quality Management: Applications and Concepts for Construction Projects
Abdul Razzak Rumane

For more information on this series, please visit: www.routledge.com/Quality-Management-and-Risk-Series/book-series/CRCQMR

Total Quality Management

Applications and Concepts for Construction Projects

Abdul Razzak Rumane

CRC Press
Taylor & Francis Group
Boca Raton London New York

CRC Press is an imprint of the
Taylor & Francis Group, an **informa** business

Designed cover image: Shutterstock – PIXEL to the PEOPLE and Abdul Razzak Rumane

First edition published 2024
by CRC Press
2385 NW Executive Center Drive, Suite 320, Boca Raton FL 33431

and by CRC Press
4 Park Square, Milton Park, Abingdon, Oxon, OX14 4RN

CRC Press is an imprint of Taylor & Francis Group, LLC

ISBN: 978-1-032-58637-3 (hbk)
ISBN: 978-1-032-58783-7 (pbk)
ISBN: 978-1-003-45146-4 (ebk)

DOI: 10.1201/9781003451464

Typeset in Times
by Newgen Publishing UK

Dedication

_____

To
My Parents
For their prayers and love
My prayers are always for my father and my mother
who encouraged and inspired me all the time.
I wish they would have been here to see this book and
give me blessings.

To
My Wife
I miss my wife, who stood with me all the time during the
writing of my earlier books.

Contents

Foreword

If you have read any of Dr. Rumane's previous books, you will know that he is able to convey his knowledge and experience in a manner that provides a foundation together with potential roadmaps and frameworks to enable adoption. This is important because the reader needs to be able to understand the fundamentals and how they could tune improvements to fit their situation and organization. Having talked with Dr. Rumane on various topics over the years, I know that his depth and breadth in quality makes him one of our foremost experts in the world. This time around, Dr. Rumane has focused on helping construction organizations adopt TQM by providing a comprehensive introduction to the subject and multiple examples of how it could be applied during a project's life cycle.

Construction organizations are complex entities in that they are organizations within organizations—unique projects within companies—delivering unique products in an environment with constant variability. This reinforces the need to instill a culture of operational excellence in order to assure profitability and a sustainable future. The American Society of Quality (ASQ) defines Total Quality Management (TQM) as "a management approach to long-term success through customer satisfaction. TQM is based on all members of an organization participating in improving processes, products, services and the culture in which they work." In other words, TQM is about leadership engaging everyone in continuously improving the organization to satisfy their customers and assure long-term profitability.

In this book, Dr. Rumane provides an understanding of TQM's roots and framework for operational excellence in construction. The roots provide the reader with an understanding of how its forefathers sought to raise the bar of organizational performance and challenge the way that things have always been done. Although the forefathers contributed elements that overlapped, the reader should note that they were united in not only chasing excellence but also in advocating that leadership and people are a critical part of making it happen.

It's important to recognize that TQM's principles are universal to any organization but how those principles are achieved could be different between industries. Dr. Rumane's book therefore provides multiple examples of how TQM could be applied within a project-based organization to provide guidance for the reader. To dive deeper into specific tactics of how to improve performance in a given scenario, I would direct the reader to Dr. Rumane's previous books on Quality Tools and other approaches.

Successful organizational and cultural change will seem hard but persevere and keep questioning and learning. As someone once said, the road to success is paved with failure!

Hasan Salih, MSc CPM, CQP MCQI.
(Chair Elect ASQ Design and Construction Division)
January 2024

Hasan is a project manager with more than 25 years of experience in the construction of mega projects across three continents with knowledge in project management, construction, commissioning, quality, safety, and manufacturing. His areas of expertise include performance leadership, change management, behavioral change, and continuous improvement strategies and tools including Kaizen and Six Sigma. He holds a Masters in Construction Project Management and is a certified Six Sigma Black Belt. He is a chartered member of the Chartered Quality Institute and the current chair elect for the ASQ Design and Construction Division.

About the Author

Abdul Razzak Rumane, PhD, is a chartered quality professional fellow of the Chartered Quality Institute (UK) and a certified consultant engineer in the field of electrical engineering and project management. He obtained a Bachelor of Engineering (Electrical) degree from Marathwada University (now Dr. Babasaheb Ambedkar Marathwada University), India, in 1972 and received his Doctor of Philosophy (PhD) from Kennedy Western University (now Warren National University), USA, in 2005. His dissertation topic was "Quality Engineering Applications in Construction Projects."

Dr. Rumane has been honored by the International University of Ministry and Education, Missouri, USA, with honorary degree of Doctor of Philosophy (D.Phil.) for expertise in quality management in the year 2021, and the Yorker International University, USA, awarded an honorary Doctorate of Engineering in the year 2007. The Albert Schweitzer International Foundation honored him with a gold medal for "Outstanding contribution in the field of Quality in Construction Projects" and the World Quality Congress awarded him with "Global Award for Excellence in Quality Management and Leadership."

Dr. Rumane is an accomplished engineer. He is associated with a number of professional organizations.

Dr. Rumane has attended many international conferences and has made technical presentations at various conferences. He is the author of eight useful books entitled *Quality Management in Construction Projects*, First Edition (2010); *Quality Tools for Managing Construction Projects* (2013); *Quality Management in Construction Projects*, Second Edition (2017); *Quality Management in Oil and Gas Projects* (2021); *Risk Management Applications Used to Sustain Quality in Projects: A Practical Guide* (2022); *Quality Management: How to Achieve Sustainability in Projects* (2023), and editor of book titled *Handbook of Construction Management: Scope, Schedule, and Cost Control* (2016). All of these books were published by CRC Press (Taylor & Francis), USA. A book entitled *Quality Auditing in Construction Projects: A Handbook* (2019) was published by Routledge (Taylor & Francis), UK. His book *Quality Management in Construction Projects* is translated into Korean.

Presently he is Treasurer of ASQ Kuwait Section and International Liaison Committee Chair ASQ (Design and Construction Division) for the year 2023–24. He served as Secretary ASQ GC for the year 2019, Secretary ASQ LMC, Kuwait, for the years 2017 and 2018, and Treasurer for the years 2022 and 2023. He was honorary chairman of the Institution of Engineers (India), Kuwait, chapter for the years 2016–17, 2013–14, and 2005–07.

Dr. Rumane's professional career exceeds 50 years, including 10 years in manufacturing industries and over 40 years in construction projects. Presently he is associated with SIJJEEL Co., Kuwait, as advisor and director of Construction Management.

Preface

Quality is a universal phenomenon which has been a matter of great concern throughout the recorded history of humankind. It was always the determination of builders and makers of the products to ensure that their products meet the customer's desire. Those were the days when products were of a "Customized nature." With the advent of globalization and competitive market, the emphasis on quality management has increased. Quality has become the most important single factor for the survival and success of the companies.

Quality has different meanings for different people. The quality has a historically important background that has evolved through different forms, periodical changes and cultural shifts such as inspection, statistical quality control, quality assurance, and quality engineering, leading to the development of Total Quality Management (TQM). Quality management resulted from the work of quality gurus and their theories. Extension of quality management gave birth to TQM.

The TQM concept was born following World War II. It was stimulated by the need to compete in the global market where higher quality, lower cost, and more rapid development are essential to market leadership. Today TQM is considered a fundamental requirement for any organization to compete, let alone lead, in its market. It is a way of planning, organizing, and understanding each activity of the process and removing all the unnecessary steps routinely followed in the organization. TQM is a philosophy that makes quality values the driving force behind leadership, design, planning, and improvement in activities.

Quality actually emerged as a dominant thinking since World War II, becoming an integral part of overall business system focused on customer satisfaction and becoming known in the twentieth century as "Total Quality Management" (TQM) with its three constitutive elements:

Total: Organization-wide
Quality: Customer satisfaction
Management: Systems of managing

TQM involves everyone in the organization in an effort to increase customer satisfaction and achieve superior performance of the products or services through continuous quality improvement. TQM helps in

- Achieving customer satisfaction
- Continuous improvement
- Developing teamwork
- Establishing vision for the employees
- Setting standards and goals for the employees
- Building motivation within the organization
- Developing corporate culture

Construction projects are mainly capital investment projects. They are customized and non-repetitive in nature. Construction projects have become more complex and technical, and the relationships and the contractual grouping of those who are involved are also more complex and contractually varied.

Construction projects have the involvement of many participants including owner, designer, contractor, and many other professionals from construction-related industries. Each of these participants is involved in implementing quality in construction projects. These participants are both influenced by and depend on each other in addition to "other players" involved in the construction process. Therefore, the construction projects have become more complex and technical, and extensive efforts are required to reduce rework and costs associated with time, materials, and engineering.

Quality in construction projects is different from that of manufacturing. Quality in construction projects is not only the quality of products and equipment used in the construction, it is the total management approach to complete the facility as per the scope of works to customer/owner satisfaction to be completed within specified schedule and within the budget to meet owner's defined purpose. This is achieved through a complex interaction of many participants in the facilities development process.

TQM is an organization-wide effort centered on quality to improve performance that involves everyone and permeates every aspect of an organization to make quality a primary strategic objective. It is a way of managing an organization to ensure satisfaction at every stage of the needs and expectations of both internal and external customers.

TQM focuses on participative management and strong operational accountability at the individual contributor level. TQM is the implementation of a quality system by an organization to satisfy the customer needs/requirements by developing a system that will make the project qualitative, competitive, and economical. A collaborative effort in the construction project is required among all the participants in order to improve alignment and misunderstanding during the execution of project. This is achieved through complex interaction of many participants in the facilities development process by application of TQM concept in the construction project.

TQM in construction projects typically involves ensuring compliance with minimum standards of material and workmanship in order to ensure the performance of the facility according to the design. TQM in a construction project is a cooperative form of doing business that relies on the talents and capabilities of both labor and management to continually improve quality. The important factor in construction projects is to complete the facility per the scope of works to customer/owner satisfaction within the specified schedule and approved budget and to complete the work to meet the owner's defined purpose. TQM will help reduce the overrun of schedule and cost.

The book contains many valuable figures and tables that make understanding of the subject easy.

For the sake of proper understanding, the book is divided into four chapters, and each chapter is divided into a number of sections covering quality-related topics that have importance or relevance for understanding TQM concept in the construction project.

Chapter 1 is about the birth of TQM. It discusses the overview of quality that has moved in different forms such as inspection, statistical quality control, quality assurance, quality engineering, quality management to the concept of TQM.

Chapter 2 is about an overview of quality in construction project. It discusses types of construction projects, types of project delivery systems, and types of contracting/pricing. It covers a brief definition of construction projects, quality of construction projects, principles of quality in construction projects, and quality management systems in construction projects.

Chapter 3 is about quality standards. It gives a brief introduction to the quality management system, ISO and ISO 9001 standards.

Chapter 4 is about the application of TQM concept. It discusses the systems engineering approach to divide lifecycle phases of construction projects to conveniently manage and control. The chapter discusses application of TQM concept that can be applied from inception of the construction project through to all phases of the project to the substantial completion of the project considering involvement of related stakeholders in each of the phases.

The book, I am certain, will meet the requirements of construction professionals, quality professionals, project owners, students, and academics and satisfy their needs.

Acknowledgments

Share the knowledge with others is the motto of this book.

Many of my colleagues and friends extended help while preparing the book by arranging reference material; many thanks to all of them for their support.

I thank the publishers and authors, whose writings are included in this book, for extending their support by allowing me to reprint their material.

I thank reviewers, from various professional organizations, for their valuable input to improve my writing. I thank members of ASQ Audit Division, ASQ Design & Construction Division, the Institution of Engineers (India), IEI, Kuwait Chapter, Kuwait Society of Engineers, and ASQ Kuwait Section, for their support to bring out this book.

I thank Mr. Hasan Salih, Chair Elect ASQ (Design and Construction Division), for his nicely worded and thought-provoking Foreword and support and best wishes all the time.

I thank Cindy Renee Carelli, Executive Editor of CRC Press, Senior Editorial Assistant, and other CRC staff for their support and contribution to make this construction-related book a reality.

I thank Mr. Collin Sutt, Chair, ASQ (Design and Construction Division), Mr. Raymond R. Crawford, Past Chair ASQ (Design and Construction Division), Mr. Cliff Moser (Former Chair, ASQ Design and Construction Division), and Dr. Adedeji Badiru of Airforce Institute of Technology for their best wishes all the time.

I thank Engr. Adel Kharafi, Former Chairman Kuwait Society of Engineers, and former President of the World Federation of Engineering Organizations (WFEO), for his good wishes all the time. I thank Engr. Ahmad Alkandari, Director, Kuwait Municipality, for his support and good wishes. I thank Engr. Ahmad Almershed, former Undersecretary, MSNA, Kuwait, for his good wishes all the time. I thank Dr. Ayed Alamri, President, Saudi Quality Council, KSA, for his support and good wishes all the time. I thank Dr. Fadel Safer, Former Minister of Public Works, Kuwait, for his support and good wishes. I thank Engr. Faisal D. Alatal, Chairman, Kuwait Society of Engineers, and President of the Federation of Arab Engineers for his support and good wishes. I thank Prof. Mohammed Aichouni, University of Hail, KSA, for his support and good wishes all the time. I thank Dr. Mohammad Ben Salamah, Chair, ASQ Kuwait Section, for his support and best wishes. I thank Dr. N. N. Murthy of Jagruti Kiran Consultants for his support and good wishes. I thank Dr. Othman Alshamrani, Imam Abdulrehman Bin Faisal University, KSA, for his support and good wishes all the time. I thank Maj. Gen. R. K. Sanan, VSM, (Retd), former Secretary Director General, the Institution of Engineers (India), for his support and best wishes all the time. I thank Ms. Rima Al Awadhi, former Chair, ASQ Kuwait Section, and Team Leader at Kuwait Oil Company for her support and good wishes. I thank Engr. Wael Aljasem of Kuwait Project Management Society for his support and good wishes. I thank Engr. Yaseen Farraj, former Director, Ministry of Public Works, Kuwait, for his support and good wishes.

I thank Mr. Bashir Ibrahim Parkar of Dar SSH International and Engr. Ateeq Mirza of KOC for their valuable input and support.

I extend my thanks to Dr. Ted Coleman, Professor and Department Chair, California State University, San Bernardino, and former Chancellor, KW University, for his everlasting support.

My special thanks to H.E. Sheikh Rakan Nayef Jaber Al Sabah for his support and good wishes.

I thank my well-wishers whose inspiration made me complete this book.

Most of the data discussed in this book is from the author's practical and professional experience and is accurate to the best of the author's knowledge and ability. However, if any discrepancies are observed in the presentation, I would appreciate communicating them to me.

The contribution of my son Ataullah, my daughter Farzeen, and daughter-in-law Masum is worth mentioning here. They encouraged me and helped me in my preparatory work to achieve the final product. I thank my brothers, sisters, and all the family members for their support, encouragement and good wishes all the time.

Abdul Razzak Rumane

Abbreviations

AACE	American Association of Cost Engineers
ASCE	American Society of Civil Engineers
ASHRAE	American Society of Heating, Refrigeration and Air-conditioning Engineers
ASQ	American Society for Quality
BIM	Building Information Modeling
BMS	Building Management System
BOM	Bill of Materials
BOQCAD	computer-aided drafting
CCS	contractor's construction schedule
CDM	Construction (Design and Management)
CEN	European Committee for Standardization
CII	Construction Industry Institute
CMAA	Construction Management Association of America
CPM	critical path method
CQCP	contractor's quality control plan
CSC	Construction Specifications, Canada
CSI	Construction Specification Institute
EPC	Engineering, Procurement, and Construction
FEED	Front End Engineering Design
FIDIC	Federation International des Ingeneurs-Counceils
GMP	Guaranteed Maximum Price
HAZID	Hazard Identification
HAZOP	Hazard and Operability
HSE	Health, Safety and Environment
ICE	Institute of Civil Engineers (UK)
ICT	Information and Communication Technology
IPD	Integrated Project Delivery
IEC	International Electrotechnical Commission
IEEE	Institute of Electrical and Electronics Engineers
IPD	Integrated Project Delivery
IoT	Internet of Things
IP	Ingress Protection
ISO	International Organization for Standardization
OH&S	Occupational Health and Safety
PCM	Planning and Control Manager
PERT	program evaluation and review technique
PHSER	Procedure for Project HSE Review
PMBOK	Project Management Book of Knowledge
PMC	Project Management Consultant
PMI	Project Management Institute
QCP	Quality Control Plan
QIS	Quality Information System
QMS	Quality Management System

QS	Quantity Surveyor
RE	Resident Engineer
SPC	Statistical Process Control
SQC	statistical quality control
TIC	Total Investment Cost
TQM	Total Quality Management
WBS	work breakdown structure

Synonyms

Consultant	Architect/Engineer (A/E), Designer, Design Professionals, Designer, Consulting Engineers, Supervision Professional, Specialist Consultant
Contractor	Constructor, Builder, EPC Contractor
Engineer	Resident Project Representative
Engineer's Representative	Resident Engineer
Owner	Client, Employer
Project Charter	Terms of Reference (TOR), Client Brief, Definitive Project Brief
Project Manager	Construction Manager
Quantity Surveyor	Cost Estimator, Contract Attorney, Cost Engineer, Cost and Works Superintendent

1 Overview of Quality

1.1 QUALITY HISTORY

Quality issues have been of great concern throughout the recorded history of humans. During the New Stone Age, several civilizations emerged, and some 4000–5000 years ago, considerable skills in construction were acquired. The pyramids in Egypt were built approximately 2589–2566 BCE. Hammurabi, the king of Babylonia (1792–1750 BCE), codified the law, according to which, during the Mesopotamian era, builders were responsible for maintaining the quality of buildings and were given the death penalty if any of their construction collapsed and their occupants were killed. The extension of Greek settlements around the Mediterranean after 200 BCE left records showing that temples and theaters were built using marble. India had strict standards for working in gold in the fourth century BCE.

China's recorded quality history can be traced back to earlier than 200 BCE. China had instituted quality control in its handicrafts during the Zhou dynasty between 1100 and 250 BCE. During this period, the handicraft industry was mainly engaged in producing ceremonial artifacts. This industry survived the long succession of dynasties that followed up to 1911 CE.

During the Middle Ages, guilds took the responsibility for quality control upon themselves. Guilds and governments carried out quality control; consumers, of course, carried out informal quality inspection throughout history.

The guilds' involvement in quality was extensive. All craftsmen living in a particular area were required to join the corresponding guild and were responsible for controlling the quality of their own products. If any of the items was found defective, then the craftsman discarded the faulty items. The guilds also initiated punishments for members who turned out shoddy products. They maintained inspections and audits to ensure that artisans followed quality specifications. The guild hierarchy consisted of three categories of workers: apprentice, journeyman, and master. The guilds had established specifications for input materials, manufacturing processes, and finished products, as well as methods of inspection and testing. They were active in managing quality during the Middle Ages until the Industrial Revolution marginalized their influence.

DOI: 10.1201/9781003451464-1

1.2   QUALITY DEFINITION

Quality has different meanings for different people. The American Society for Quality (ASQ) glossary defines quality as a subjective term for which each person has his or her own definition. In technical usage, quality can have two meanings:

1. The characteristics of a product or service that bear on its ability to satisfy stated or implied needs.
2. A product or service free of deficiencies.
   It further states that it is
   - Based on customers' perceptions of a product's design and how well the design matches the original specifications.
   - The ability of a product and service to satisfy stated or implied needs.
   - Achieved by conforming to established requirements within an organization.

The International Organization for Standardization (ISO, 1994a) defines quality as "the totality of characteristics of an entity that bears on its ability to satisfy stated or implied needs."

The quality definition is further be based on the following criteria:

1. Product based
2. User based
3. Manufacturing based
4. Value based
5. Customer's perception based

The above definitions can further be summarized under the names of those contributors to the quality movement whose philosophies, methods, and tools have been proven useful in quality practices. They are called the "quality gurus." Their definitions of quality are as follows:

1. *Philip B. Crosby*—Conformance to requirements not as "goodness" nor "elegance."
2. *W. Edward Deming*—Quality should be designed into both the product and the process.
3. *Armand V. Feigenbaum*—Best for customer use and selling price.
4. *Kaoru Ishikawa*—Quality of the product as well as after-sales services, quality of management, the company itself, and the human being.
5. *Joseph M. Juran*—Quality is fitness for use.
6. *John S. Oakland*—Quality is meeting customer's requirements.

Based on these definitions, it is possible to evolve a common definition of quality, which is mainly related to the manufacturing, processes, and service industries as follows:

- Meeting the customer's need
- Fitness for use
- Conforming to requirements

1.3   EVOLUTION OF QUALITY MANAGEMENT SYSTEM

The Industrial Revolution began in Europe in the mid-nineteenth century. It gave birth to factories, and the goals of the factories were to increase productivity and reduce costs. Prior to the Industrial Revolution, items were produced by individual craftsman for individual customers, and it was possible for workers to control the quality of their products. Working conditions then were more conducive to professional pride. Under the factory system, the tasks needed to produce a product were divided among several or many factory workers. Under this system, large groups of workmen were performing similar types of work, and each group was working under the supervision of a foreman who also took on the responsibility of controlling the quality of the work performed. Quality in the factory system was ensured by means of skilled workers, and the quality audit was done by inspectors.

The broad economic result of the factory system was mass production at low costs. The Industrial Revolution changed the situation dramatically with the introduction of a new approach to manufacturing.

The beginning of the twentieth century marked the inclusion of process in quality practices. During World War I, the manufacturing process became more complex. Production quality was the responsibility of quality control departments. The introduction of mass production and piecework created quality problems as workmen were interested in earning more money by the production of extra products, which in turn led to bad workmanship. This situation made factories introduce full-time quality inspectors, which marked the real beginning of inspection quality control and thus the introduction of quality control departments headed by superintendents. Walter Shewhart introduced statistical quality control (SQC) in the process. His concept was that quality is not relevant to the finished product but to the process that created the product. His approach to quality was based on continuous monitoring of process variation. The SQC concept freed manufacturers from the time-consuming 100% quality control system because it accepted that variation is tolerable up to certain control limits. Thus, the quality control focus shifted from the end of line to the process.

The systematic approach to quality in industrial manufacturing started during the 1930s when some attention was given to the cost of scrap and rework. With the impact of mass production, which was required during World War II, it became necessary to introduce a more stringent form of quality control. This was instituted by manufacturing units and was identified as SQC. SQC made a significant contribution in that it provided a sampling rather than 100% product inspection. However, SQC was instrumental in exposing the underappreciation of the engineering aspect of product quality.

From the foregoing writings and many others on the history of quality, it is evident that the quality system in its different forms has moved through distinct quality eras such as

1. Quality inspection
2. Quality control
3. Quality assurance
4. Quality engineering
5. Quality management

Table 1.1 summarizes periodical changes in the quality system.

TABLE 1.1
Periodical Changes in Quality System

Period	System
• Middle Ages (1200–1799)	• Guilds-skilled craftsmen were responsible for controlling their own products.
• Mid-eighteenth century (Industrial Revolution)	• Establishment of factories. Increase in productivity. Mass production. Assembly lines. Several workers were responsible for producing a product. Production by skilled workers and quality audit by inspectors.
• Early nineteenth century	• Craftsmanship model of production.
• Late nineteenth century (1880s)	• Fredrick Taylor and "Scientific Management." Quality management through inspection.
• Beginning of twentieth century (1920s)	• Walter Shewhart introduced "statistical process control". Introduction of full-time Quality Inspection and Quality Control Department. Quality management.
• 1930s	• Introduction of sampling method.
• 1950s	• Introduction of Statistical Quality Process in Japan.
• Late 1960s	• Introduction of Quality Assurance.
• 1970s	• Total Quality Control.
	• Quality Management.
• 1980s	• Total Quality Management (TQM).
• Beginning of twenty-first century	• Integrated Quality Management (IQM).

Source: Abdul Razzak Rumane. (2017). *Quality Management in Construction Projects*, Second Edition. Reprinted with permission from Taylor & Francis Group.

1.3.1 QUALITY INSPECTION

Prior to the Industrial Revolution, items were produced by an individual craftsman, who was responsible for material procurement, production, inspection, and sales. In case any quality problems arose, the customer would take up issues directly with the producer. The Industrial Revolution provided the climate for continuous quality improvement. In the late nineteenth century, Fredrick Taylor's system of Scientific Management was born. It provided the backup for the early development of quality management through inspection. At the time when goods were produced individually by craftsmen, they inspected their own work at every stage of production and discarded faulty items. When production increased with the development of technology, scientific management was born out of a need for standardization rather than craftsmanship. This approach required each job to be broken down into its component tasks. Individual workers were trained to carry out these limited tasks, making craftsmen redundant in many areas of production. The craftsmen's tasks were divided among many workers. This also resulted in mass production at lower cost, and the concept of standardization started resulting in interchangeability of similar types of bits and pieces of product assemblies. One result of this was a power shift away from workers and toward management.

With this change in the method of production, inspection of the finished product became the norm rather than inspection at every stage. This resulted in wastage because defective goods were not detected early enough in the production process. Wastage added costs that were reflected either in the price paid by the consumer or in reduced profits. Due to the competitive nature of the market, there was pressure on manufacturers to reduce the price for consumers, which in turn required cheaper input prices and lower production costs. In many industries, emphasis was placed on automation to try to reduce the costly mistakes generated by workers. Automation led to greater standardization, with many designs incorporating interchanges of parts. The production of arms for the 1914–1918 war accelerated this process.

An inspection is a specific examination, testing, and formal evaluation exercise and overall appraisal of a process, product, or service to ascertain if it conforms to established requirements. It involves measurements, tests, and gauges applied to certain characteristics with regard to an object or an activity. The results are usually compared to specified requirements and standards for determining whether the item or activity is in line with the target. Inspections are usually nondestructive. Some of the nondestructive methods of inspection are:

- Visual
- Liquid dyed penetrant
- Magnetic particle
- Radiography
- Ultrasonic
- Eddy current
- Acoustic emission
- Thermography

The degree to which inspection can be successful is limited by the established requirements. Inspection accuracy depends on:

1. Level of human error
2. Accuracy of the instruments
3. Completeness of the inspection planning

Human errors in inspection are mainly due to:

- Technique errors
- Inadvertent errors
- Conscious errors
- Communication errors

Most construction projects specify that all the contracted works are subject to inspection by the owner/consultant/owner's representative.

1.3.2   QUALITY CONTROL

The quality control era started at the beginning of the twentieth century. The Industrial Revolution had brought about the mechanism and marked the inclusion of process in quality practices. The ASQ termed the quality control era as a process orientation that consists of product inspection and SQC.

The control process involves observing actual performances, comparing them with some standards, and then taking action if the observed performance is significantly different from the standard. Control process involves a universal sequence of steps as follows:

1. Performance standard
2. Measurement standard
3. Project/product goal
4. Process goal
4. Compare actual measured performance against standards
6. Corrective action on the difference observed/recorded

From the foregoing, quality control can be defined as a process of analyzing data collected through statistical techniques to compare with actual requirements and goals to ensure compliance with some standards.

Quality control in construction projects is performed at every stage through the use of various control charts, diagrams, checklists, and so on and can be defined as

- Checking of executed/installed works to confirm that works have been performed/executed as specified, using specified/approved materials, installation methods, and specified references, codes, and standards to meet the intended use
- Planning, monitoring, and controlling project schedule
- Controlling budget

A control chart is a graphical representation of the mathematical model used to detect changes in a parameter of the process. Charting statistical data is a test of the null hypothesis that the process from which the sample came has not changed. A control chart is employed to distinguish between the existence of a stable pattern of variation and the occurrence of an unstable pattern. If an unstable pattern of variation is detected, action may be initiated to discover the cause of the instability. Removal of the assignable cause should permit the process to return to a stable state.

There are a variety of methods, tools, and techniques that can be applied for quality control and the improvement process. These are used to create an idea, engender planning, analyze the cause, analyze the process, foster evaluation, and create a wide variety of situations for continuous quality improvement. These tools can also be used during various stages of a construction project.

Figure 1.1 illustrates a control chart for the air handling unit in an air distribution system.

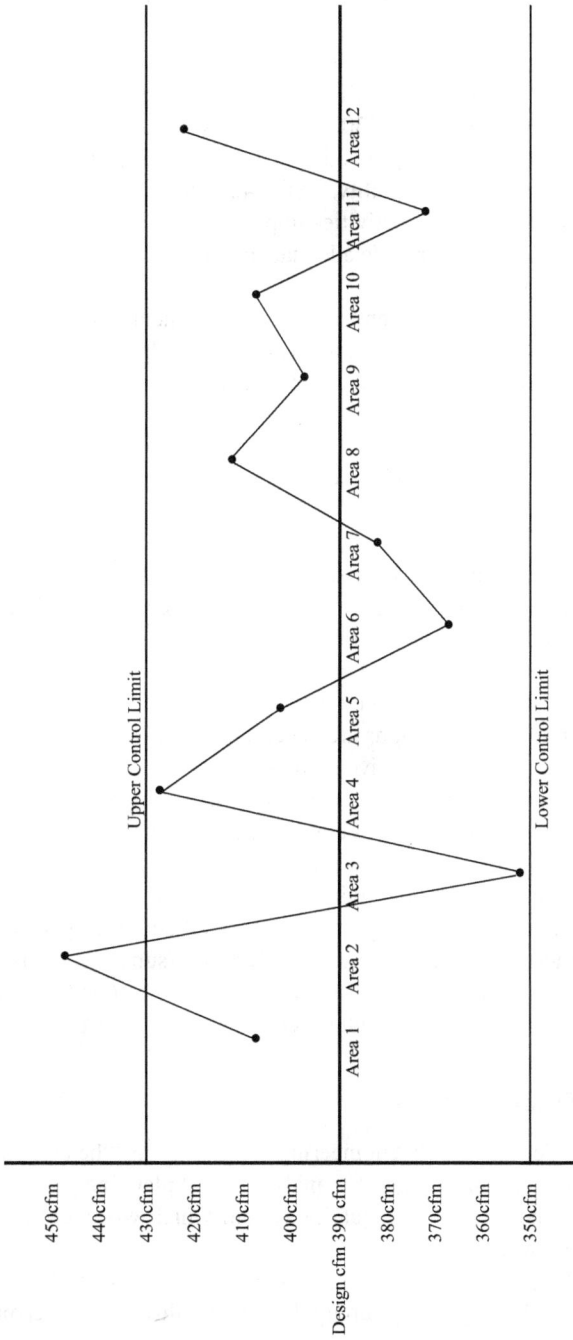

FIGURE 1.1 Control chart for air handling unit air distribution (CFM). (Abdul Razzak Rumane. (2013). *Quality Tools for Managing Construction Projects*, Reprinted with permission from Taylor & Francis Group.)

1.3.3   QUALITY ASSURANCE

Quality assurance is the third era in the quality management system.

The ASQ defines quality assurance as "all the planned and systematic activities implemented within the quality system that can be demonstrated to provide confidence that a product or service will fulfill requirements for quality."

According to ISO 9000 (or BS 5750), quality assurance is "those planned and systematic actions necessary to provide adequate confidence that product or service will satisfy given requirements for quality." ISO 8402-1994 defines quality assurance as "all the planned and systematic activities implemented within the quality system, and demonstrated as needed, to provide adequate confidence that an entity will fulfill requirements for quality."

The third era of quality management saw the development of quality systems and their application principally to the manufacturing sector. This was due to the impact of the following external environment upon the development take-up of quality systems at this time:

- Growing and, more significantly, maturing populations
- Intensifying competition

These converging trends contributed greatly to the demand for more, cheaper, and better-quality products and services. The result was the identification of quality assurance schemes as the only solution to meet this challenge.

Quality assurance is the activity of providing evidence to establish confidence among all concerned that quality-related activities are being performed effectively. All these planned or systematic actions are necessary to provide adequate confidence that a product or service will satisfy given requirements for quality.

Quality assurance covers all activities from design, development, production/construction, installation, and servicing to documentation and also includes regulations of the quality of raw materials, assemblies, products, and components; services related to production; and management, production, and inspection processes.

Quality assurance in construction projects covers all activities performed by the design team, contractor, and quality controller/auditor (supervision staff) to meet owners' objectives as specified and to ensure that the project/facility is fully functional to the satisfaction of the owners/end users.

1.3.4   QUALITY ENGINEERING

Feigenbaum (1991) defines quality engineering technology as "the body of technical knowledge for formulating policy and for analyzing and planning product quality in order to implement and support that quality system which will yield full customer satisfaction at minimum cost."

Feigenbaum (1991) has further elaborated the entire range of techniques used in quality engineering technology by grouping them under three major headings:

1. Formulating of quality policy
2. Product-quality analysis
3. Quality operations planning

1.3.5 QUALITY MANAGEMENT

The ASQ glossary defines quality management as "the application of quality management system in managing a process to achieve maximum customer satisfaction at the lowest overall cost to the organization while continuing to improve the process."

Quality actually emerged as a dominant thinking only since World War II, becoming an integral part of the overall business system focused on customer satisfaction, and becoming known in recent times as "Total Quality Management," with its three constitutive elements:

- Total: Organization-wide
- Quality: Customer satisfaction
- Management: Systems of managing

1.3.6 TOTAL QUALITY MANAGEMENT

The Total Quality Management (TQM) concept was born following World War II. It was stimulated by the need to compete in the global market where higher quality, lower cost, and more rapid development are essential to market leadership. Today TQM is considered a fundamental requirement for any organization to compete, let alone lead, in its market. It is a way of planning, organizing, and understanding each activity of the process and removing all the unnecessary steps routinely followed in the organization. TQM is a philosophy that makes quality values the driving force behind leadership, design, planning, and improvement in activities.

1.3.7 CHANGING VIEWS OF QUALITY

The failure to address the culture of an organization is frequently the reason for management initiatives either having limited success or failing altogether. To understand the culture of the organization and using that knowledge to implement cultural change is an important element of TQM. The culture of good teamwork and cooperation at all levels in an organization is essential to the success of TQM. Table 1.2 describes cultural changes needed in an organization to meet TQM.

1.4 QUALITY GURUS AND TQM PHILOSOPHIES

The TQM approach was developed immediately after World War II. There are prominent researchers and practitioners whose work has dominated this movement. Their ideas, concepts, and approaches in addressing specific quality issues have become part of the accepted wisdom in TQM, resulting in a major and lasting impact within the field. These persons have become known as "quality gurus." They all

TABLE 1.2
Cultural Changes Required to Meet Total Quality Management

From	To
• Inspection orientation	• Defect prevention
• Meet the specification	• Continuous improvement
• Get the product out	• Customer satisfaction
• Individual input	• Cooperative efforts
• Sequential engineering	• Team approach
• Quality control department	• Organizational involvement
• Departmental responsibility	• Management commitment
• Short-term objective	• Long-term vision
• People as cost burden	• Human resources as an assets
• Purchase of products or services on price-alone basis	• Purchase on total cost minimization basis
• Minimum cost suppliers	• Mutual beneficial supplier relationship

Source: Abdul Razzak Rumane. (2017). *Quality Management in Construction Projects*, Second Edition. Reprinted with permission from Taylor & Francis Group.

emphasize the involvement of organizational management in the quality efforts. These philosophers are

1.   Philip B. Crosby
2.   W. Edwards Deming
3.   Armand V. Feigenbaum
4.   Kaoru Ishikawa
5.   Joseph M. Juran
6.   John S. Oakland
7.   Shigeo Shingo
8.   Genichi Taguchi

A brief summary of their philosophies and approaches is given in the following sections.

1.4.1   PHILIP B. CROSBY

Crosby's philosophy is seen by many to be encapsulated in his five "Absolute Truths of Quality Management." These are

1.   Quality is defined as conformance to requirement, not as "goodness" or "elegance."
2.   There is no such thing as a quality problem.
3.   It is always cheaper to do it right the first time.
4.   The only performance measurement is the cost of quality.
5.   The only performance standard is zero defects.

Crosby's perspective on quality has three essential beliefs:

1. A belief in qualification
2. Management leadership
3. Prevention rather than cure

His main emphasis is on the quantitative aspect, that is, the performance standard of "zero defects."

1.4.2 W. Edwards Deming

Deming was perhaps the best-known figure associated with the quality field and is considered its founding father. His philosophy is based on four principal methods:

1. The Plan–Do–Check–Act (PDCA) cycle
2. Statistical process control
3. The 14 principles of transformation
4. The seven-point action plan

1.4.2.1 PDCA Cycle

PDCA cycle is an iterative, four-step management method used for the continuous improvement of business processes and products. Once it has been completed, it recommences without ceasing. The approach is seen as reemphasizing the responsibility of management to be actively involved in the organization's quality program. The PDCA cycle is also known as the Plan–Do–Study–Act (PDSA) cycle.

The PDCA or PDSA cycle consists of a four-step model for carrying out change. Just as a circle has no end, the PDCA cycle should be repeated again and again for continuous improvement. PDCA is a basic model that can be compared to the continuous improvement process, which can be applied on a small scale. PDCA cycle is used for

- Continuous improvement
- Designing a new process, product, or project
- Defining repetitive processes in a project
- Development of new project

PDCA cycle is developed by adopting the following procedure:

1. Plan. Recognize an opportunity and plan the scope.
2. Do. Implement the activities to perform the required results.
3. Check. Review and monitor the results.
4. Act. Take action for successful implementation of required results. If the results are successful, incorporate the same in the repetitive cycle.

PDCA is mainly used for continuous process improvement. The PDCA cycle, when used as a process improvement tool for design improvement/design conformance

● CORRECTIVE OR PREVENTIVE ACTION   ● IDENTIFY GOALS AND OBJECTIVES

●
   IMPLEMENT APPROVED CHANGES   ● ESTABLISH REQUIREMENTS

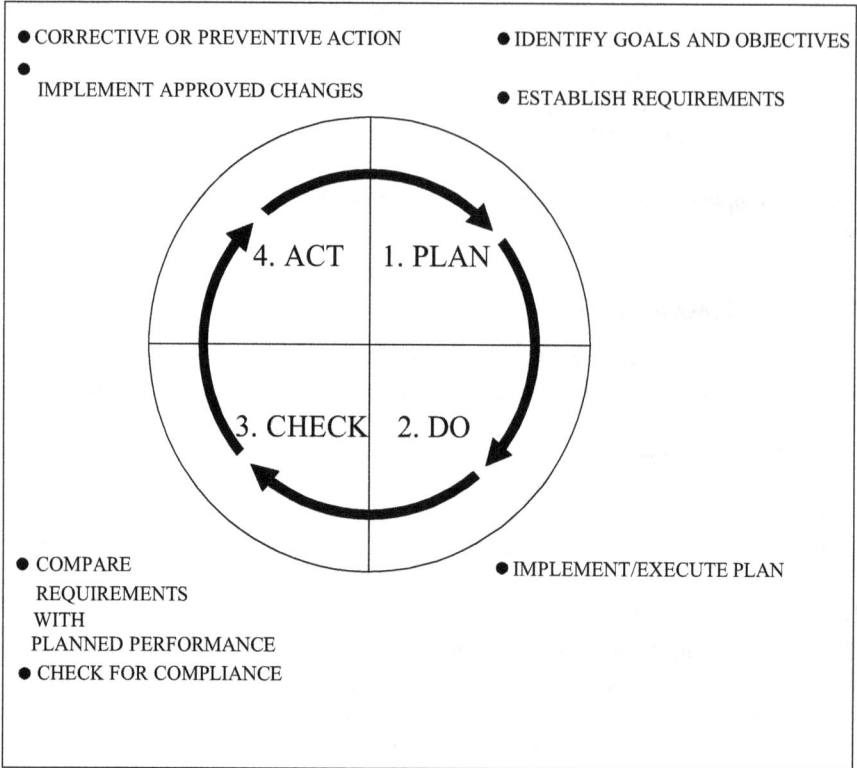

4. ACT	1. PLAN
3. CHECK	2. DO

● COMPARE   ● IMPLEMENT/EXECUTE PLAN
   REQUIREMENTS
   WITH
   PLANNED PERFORMANCE
● CHECK FOR COMPLIANCE

FIGURE 1.2   Development of PDCA cycle.

in construction projects to meet owner's requirements, shall indicate the following actions:

Plan: Establish scope.
Do: Develop design.
Check: Review and compare.
Act: Implement comments, take corrective action, and/or release contract documents to construct/build the project/facility.

Figure 1.2 illustrates basic concepts to develop the PDCA cycle. Figure 1.3 illustrates the PDCA cycle model for conformance of construction projects designed to owner requirements/scope of work.

1.4.2.2   Statistical Process Control

Statistical Process Control (SPC) is a quantitative approach based on the measurement of process control. Deming believed in the use of SPC charts as the key

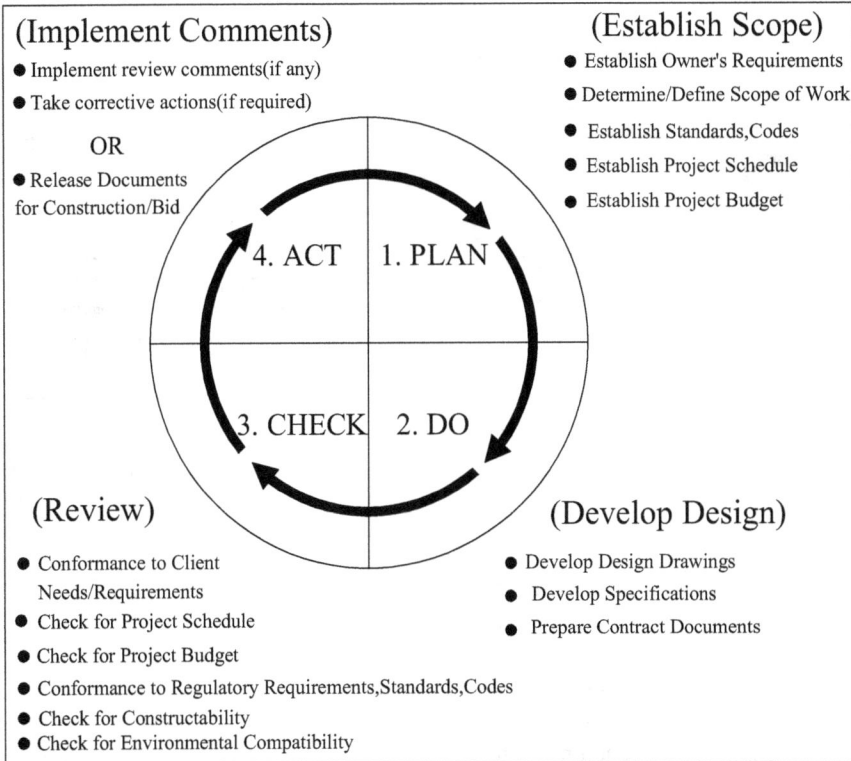

(Implement Comments)
- Implement review comments(if any)
- Take corrective actions(if required)

OR
- Release Documents for Construction/Bid

(Establish Scope)
- Establish Owner's Requirements
- Determine/Define Scope of Work
- Establish Standards,Codes
- Establish Project Schedule
- Establish Project Budget

4. ACT | 1. PLAN

3. CHECK | 2. DO

(Review)
- Conformance to Client Needs/Requirements
- Check for Project Schedule
- Check for Project Budget
- Conformance to Regulatory Requirements,Standards,Codes
- Check for Constructability
- Check for Environmental Compatibility

(Develop Design)
- Develop Design Drawings
- Develop Specifications
- Prepare Contract Documents

FIGURE 1.3 PDCA cycle for construction projects (design phases).

method for identifying special and common causes and assisting diagnosis of quality problems. His aim was to remove "outliers," that is, quality problems relating to the special causes of failure. This was achieved through training, improved machinery and equipment, and so on. SPC enabled the production process to be brought "under control."

The remaining quality problems were considered to be related to common causes, that is, they were inherent in the design of the production process. Eradication of special causes enabled a shift in focus to common causes, thereby improving quality.

There are two categories of control charts based on the type of data collected. These are

1. Variable control charts
2. Attributes control charts

Figure 1.4 illustrates the categories of control chart.

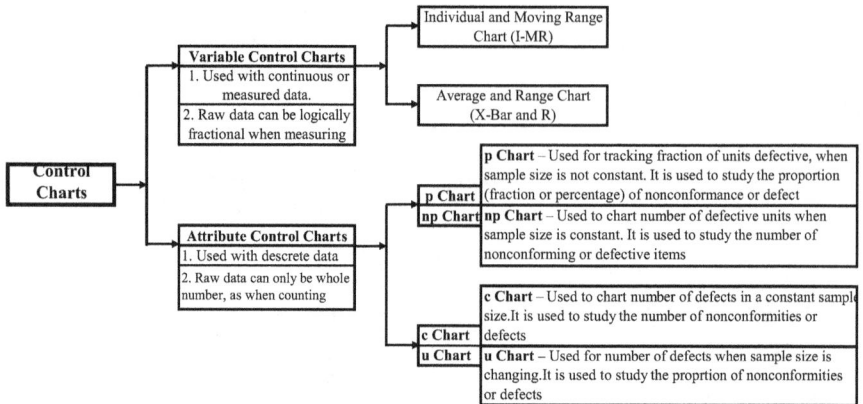

FIGURE 1.4 Categories of Control Charts. (Abdul Razzak Rumane. (2017). *Quality Tools for Managing Construction Projects*, Second Edition. Reprinted with permission from Taylor & Francis Group.)

1.4.2.3   Principles for Transformation

Deming's 14 principles for transformation mainly relate to

1. Product, process improvement
2. Adoption of new philosophy
3. Sampling inspection method
4. End low price awarding system
5. Continuous improvement
6. On-the-job training
7. Organizational leadership
8. Involvement/participation of all staff to work collectively without any fear
9. Research and development
10. Team approach
11. Integrated leadership
12. Involvement/participation of all staff
13. Education and self-improvement
14. Participation of everyone in the company to accomplish transformation

1.4.3   ARMAND V. FEIGENBAUM

Feigenbaum defines quality as "best for the customer use and selling price," and quality control as an effective method for coordinating the quality maintenance and quality improvement efforts across various groups in an organization, so as to enable production at the most economical levels that allow for full customer satisfaction. Feigenbaum's philosophy of quality has a four-step approach. These are:

Step 1. Set quality standards
Step 2. Appraise conformance to standards
Step 3. Act when standards are not met
Step 4. Plan to make improvements

1.4.4 KAORU ISHIKAWA

The founding philosophy of Ishikawa's approach is "companywide quality control." He has identified 15 effects of companywide quality control. Ishikawa's approach deals with organizational aspects and is supported by the "quality circles" technique and the "seven tools of quality control." Quality circles are Ishikawa's principal method for achieving participation, composed of 5 to 15 workers from the same area of achieving and led by a foreman or supervisor who acts as a group leader to liaison between the workers and the management. The function of quality circles is to identify local problems and recommend solutions. The aim of quality circles is to

- Contribute to the improvement and development of the enterprise
- Respect human relations and build a happy workshop offering job satisfaction
- Deploy human capabilities fully and draw out infinite potential

Quality circles are small groups of employees who meet frequently to help resolve company quality problems and provide recommendations to management. Quality circles were initially developed in Japan and only recently have achieved some degree of success in the United States. The employees involved in quality circles meet frequently either at someone's home or at the plant before the shift begins. The group identifies problems, analyzes data, recommends solutions, and carries out management-approved changes. The success of quality circles is heavily based upon management's willingness to employ recommendations.

The key elements of quality circles are as follows:

- They are completely voluntary.
- Employees are trained in group dynamics, motivation, communications, and problem-solving.
- Members rely upon each other for help.
- Management support is achieved but as needed.
- Creativity is encouraged.
- Management listens to recommendations.

The quality circles:

- Improve quality of products and services
- Improve organizational communications
- Improved worker performance
- Improved morale of workers

Ishikawa emphasized four points to be considered for the formation of the quality circles. These are

1. **Voluntarism:** Circles are to be created on a voluntary basis, and not by a command from above. Begin circle activities with people who wish to participate.
2. **Self-development:** Circle members must be willing to study.
3. **Mutual development:** Circle members must aspire to expand their horizons and cooperate with other circles.
4. **Eventual total participation:** Circles must establish their ultimate goal of full participation of all workers in the same workplace.

The quantitative techniques of the Ishikawa approach are referred to as "Ishikawa's Seven Tools of Quality Control" and is listed in Table 1.3. The approach includes both quantitative and qualitative aspects, which, taken together, focus on achieving companywide quality.

These form a set of pictures of quality, representing in diagrammatic or chart form the quality status of the operation of the process being reviewed. Ishikawa considered that all staff should be trained in these techniques as they have a useful role to play in managing quality.

Ishikawa developed a technique of graphically displaying the causes of any quality problem. His method is called by several names, such as the Ishikawa diagram, fishbone diagram, and cause-and-effect diagram. The Ishikawa diagram is essentially an end or goal-oriented picture of a problem situation. It is based around a set of "M" causes such as Manpower (personnel), Machine (plant and equipment), Material (raw material and parts), Method (techniques and technology), Measurement (sampling, instrumentation), and Mother Nature (environment). Figure 1.5 illustrates the Ishikawa "fishbone" diagram.

TABLE 1.3
Seven Tools of Quality Control by Kaoru Ishikawa

Sr. No.	Name of Quality Tool	Usage
Tool 1	Pareto Charts	Used to identify the most significant cause or problem.
Tool 2	Ishikawa/Fishbone diagrams	Charts of cause and effect in processes.
Tool 3	Stratification	Layer charts which place each set of data successively on top of the previous one. How is the data made up?
Tool 4	Check sheets	To provide a record of quality. How often it occurs?
Tool 5	Histogram	Graphs are used to display frequency of various ranges of values of a quantity.
Tool 6	Scatter diagram	To determine whether there is a correlation between two factors.
Tool 7	Control Charts	A device in Statistical Process Control to determine whether or not the process is stable.

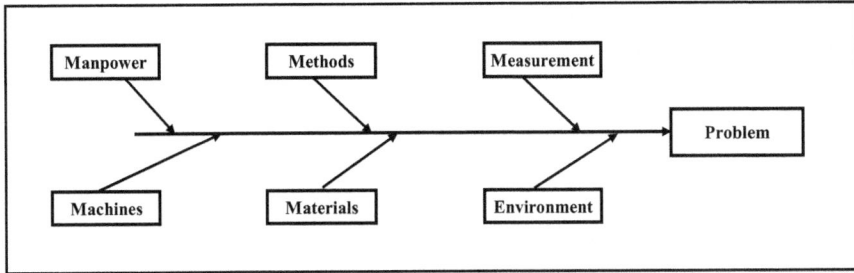

FIGURE 1.5   The Ishikawa "Fishbone" diagram.

The figure shows the basics of Ishikawa diagram.
Six steps are used to perform a cause-and-effect analysis. These are

Step 1. Identify the problem to analyze its technical cause.
Step 2. Select an interdisciplinary brainstorming team.
Step 3. Draw a problem box and prime arrows.
Step 4. Specify major categories contributing to the problem.
Step 5. Identify a defect cause.
Step 6. Identify corrective action and perform the analysis in the same manner as for the cause-and-effect analysis.

1.4.5   JOSEPH M. JURAN

Juran's philosophy is perhaps best summed as "Quality does not happen by accident; it has to be planned."

The emphasis of Juran's work is on planning organizational issues, management's responsibility for quality, and the need to set goals and targets for improvement. Juran's definition of quality is "Fitness for use or purpose." His thinking on quality is an operational framework of three quality processes. These are

1. Quality planning
2. Quality control
3. Quality improvement

These are best known as Juran's quality trilogy.

1.4.6   JOHN S. OAKLAND

Oakland's philosophy of quality is "We cannot avoid seeing how quality has developed into the most important competitive weapon, and many organizations have realized that TQM is the [sic] way of managing for the future" (Oakland, 1993). He gives absolute importance to the pursuit of quality as the cornerstone of organizational success.

Oakland's view is that "quality starts at the top," with quality parameters inherent in every organizational decision. He offers his own overarching approach for TQM on the many well-established methods, tools, and techniques for achieving quality, and some new insight.

1.4.7 SHIGEO SHINGO

Shingo's early philosophy embraced the "scientific management ideas" originated by Fredrick Taylor in the early part of the twentieth century. Shingo believed that statistical methods detect errors too late in the manufacturing process. He suggested that, instead of detecting errors, it was better to engage in preventative measures aimed at eliminating error sources. Shingo continues to believe in mechanizing the monitoring of error, considering that human assessment was "inconsistent" and prone to error and the introduction of controls within a process. He used people to identify underlying causes and produce preventative solutions. Shingo has a clear belief, like Crosby, in a "zero-defects" approach. His approach emphasizes zero defects through good engineering and process investigation and rectification.

Shingo is strongly associated with the "Just-in-Time" manufacturing philosophy. He was the inventor of the Single-Minute Exchange of Die system that drastically reduced the equipment setup time from hours to minutes. Just-in-Time is an integrated set of activities designed to achieve high-level volume production, with minimal inventories of parts that arrive at the workstation when they are needed.

Shingo is also associated with the Poka-Yoke system to achieve zero defects (failsafe procedures). The Poka-Yoke system includes checklists that (1) prevent workers from making an error that leads to defects before starting or (2) give rapid feedback of abnormalities in the process to the worker in time to correct them.

1.4.8 GENICHI TAGUCHI

Taguchi's two founding ideas of quality work are essentially quantitative. The first is a statistical method to identify and eradicate quality problems. The second rests on designing products and processes to build in quality right from the outset.

Taguchi's prime concern is with customer satisfaction and with the potential for "loss of reputation and goodwill" associated with failure to meet customer expectations. Such a failure, he considered, would lead the customer to buy elsewhere in the future, damaging the prospects of the company, its employees, and society. He saw that loss not only occurred when a product was outside its specification but also when it varied from its target value. Taguchi recognized the organization as "open system," interacting with its environment. The principal tools and techniques espoused by Taguchi center on the concept of continuous improvement and eradicating, as far as possible, potential causes of "non-quality" at the outset. His concept of product development has three stages:

1. System design stage
2. Parameter stage
3. Tolerance design stage

The first stage is concerned with system design reasoning involving both product and process. This framework is carried on to the second stage—parameter design. The third stage, tolerance design, enables the recognition of factors that may significantly affect the variability of the product.

1.4.9 SUMMARY OF PHILOSOPHIES

Although there are differences in certain areas among these philosophers, all of them generally advocate the same steps. Their emphasis is on customers' satisfaction, management leadership, teamwork, continuous improvement, and minimizing defects.

Based on these, the common features of their philosophies can be summarized as follows:

1. Quality is conformance to the customer's defined needs.
2. Senior management is responsible for quality.
3. Institute continuous improvement of processes, products, and services through the application of various tools and procedures to achieve a higher level of quality.
4. Establish performance measurement standards to avoid defects.
5. Take a team approach by involving every member of the organization.
6. Provide training and education to everyone in the organization.
7. Establish leadership to help employees perform a better job.

Thus, their concept of quality forms the basic tenets of TQM.

1.5 BIRTH OF TOTAL QUALITY MANAGEMENT

The TQM approach was developed immediately after World War II. There are prominent researchers and practitioners whose works have dominated quality movement. Their ideas, concepts, and approaches in addressing specific quality issues have become part of the accepted wisdom in the field of quality resulting in a major and lasting impact on the business. These persons are known as "quality gurus." They all emphasize upon involvement of organizational management in the quality efforts. Their philosophies are already discussed in Section 1.4.

Their approach to quality emphasizes on customer satisfaction, management leadership, teamwork, continuous improvement, and minimizing defects.

The TQM was stimulated by the need to compete in the global market where higher quality, lower cost, and more rapid development are essential to market leadership. TQM was/is considered a fundamental requirement for any organization to compete, let alone lead its market. It is a way of planning, organizing, and understanding each activity of the process and removing all the unnecessary steps routinely followed in the organization. TQM is a philosophy that makes quality values the driving force behind leadership, design, planning, and improvement in activities. It places quality as a strategic objective and focuses on the continuous improvement of products, processes, services, and cost, to compete in the global market by minimizing rework and maximizing profitability to achieve market leadership and customer satisfaction. It is a way of managing people and business processes to meet customer satisfaction.

TQM involves everyone in the organization in an effort to increase customer satisfaction and achieve superior performance of the products or services through continuous quality improvement. TQM helps in

- Achieving customer satisfaction
- Developing corporate culture
- Building motivation within organization
- Developing teamwork
- Establishing vision for the employees
- Setting standards and goals for the employees
- Continuous improvement

TQM was considered a fundamental requirement for any industry to compete, let alone lead, in its market. The TQM concept was there till the end of twentieth century.

TQM is a management approach that strives for the following in any business environment:

- Under strong top management leadership established clear mid- and long-term vision and strategies.
- Properly utilize the concepts, values, and scientific methods of TQM.
- Regard human resources and information as vital organizational infrastructures.
- Under an appropriate management system, effectively operate a quality assurance system and other cross-functional management systems such as cost, delivery, environment, and safety.
- Supported by fundamental organizational powers, such as core technology, speed, and vitality, ensure sound relationship with customers, employees, society suppliers, and stockholders.
- Continuously release corporate objectives in the form of achieving an organization's mission, building an organization with a respectable presence, and continuously securing profits.

TQM focuses on participative management and strong operational accountability at the individual contributor level. Total quality involves not just managers but everyone in the organization in a complete transformation of the prevailing culture. It is a change to the way people do things and relies on trust between managers and staff. TQM is applicable to all kinds of organizations, both in the public and private sectors. It is also applicable to those providing services as well as those involved in producing goods or manufacturing activities.

The quality actually emerged as a dominant thinking since World War II, becoming an integral part of the overall business system focused on customer satisfaction, and becoming known in the twentieth century as "Total Quality Management" (TQM) with its three constitutive elements:

Total: Organization-wide
Quality: Customer satisfaction
Management: Systems of managing

FIGURE 1.6   Concept of Total Quality Management.

Figure 1.6 illustrates the concept of TQM.

According to Construction Industry Institute (CII) source document 74: "Total Quality Management is often termed a journey, not a destination." This is because of its nature as a collection of improvement-centered processes and techniques that are performed in a transformed management environment. The concept of "continuous improvement" holds that this environment must prevail for the life of the enterprise and that the methods will routinely be used on a regular, recurring basis. The improvement process never ends; therefore, no true destination is ever reached. Figure 1.7 describes the phases of the TQM journey.

The document further states (p. 1) that from the viewpoint of the individual company, the strategic implications of TQM include

- Survival in an increasingly competitive world
- Better service to the customer
- Enhancement of the organization's "shareholder value"
- Improvement of the overall quality and safety of our facilities
- Reduced project duration and costs
- Better utilization of talents of the people

Exploration and Commitment	Planning and Preparation	Implementation	Sustaining
Perceived need for change	Strategic Quality deployment process	Management oversight structure	Absorption of TQM infrastructure into regular mangement system
Investigation of approaches	Initial development of Quality infrastructure	Realignment of reward system	
		Formation of teams	Long-range planning
Engagement of consultant	Expansion of training '- More people '-More subjects '- Mgmt. Role modeling	Teams-skills training Pilot improvement projects	Focus on process and customers Ongoing training
Top management basic training	Team management system	Implementation of results	Ongoing improvement efforts
Confirmation of TQM commitment	Team-process models	Company-wide expansion Vendor / supplier process	Management for Continous Improvement

Time

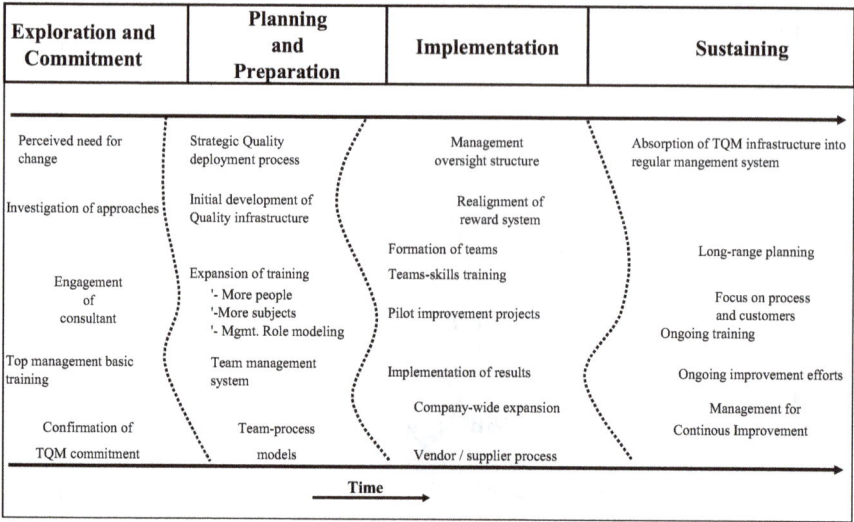

FIGURE 1.7 Phases of the TQM journey. (CII Source document 74. Reprinted with permission from CII, University of Texas.)

The ISO document has listed eight quality management principles on which the quality management system standards of the revised ISO 9000:2000 series are based. These are as follows:

Principle 1—Customer focus
Principle 2—Leadership
Principle 3—Involvement of people
Principle 4—Process approach
Principle 5—System approach to management
Principle 6—Continual improvement
Principle 7—Factual approach to design making
Principle 8—Mutual beneficial supplier relationship

These principles were further revised in ISO 9001:2015. Accordingly, there are seven quality management principles. These are as follows:

Principle 1—Customer focus
Principle 2—Leadership
Principle 3—Involvement of people/Engagement of people
Principle 4—Process-driven approach to quality standards/System approach to management
Principle 5—Continual improvement
Principle 6— Evidence-based decision making/Factual approach to design making
Principle 7—Relationship management of suppliers, customers, and regulators

TQM is widely accepted as the basis for achieving sustainable competitive advantage. It is required to achieve customer satisfaction and ensure survival in the competitive world because of the global nature of competition, which never rests, and there is no end to product or service improvement. TQM is based on the quality principles discussed above.

Pursuing TQM in construction projects results in the following strategic benefits:

- Survive in an increasingly competitive world
- Improve market share and profitability
- Better serve the needs of customers
- Improve quality and safety of the project/facility
- Reduce duration of project execution
- Reduce project cost
- Fully utilize the talents of employees

TQM focuses on participative management and strong operational accountability at the individual contributor level. Total quality involves not just managers but everyone in the organization in a complete transformation of the prevailing culture. It is a change to the way people do things and relies on trust between managers and staff. TQM is applicable to all kinds of organizations, both in the public and private sectors. It is also applicable to those providing services as well as those involved in producing goods or manufacturing activities.

2 Overview of Quality in Construction Projects

2.1 CONSTRUCTION PROJECTS

A project is a plan or program performed by the people with assigned resources to achieve an objective within a finite duration. A project is a temporary endeavor undertaken to create a unique product or service. "Temporary" means that every project has a definite beginning and a definite end. Projects have specified objectives to be completed within certain specifications and funding limits. Projects are often critical components of the performing organization's business strategy. Examples of projects include

- Developing a new product or service
- Effecting a change in structure, staffing, or style of an organization
- Designing a new transportation vehicle/aircraft
- Developing or acquiring a new or modified information system
- Running a campaign for political office
- Implementation of a new business procedure or process
- Constructing a building or facilities

The duration of a project is finite; projects are not ongoing efforts, and the project ceases when its declared objectives have been attained. Some of the characteristics of projects, for example, are

1. Performed by people
2. Constrained by limited resources
3. Planned, executed, and controlled

Construction is the translation of owner's goals and objectives into a facility built by the contractor/builder as stipulated in the contract documents, plans, and specifications on schedule and within budget. Construction projects are a custom rather than a routine, repetitive business and differ from manufacturing. Construction projects work against a defined scope, schedule, and budget to achieve the specified result.

 DOI: 10.1201/9781003451464-2

Construction projects have the involvement of many participants comprising owner, designer, contractor, and many other professionals from construction-related industries. Each of these participants is involved in implementing quality in construction projects. These participants are both influenced by and depend on each other in addition to "other players" involved in the construction process. Therefore, the construction projects have become more complex and technical, and extensive efforts are required to reduce rework and costs associated with time, materials, and engineering, and so forth.

Construction projects are mainly capital investment projects. They are customized and non-repetitive in nature. Construction projects have become more complex and technical, and the relationships and the contractual grouping of those who are involved are also more complex and contractually varied. The products used in construction projects are expensive, complex, immovable, and long-lived. Generally, a construction project is composed of building materials (civil), electro-mechanical items, finishing items, and equipment. For oil and gas projects, there are other items such as processing equipment/machines, pumps, piping material, values, storage tanks, and instrumentation. These are normally produced by other construction-related industries/manufacturers. These industries produce products as per their own quality management practices complying with certain quality standards or against specific requirements for a particular project. The owner of the construction project or his representative has no direct control over these companies unless he/his representative/appointed contractor commits to buy their product for use in their facility. These organizations may have their own quality management program. In manufacturing or service industries, the quality management of all in-house manufactured products is performed by manufacturer's own team or under the control of the same organization having jurisdiction over their manufacturing plants at different locations. Quality management of vendor-supplied items/products is carried out as stipulated in the purchasing contract as per the quality control specifications of the buyer.

2.1.1 Types of Construction Projects

There are numerous types of construction projects.

A. Process-Type Projects
   1. Liquid Chemical Plants
   2. Liquid/Solid Plants
   3. Solid Process Plants
   4. Petrochemical Plants
   5. Petroleum Refineries
B. Non-Process-Type Projects
   1. Power Plants
   2. Manufacturing Plants
   3. Support Facilities
   4. Miscellaneous (R&D) Projects
   5. Civil Construction Projects **
   6. Commercial/A&E Projects **

** Civil Construction Projects and Commercial/A&E Projects can further be categorized into four somewhat arbitrary but generally accepted major types of construction. These are

   Residential Construction
   Building Construction (Institutional and Commercial)
   Industrial Construction
   Heavy Engineering Construction

Table 2.1 illustrates the types of construction projects.

Construction projects are custom-oriented and custom designed having specific requirements set by the customer/owner to be completed within a finite duration and budget. Every project has elements that are unique. No two projects are alike. It is always the owner's desire that his project should be unique and better. To a great extent, each project has to be designed and built to serve a specified need. Construction projects are more custom than a routine and repetitive business. Construction projects differ from manufacturing or production. Construction projects involve many participants including owners, designers, contractors, and many other professionals from construction-related industries. These participants are both influenced by and depend on each other in addition to "other players" who are part of the construction process. Therefore, construction projects have become more complex and technical, and extensive efforts are required to reduce rework and costs associated with time, materials, engineering, and so forth.

Construction projects comprised of a cross-section of many different participants. These participants both influence and depend on each other in addition to "other players" involved in the construction process. Figure 2.1 illustrates the concept of traditional construction project organization.

Traditional construction projects have involvement of three main groups. These are

1. Owner—A person or an organization who initiates and sanctions a project. He/she requests the need of the facility and is responsible for arranging the financial resources for the creation of the facility.
2. Designer (A/E)—This consists of architects or engineers or consultants. They are the owner's appointed entity accountable for converting the owner's conception and need into a specific facility with detailed directions through drawings and specifications within the economic objectives. They are responsible for the design of the project and in certain cases supervision of the construction process.
3. Contractor—A construction firm engaged by the owner to complete the specific facility by providing the necessary staff, workforce, materials, equipment, tool, and other accessories to the satisfaction of the owner/end user in compliance with the contract documents. The contractor is responsible for implementing the project activities and to achieve the owner's objectives.

Construction projects are executed based on a predetermined set of goals and objectives. Under traditional construction projects, the owner heads the team, designating as a project manager. The project manager is a person/member of the owner's

TABLE 2.1
Types of Construction Projects

1	**Process-type projects**		
1.1	Liquid chemical plants		
1.2	Liquid/solid plants		
1.3	Solid process plants		
1.4	Petrochemical plants		
1.5	Petroleum refineries		
2	**Non-process-type projects**		
2.1	Power plants		
2.2	Manufacturing plants		
2.3	Support facilities		
2.4	Miscellaneous (R&D) projects		
2.5	Civil construction projects	Residential construction	Family homes, multi-unit town houses, gardens, apartments, condominiums, high-rise apartments, villas
2.6	Commercial A/E projects	Building construction (institutional and commercial)	Schools, universities, hospitals, commercial office complexes, shopping malls, banks, theaters, stadiums, government buildings, warehouses, recreation centers, amusement parks, holiday resorts, neighborhood centers
		Industrial construction	Petroleum refineries, petroleum plants, power plants, heavy manufacturing plants, steel mills, chemical processing plants
		Heavy engineering	Dams, tunnels, bridges, highways, railways, airports, urban rapid transit systems, ports, harbors, power lines, and communication network
		Environmental	Water treatment and clean water distribution, sanitary and sewage system, waste management

Categories of civil construction projects and commercial A/E projects

Source: Abdul Razzak Rumane. (2013). *Quality Tools for Managing Construction*, CRC Press, Florida. Projects. Reprinted with permission of Taylor & Francis Group.

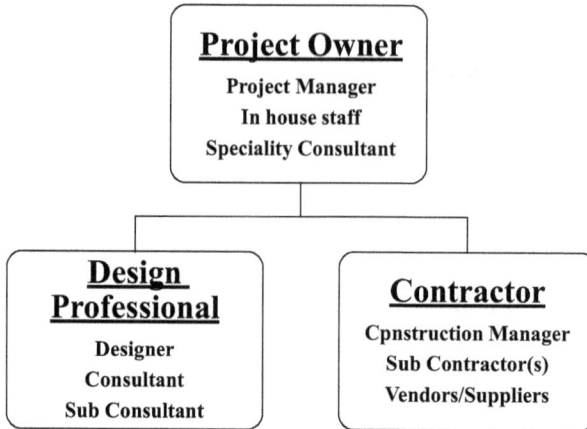

FIGURE 2.1 Traditional construction project.

staff or independently hired person/firm who has overall or principal responsibility for the management of the project as a whole.

2.2 PROJECT DELIVERY SYSTEMS IN CONSTRUCTION PROJECTS

A project delivery system is defined as the organizational arrangement among various participants comprising owner, designer, contractor, and many other professionals involved in the design and construction of a project/facility to translate/transform the owner's needs/goals/objectives into a finished facility/project to satisfy the owner's/ end user's requirements.

The project delivery system

- Establishes scope and responsibility for how the project is delivered to the owner.
- Includes project design and construction.
- Defines responsibility/obligations each of the participants are expected to perform, such as scheduling, cost control, quality management, safety management, and risk management during various phases of the construction project life cycle (concept design, schematic design, detailed design, construction, and testing, commissioning, and handover).
- Is the approach by which the project is delivered to the owner but is separate and distinct from the contractual arrangements for financial compensation.
- Establishes procedures, actions, and sequence of events to be carried out.

The owner/client selects a particular type of project delivery system as per the project procurement strategy of the organization. The type of project delivery system varies from project to project taking into consideration the objectives of the project. Project delivery method is a system to achieve satisfactory completion of construction projects from inception to occupancy.

2.2.1 Types of Project Delivery Systems

There are several types of contract delivery systems. Table 2.2 illustrates the most common project delivery systems followed in construction projects.

TABLE 2.2
Categories of Project Delivery Systems

Serial Number	Category	Classification	Sub-Classification
1	Traditional System (Separated and Cooperative)	Design–Bid–Build Variant of Traditional System	Design–Bid–Build Sequential Method Accelerated Method
2	Integrated System	Design–Build	Design–Build
		Design–Build	Joint Venture (Architect and Contractor)
		Variant of Design–Build System	Package Deal
		Variant of Design–Build System	Turnkey Method (EPC)- (Engineering, Procurement, Construction)
		Variant of Design–Build System (Turnkey)	Build–Operate–Transfer (BOT)
			Build–Own–Operate–Transfer (BOOT)
			Build–Transfer–Operate (BTO)
			Design–Build–Operate– Maintain (DBOM)
		Variant of Design–Build System (Funding Option)	Lease–Develop–Operate (LDO)
			Wraparound (Public–Private Partnership)
		Variant of Design–Build System	Build–Own–Operate (BOT) Buy–Build–Operate (BBO)
3	Management- Oriented System	Management Contracting Construction Management	Project Manager (Program Management) Agency Construction Manager Construction Manager-at-Risk
4	Integrated Project Delivery System	Integrated Form of Contract	

Source: Abdul Razzak Rumane. (2013). *Quality Tools for Managing Construction Projects*. Reprinted with permission of Taylor & Francis Group.

2.2.1.1 Design–Bid–Build

In this method, the owner contracts design professional(s) to prepare detailed design and contract documents. These are used to receive competitive bids from the contractors. A design–build–bid–build contract has well-defined scope of work. This method involved three steps:

1. Preparation of complete detailed design and contract documents for tendering
2. Receiving bids from prequalified contractors
3. Award of contract to successful bidder

In this method, two separate contracts are awarded: one to the designer/consultant and one to the contractor. In this type of contract structure, design responsibility is primarily that of the architect or engineer employed by the client, and the contractor(s) is primarily responsible for construction only. In most cases the owner contracts a designer/consultant to supervise the construction process. These types of contracts are lump sum fixed price contracts. Any variation or changes during the construction need prior approval from the owner. Since a complete design is prepared before construction, the owner knows the cost of the project, the time of completion of project, and knows what the configuration of the project is. The client through the architect or engineer retains control of design during construction. This type of contracting system requires considerable time; each step must be completed before starting the next step. Table 2.3 illustrates the main aspects, advantages, and disadvantages of design–bid–build type (traditional type) of project delivery system whereas Figure 2.2 illustrates the contractual relationship.

2.2.1.1.1 Multiple-Prime Contracting

Multiple-Prime Contracting is a variation of design–bid–build type of project delivery system. In this type of contracting system, the owner holds separate contracts with contractors of various disciplines and retains control over the project. This system can result in lower cost to the owner because it avoids the compounded profit and overhead margins that are common to the single contract method. Since the project work is divided into a number of packages, more firms take part in the bidding which results in lower prices and also reduces liability for delays by postponing the bidding of follow-on work. With this type of system, "fast tracking" of project is possible. In this type of project delivery system, the owner normally engages CM to manage the project. Figure 2.3 illustrates the contractual relationship in Multiple-Prime Contracting system.

2.2.1.2 Design–Build

In design–build contract, the owner contracts a firm singularly responsible to design and build the project. In this type of contracting system, the contractor is appointed based on an outline design or design brief to understand the owner's intent of the project. The owner has to clearly define his/her needs, performance requirements, and comprehensive scope of work prior to signing the contract. It is a must that project definition is understood by the contractor to avoid any conflict in future as the contractor is responsible for the detailed design and construction of the project. A design–build type of contract is often used to shorten the time required to complete a project. Since the contract with the design–build firm is awarded before starting any design

TABLE 2.3
Design–Bid–Build

Project Delivery System	Main Aspects	Advantages	Disadvantages
Traditional Contracting System (Design–Bid–Build)	In this method the owner contracts design professional(s) to prepare detailed design and contract documents. These are used to receive competitive bids from the contractors. A design–bid–build contract has well-defined scope of work. This method involved three steps: 1. Preparation of complete detailed design and contract documents for tendering. 2. Receiving bids from prequalified contractors. 3. Award of contract to successful bidder. In this method two separate contracts are awarded, one to the designer/consultant and one to the contractor. In this type of contract structure, design responsibility is primarily that of the architect or engineer employed by the client and the contractor(s) is primarily responsible for construction only.	• Traditional, well-known delivery system. • Suitable for all types of clients. • Client has control over design process and quality. • Project scope is defined. • Project schedule is known. • Cost certainty at the time of awarding the contract. • The client has direct contractual control over designer (consultant) and contractor. • Low risk for client. • Contractual roles and responsibilities of all parties are clearly defined and known and understood by each participant. • Well-defined relationship among all the parties. • Higher level of competition resulting in low bid cost.	• Long project life cycle. • Client has no control over selection of contractor/subcontractor. • Client must have the resources and expertise to administer the contracts of designer and contractor. • Higher level of inspection by supervision agency. • Lack of input from constructors may result in design and constructability issues. • Project cost may be higher than estimated by the designer. • If contract documents are unclear, it raises the unexpected costs drastically. • Project likely to be delayed as it may not be possible to complete as per estimated schedule. • Exposure to risk where unreasonable time is set to complete the design phase. • No incentives to the contractor. • Not suitable for projects which are change-sensitive.

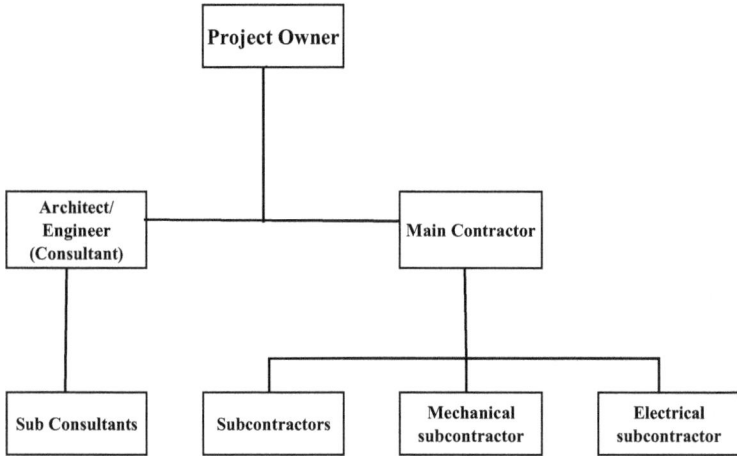

FIGURE 2.2 Design–bid–build (traditional contracting system) contractual relationship.

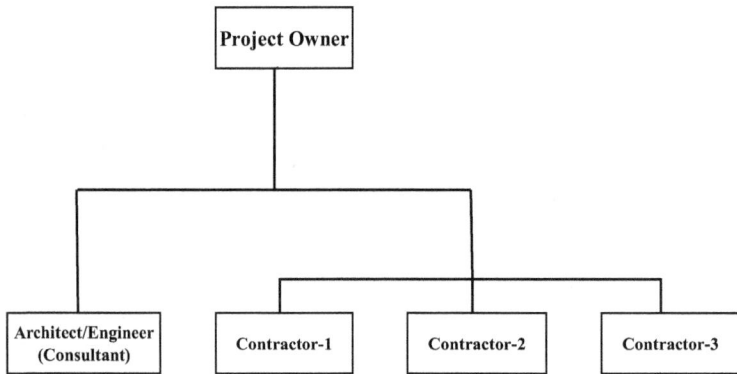

FIGURE 2.3 Multiple-prime contractor contractual relationship.

or construction, a cost plus contract or reimbursable arrangement is normally used instead of lump sum, fixed-cost arrangement. This type of contract requires extensive involvement of the owner during the entire life cycle of the project. He/she has to be involved for taking decisions during the selection of design alternatives and the monitoring of costs and schedules during construction, and therefore the owner has to maintain/hire a team of qualified professionals to perform these activities. Design/ build contracts are used for relatively straightforward work, where no significant risk or change is anticipated and when the owner is able to specify precisely what is required. Table 2.4 illustrates the main aspects, advantages, and disadvantages of design–build type of project delivery system whereas Figure 2.4 illustrates the contractual relationship.

In case of design–bid–build type of project delivery system, the contractor is contracted after completion of design based on successful bidding whereas in

TABLE 2.4
Design–Build

Project Delivery System	Main Aspects	Advantages	Disadvantages
Design–Build	In design–build contract, owner contracts a firm (contractor) singularly responsible for design and construction of the project. In this type of contracting system, the contractor is appointed based on an outline design or design brief to understand the owner's intent of the project. The owner has to prepare a comprehensive scope of work and has to clearly define his/her needs and performance requirements/ specifications prior to signing of the contract. It is a must that project definition is understood by the contractor to avoid any conflict in future as the contractor is responsible for detailed design and construction of the project. Owner has to involve for taking decisions during the selection of design alternatives and the monitoring of costs and schedules during construction and therefore the owner has to maintain/hire a team of qualified professionals to perform these activities.	• Reduce overall project time because construction begins before completion of design. • Singular responsibility, contractor takes care of the design, schedule, construction services, quality, methods, and technology. • For owner/client, the risk is transferred to design/build contractor. • Project cost is defined in early stage and has certainty. • Early involvement of contractor assists constructability. • Suitable for straightforward projects where significant changes or risks are not anticipated and owner is able to precisely specify the objectives/ requirements. • Risk management is better than Design/ Bid/Build.	• Not suitable for complex projects. • Owner has reduced control over design quality. • Extensive involvement to ensure design deliverable meets project performance requirements. • Extensive involvement of owner during entire life cycle of project. • Real price for a contract cannot be estimated by the owner/client in the beginning. • Changes by owner can be expensive and may result in heavy cost penalties to the owner. • Poor identification of owner needs and wrong understanding of project brief/ concept can cause main problem during the project realization. • Project quality cannot be assured if it is not monitored properly by the owner. • For contractor, more risks in this type of contract. • Not suitable for renovation projects.

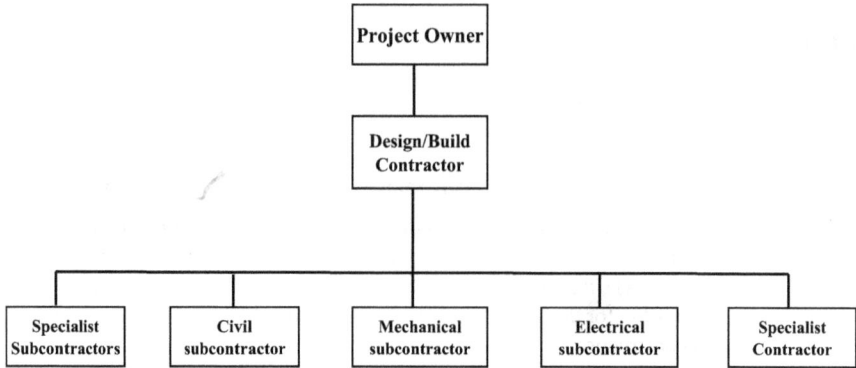

FIGURE 2.4   Design–build delivery system contractual relationship.

design–build type of deliverable system, the contractor is contracted right from the early stage of the construction project and is responsible for the design development of the project.

Figure 2.5 illustrates a typical logic flow diagram for design–bid–build type of construction project, and Figure 2.6 illustrates a typical logic flow diagram for design–build type of contracting system.

2.2.1.3   The Turnkey Contract (EPC)

The turnkey contract (EPC) are the types of contracts where, on completion, turns a key in the door and everything is working to full operating standards. In this type of method, the owner employs a single firm to undertake design, procurement, construction, and commissioning of the entire work. The firm is also involved in the management of project during the entire process of the contract. The client is responsible for preparation of their statement requirements which becomes a strict responsibility of the contractor to deliver. This type of contract is used mainly for the process type of projects and is sometimes called Engineering, Procurement, and Construction (EPC). Table 2.5 illustrates differences between design–build and EPC type of project delivery system.

2.2.1.4   Build–Own–Operate–Transfer

This type of method is generally used by governments to develop public infrastructure by involving private sector to finance, design, operate, and manage the facility for a specified period and then transfer the same to the same government free of charge. The terms BOOT and BOT are used synonymously.

Examples of BOT projects include

- Airports
- Bridges
- Motorways/Toll roads
- Parking Facilities
- Tunnels

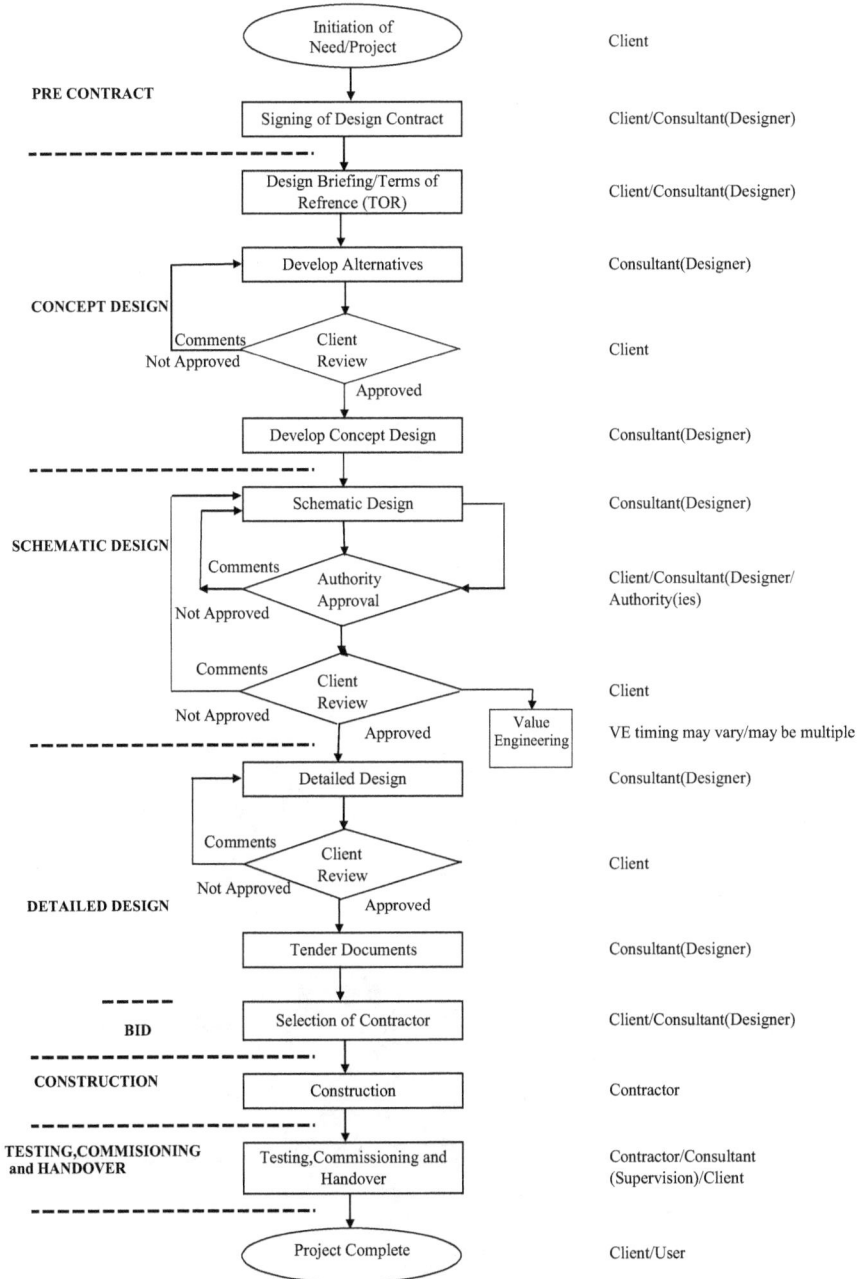

FIGURE 2.5 Logic flow diagram for construction projects: design–bid–build system. (Abdul Razzak Rumane (2017), *Quality Management in Construction Projects*, Second Edition. Reprinted with permission from Taylor & Francis Group.)

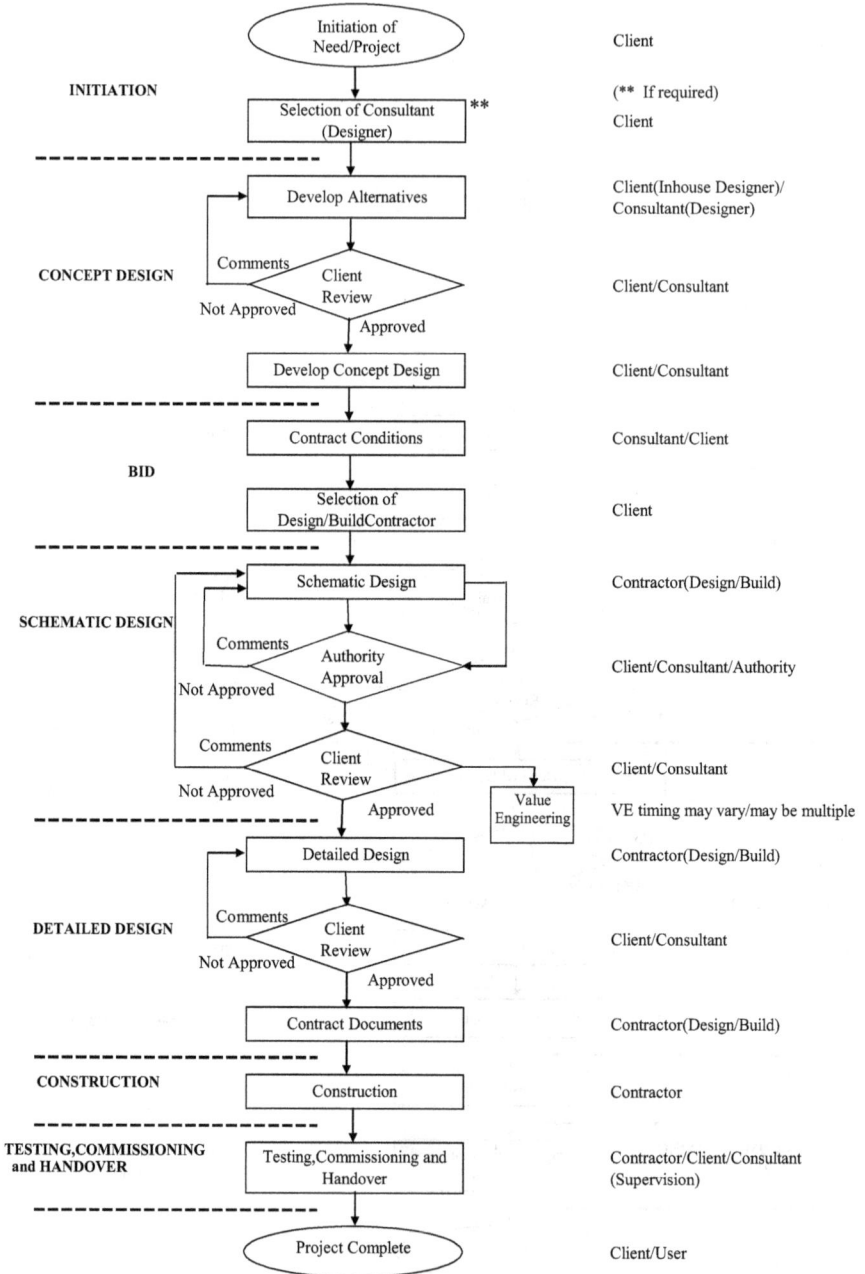

FIGURE 2.6 Logic flow diagram for construction projects: design–build system. (Abdul Razzak Rumane (2017), *Quality Management in Construction Projects*, Second Edition. Reprinted with permission from Taylor & Francis Group.)

TABLE 2.5
Difference between Design–Build and EPC

Design–Build	EPC
• Client/owner has an input into the outline design or design brief to understand owner's intent of the project	• Client has an input in the output (operating capacity)
• Client/owner contract a designer/consultant (in case in-house facility is not available) to decide design outline	• EPC (Engineering–Procurement–Construction) is a direct contract between client/owner to build a complete project to meet agreed upon output
• Detailed design is carried out by the Design–Build contractor (singular responsibility)	• All project activities are carried out by EPC contractor
• Client/owner has to employ a team of professionals to perform supervision/management of detail design, monitor quality, schedule, and cost	• Client/owner is not involved with detailed design process, except in the event of variation and quality procedures
• Claims risk is higher	• Claims risk is lower
• Mainly suitable for building projects	• Mainly used for projects such as power plants, process industry, oil and gas sector

Certain countries allow the private sector to develop commercial and recreational facilities on government land through BOT scheme.

2.2.1.5 Project Manager

A project manager contract is used by the owner when the owner decides to turn over the entire project management to a professional project manager. In project manager type of contract, the project manager is owner's representative and is directly responsible to the owner. The project manager is responsible for planning, monitoring, and management of the project. In its broadest sense, the project manager has responsibility for all the phases of the project from inception of the project till the completion and handing over of the project to the owner/end user. Project manager is involved in giving advice to the owner and is responsible to appoint design professional(s), consultant, and supervision firm, and select the contractor to construct the project. Table 2.6 illustrates the main aspects, advantages, and disadvantages of project manager type of project delivery system whereas Figure 2.7 illustrates the contractual relationship.

2.2.1.6 Construction Management

In this method, the owner contracts a construction management firm to coordinate the project for the owner and provide construction management services. There are two general forms of construction management. These are

1. Agency Construction Management (agency CM)
2. Construction Management-at-Risk (CM-at-risk)

TABLE 2.6
Project Manager Delivery System

Project Delivery System	Main Aspects	Advantages	Disadvantages
Project Manager (PM)	A project manager type delivery system is used by the owner when the owner decides to turn over the entire project management to a professional project manager. In project manager type of contract, project manager is owner's representative and is directly responsible to the owner. The project manager acts as a management consultant on behalf of the owner. The project manager is responsible for planning, monitoring, and management of the project. In its broadest sense, the project manager has responsibility for all the phase of the project from inception of the project till the completion and handing over of the project to the owner/end user. Project manager is involved in giving advice to the owner and is responsible to appoint design professional(s), consultant, supervision firm, and select the contractor/package contractor to construct the project.	• Owner/client retains full control of design. • Provide the opportunity for "Fast Track" or overlapping design and construction phases. • Reduces the owner's general management and oversight responsibilities. • Changes can be accommodated in unlet packages as long as there is no impact on time and cost. • Expert opinion and independent view toward constructability, cost, value engineering, and team member selection. • Multi-prime type of delivery system is possible.	• Added project management cost to the owner. • Owner must have resources and expertise to deal with PM. • Owner cedes much of day-to-day control over the project to the PM. • Project manager is not at risk to the cost. • Owner continues to hold construction contracts and retains contractual liability.

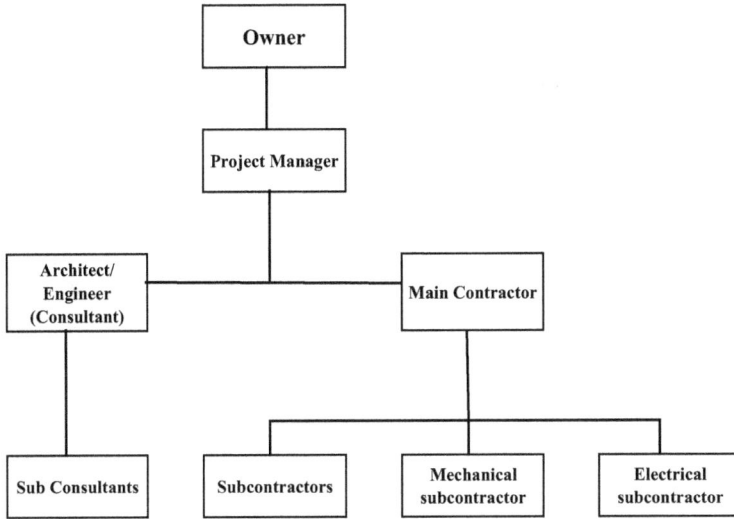

FIGURE 2.7 Project manager type delivery system contractual relationship.

The agency CM type is a management process type of contract system having four-party arrangement involving the owner, designer, construction management firm, and contractor. The construction manager provides advice to the owner regarding cost, time, safety, and the quality of materials/products/systems to be used on the project. The agency CM firm performs no design or construction but assists the owner in selecting design firm(s) and contractor(s) to build the project.

Agency CM could be implemented in conjunction with any type of project delivery system.

The basic concept of construction management type of contract is that the firm is knowledgeable and capable of coordinating all aspects of the project to meet the intended use of the project by the owner. Agency construction manager acts as a principal agent to advise the owner/client, whereas CM-at-risk is responsible for on-site performance and actually performs some of the project works. CM-at-risk type of contract has two stages. The first stage encompasses pre-construction services and during second stage the CM-at-risk is responsible for performing the construction work. CM-at-risk project delivery system is also known as construction manager/general contractor (CM/GC).

Table 2.7 illustrates the main aspects, advantages, and disadvantages of CM-at-risk type of project delivery system whereas Figure 2.8 illustrates the contractual relationship for agency CM, and Figure 2.9 illustrates the contractual relationship for CM-at-risk.

2.2.1.7 Integrated Project Delivery

In this delivery method the owner, designer, and contractor are contractually required to collaborate among themselves so that the risk responsibility and liability for the project delivery are collectively managed and appropriately shared. Table 2.8 illustrates

TABLE 2.7
Construction Management

Project Delivery System	Main Aspects	Advantages	Disadvantages
Construction Management	In this method, owner contracts a construction management firm to coordinate the project for the owner and provide construction management services. There are two general types of construction management type of contracts. These are 1. Agency construction management 2. Construction management-at-risk The agency construction management is a management process system which can be implemented regardless of project delivery method. Agency CM has four-party arrangement involving owner, designer, construction management firm, and the contractor. CM-at-Risk type of delivery system has two contracts, one between owner and designer and other one between owner and the CM-at-risk. CM-at-risk is selected on qualification basis. The construction manager, a person, or a firm provides advice to the owner regarding cost, time, safety, and about the quality of materials/products/ systems to be used on the project. The architect/engineer is responsible for design of the project. The basic concept of construction management type of contract is that the firm is knowledgeable and capable of coordinating all aspects of the project to meet the intended use of the project by the owner. Agency construction manager acts as an advisor to the owner/client, whereas construction manager-at-Risk is responsible for on-site performance and actually performs some of the project works while during the design phase construction manager acts as consultant to the owner and offer pre-construction services. The agency CM performs no design or construction but assists the owner in selecting design firm(s), and contractor(s) to build the project.	• Client retains full control of project. • Agency CM helps owner providing advice during the design phase, bid evaluation, overseeing construction, managing project schedule, cost, and quality. • Suitable for large and complex projects with multiple phases and contract packages. • Since construction manager is an expert entity, it provides an independent view regarding constructability, cost, and value engineering. • Shorter project schedule. • Changes can be accommodated in unlet packages.	• Construction manager takes control of packages and interaction with contractors/ subcontractors but has no contractual role. • No cost certainty till final package is let. • Client retains all the risk. • Client has to manage the contractual agreements with each package contractor. • Client has to carry majority risk.

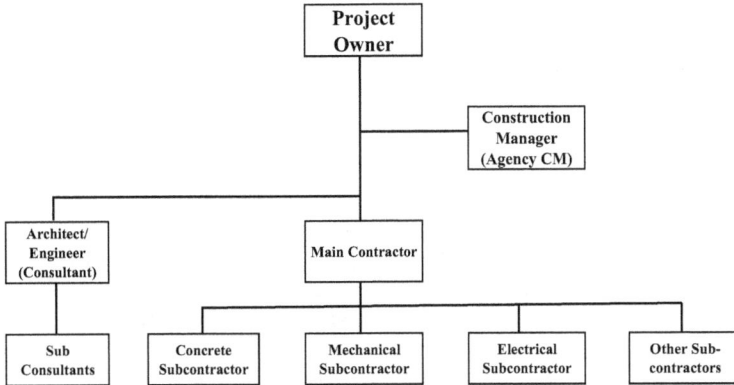

FIGURE 2.8 Agency construction management contractual relationship.

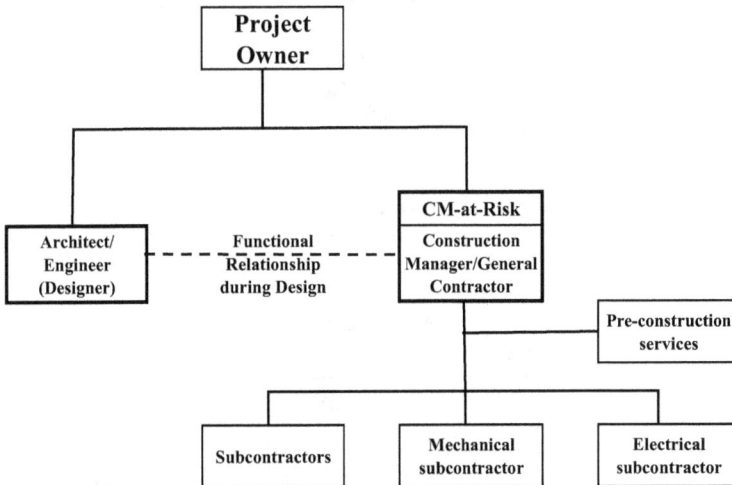

FIGURE 2.9 Construction manager-at-risk contractual relationship (CM-at-risk).

the main aspects, advantages, and disadvantages of Integrated Project Delivery (IPD) type of project delivery system, and Figure 2.10 illustrates the contractual relationship between various parties.

In all types of project delivery system, there are mainly three participants:

1. Owner
2. Designer (A/E)
3. Contractor (known as main contractor or general contractor)

TABLE 2.8
Integrated Project Delivery System

Project Delivery System	Main Aspects	Advantages
Integrated Project Delivery System	Integrated project delivery system is an alternative project delivery method distinguished by a contractual agreement at a minimum between owner, design professional, and contractor where the risks and rewards are shared and stakeholder's success is dependent on project. It is a relational contracting which sets performance and results expectations. It creates an integrated team which results optimization of the project. Value analysis throughout the design process, not at the end. It has full transparency wherein the information is shared openly and only consensus decisions are implemented.	• Integrated form of agreement between owner, architect/engineer, and contractor. • Owner's business plan as the basic design criteria. • Design and cost of the project stay in balance. • Common understanding of project parameters, objectives. and constraints. • Better relationship among project teams. • Better collaboration. • More certainty of outcome within the agreed upon target cost, schedule, and program. • Superior risk management. • Establishes a collaborative governance structure. • Incorporates lean project delivery.

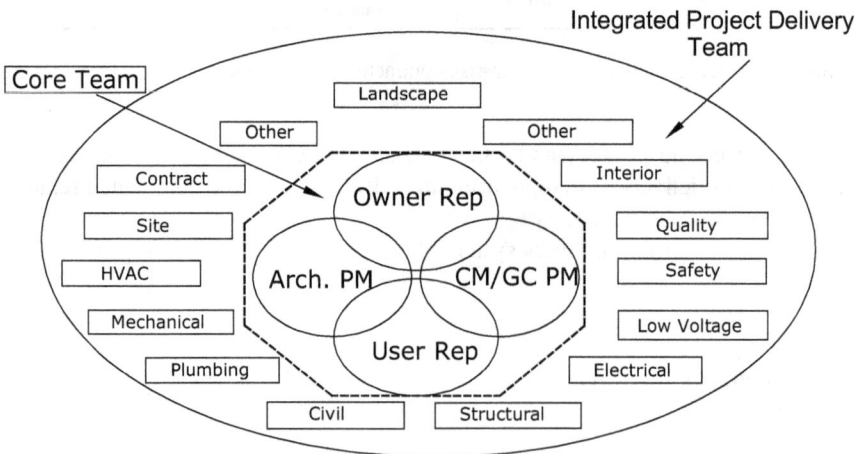

FIGURE 2.10   Integrated project delivery system.

Serial Number	Category	Classification	Sub-Classification	Owner/ Designer	Owner/ Contractor	Designer/ Contractor	Owner/ CM	CM/ Designer	CM/ Contractor
1	Traditional System	Design–Bid–Build	Design–Bid–Build	C	C	F	--	--	--
2	Integrated System	Design–Build	Design–Build	C	C	I	--	--	--
				Contractual Relationship					
3	Management-Oriented System	Construction Management	Agency Construction Manager	C	C	--	C	F	F
			Construction Manager-at-Risk	C	--	C	C	F	C

Note: C – Contractual Relationship, F – Functional Relationship, I – Integral Part of Contract.

FIGURE 2.11 Relationship among project participants.

However, if the owner engages a construction manager or project manager then it becomes four-party contract. Figure 2.11 summarizes the relationship among various project participants.

2.2.2 TYPES OF CONTRACTING/PRICING

A contract is a formalized means of agreement, enforceable by law, between two or more parties to perform certain works or provide certain services against an agreed upon financial incentives to complete the works/services. Regardless of the type of project delivery system selected, the contractual arrangement by which the parties are compensated must also be established. In construction projects, determining how to procure the product is as important as determining what and when. In most procurement activities, there are several options available for purchase, or subcontracts. The basis of compensation type relates to the financial arrangement among the parties as to how the designer or contractor is to be compensated for their services. Following are the most common types of contract/compensation methods:

1. Firm fixed price or lump sum contract
   a. Firm fixed price
   b. Fixed price incentive fee
   c. Fixed price with economic adjustment price

Table 2.9 illustrates the advantages and disadvantages of fixed price or lump sum price contract.

2. Unit price
   a. Specific price for a particular task or unit of work performed by the contractor
   b. In this method each unit of work is to be precisely defined

TABLE 2.9
Fixed Price/Lump Sum Contracts

Serial Number	Project Contracting Type	Main Aspects	Advantages	Disadvantages
1	Fixed price or lump sum	With this type of contract, the contractor agrees to perform the specified work/ services with a fixed price (it is also called lump sum) for the specified and contracted work. Any extra work is executed only upon receipt of instruction from the owner. Fixed price contracts are generally inappropriate for work involving major uncertainties, such as work involving new technologies.	• Low financial risk to the owner. • Total cost of project is known before construction. • Project viability is known before a commitment is made. • Suitable for projects which can be completely designed and whose quantities are definable. • Minimal owner supervision. • Contractor will usually assign its best personnel to the work. • Maximum financial motivation of contractor. • Contractor has to solve his own problems and do so quickly. • Contractor selection by competitive bidding is fairly easy, apart from the deliberate low price.	• High financial risk to the contractor. • Owner bears the risk of poor quality from the contractor trying to maximize the profit within fixed cost. • Contractor's price may include high contingency. • Variation can be time consuming and costly. • Variations (changes) are difficult and costly. • More time taken for bidding and for developing a good design basis. • Contractor will tend to choose the cheapest and quickest solutions, making technical monitoring and strict quality control by the owner essential.

TABLE 2.10
Unit Price Contracts

Serial Number	Project Contracting Type	Main Aspects	Advantages	Disadvantages
1	Unit Price	The owner and contractor agree to structure the contract on specified unit price for the estimated quantities of the work. In unit price contract, the work to be performed is broken into various parts and a fixed price is established for each unit of work. This type of contract is well-suited for repetitive or easily quantifiable tasks. A unit price contract provides benefits to both owners and contractors.	• Complete design definition is not essential for the bidding process. • Suitable for competitive bidding and relatively easy contractor selection subject to sensitivity evaluation. • Typical drawings can be used for bidding. • Bidding is speedy and inexpensive, and an early start is possible. • Flexibility depending on the contract conditions; the scope and quantity of work can be varied and easily adjustable.	• The final cost is not known at the outset. • Additional site staff are needed to measure, control, and report on the cost and the status of the work. • Biased bidding and front-end loading may not be detected.

Table 2.10 illustrates the advantages and disadvantages of unit price contracts.

3. Cost reimbursement contract (cost plus)
   a. Cost plus percentage fee
   b. Cost plus fixed fee
   c. Cost plus incentive fee
   d. Cost plus award fee

Table 2.11 illustrates the advantages and disadvantages of cost reimbursement contracts.

4. Remeasurement contract
   a. Bill of quantities
   b. Schedule of rates
   c. Bill of materials

TABLE 2.11
Cost Reimbursement Contracts

Serial Number	Project Contracting Type	Main Aspects	Advantages	Disadvantages
1	Cost reimbursement contract (Cost plus)	It is a type of contract in which contractor agrees to do the work for the cost related to the project plus an agreed upon amount of fee that covers profits and non-reimbursable overhead costs. Following are different types of cost plus contracts: 1. Cost plus % fee contract 2. Cost plus fixed fee contract 3. Cost plus incentive fee contract	• Suitable to start construction concurrently while design is in progress • Suitable for renovation type of projects • Suitable for projects expecting major changes • Suitable for projects with possible introduction of latest technology • Flexibility in dealing with changes • An early start can be made • Useful where site problems such as trade union actions like delays or disruptions may be encountered • Owner can control all aspects of the work	• Possibility of overspending by the contractor • Difficult to predict final cost. • Project quality is likely to be affected as the fee will be same no matter how low the costs are • Final cost is unknown • Difficulties in evaluating proposals, e.g. L1 may not result in selecting a contractor in achieving lowest project cost • Contractor has little incentive for early completion or cost economy • Contractor may assign its "second division" personnel, make excessive use of agency personnel, or use the job as a training vehicle for new personnel • Biased bidding of fixed fees and reimbursable rates may not be detected

TABLE 2.12
Remeasurement Contracts

Serial Number	Project Contracting Type	Main Aspects	Advantages	Disadvantages
1	Remeasurement contract	It is a contract in which contractor is paid as per the actual quantities of the work done. In this type of contract, contractor agrees to do the work based on one of the following criteria: 1. Bill of quantities, or 2. Schedule of rates, or 3. Bill of material. In this type of contract, payment is linked to measured work completion.	• Suitable for competitive bidding. • Fair competition. • For contractor, low risk. • Suitable for projects where quantities of work cannot be determined in advance of construction. • Fair basis for competition.	• Require adequate breakdown and design definition of work unit. • Final cost not known with certainty until the project is complete. • Additional requirements of administrative staff to measure, control, and report.

Table 2.12 illustrates the advantages and disadvantages of re-imbursement contract.

5. Target price contract

Table 2.13 illustrates the advantages and disadvantages of target price contract.

6. Time and material contract

Table 2.14 illustrates the advantages and disadvantages of time and material contract.

7. Guaranteed maximum price (GMP)
   a. Cost plus fixed fee GMP contract
   b. Cost plus fixed fee GMP and bonus contract
   c. Cost plus fixed fee GMP with arrangement for sharing any cost-saving type contract

TABLE 2.13
Target Price Contracts

Serial Number	Project Contracting Type	Main Aspects	Advantages	Disadvantages
1	Target price contract	A target cost contract is based on the concept of top-down approach which provides a fixed price for an agreed range of out-turn cost around the target. In this type of contract over run or under spend are shared by the owner and the contractor at a predetermined agreed upon percentages.	• Final cost is known. • Contractor may share the savings as per agreed upon percentages. • Flexibility in controlling the work. • Almost immediate start on the work, even without a scope definition. • Encourages economic and speedy completion (up to a point) • Contractor is rewarded for superior performance.	• Tight cost control is required. • Difficult to adjust major variations or cost inflation. • No opportunity to competitively bid the targets. • Difficulty in agreeing on an effective target for superior performance. • Variations are difficult and costly once the target has been established. • If the contractor fails to achieve the targets, it may attempt to prove the owner's fault.

TABLE 2.14
Time and Material Contracts

Serial Number	Project Contracting Type	Main Aspects	Advantages	Disadvantages
1	Time and material contract	A time and material contract has elements of unit price and cost plus type of contract. This type of contract is mainly used for maintenance contract and small projects.	• The owner pays the contractor based on actual cost for material and time as per pre-agreed rates. • Suitable for design services where it is difficult to determine total expected efforts in advance.	• Contractor must be accurate in estimating the price, which normally includes indirect and overhead cost.

2.2.2.1   Guaranteed Maximum Price

In this method, the contractor is compensated for actual costs incurred in connection with design and construction of the project, plus a fixed fee-all subject, however, to a ceiling above which the client is not obligated to pay. A GMP contract specifies a target profit (or fee), a price ceiling (but not for profit ceiling or floor), and a profit (or fee) adjustment formula. These elements are all negotiated at the outset. GMP combines construction management with design/build. With GMP contract, amounts below the maximum are typically shared between the client and the contractor, while the contractor is responsible to absorb the cost above the maximum. Any changes which result from the specific instructions of the owner fall outside the guaranteed price. Cost plus GMP, as it is also known, type of contract is typically used

- When time pressure requires letting of the contract before design development is sufficiently advanced to allow a conventional lump sum type of contract to be fixed
- If this type of contract is likely to be less costly than other types
- Where financing or other constraints preclude the use of alternatives such as two-stage contract of construction management
- If it is impractical to obtain certain types of services with improved delivery or technical performance, or quality without the use of this type of contract.

In this method, the contractor and owner have the knowledge that the drawings and specifications are not complete, and the contractor and the owner agree to work together to complete the drawings and specifications as provided in the contract agreement. This type of contract is weighted heavily in favor of owner. The contractor takes on all the risks in this type of contracting system. Value engineering studies are conducted to identify design alternatives to help the project maintain the budget and schedule. This type of contract needs

- Adequate cost pricing information for establishing a reasonable firm target price at the time of initial contract negotiation
- Contractor's tendering/bidding department should have adequate information to provide necessary data to support the negotiation of final cost and incentive price revision
- High administration cost from owner side to monitor what the contractor is actually spending to get the benefit of under-spending
- Evaluation of minimum two or three proposals for any major subcontract work.

In certain GMP contracts, the owner monitors and controls the contractor's expenses toward the project resources such as construction equipment, machinery, manpower, and staff on a monthly basis by fixing the basic price and profit percentage agreed at the initial stage.

Table 2.15 illustrates the advantages and disadvantages of GMP type of contracts.

TABLE 2.15
Cost Plus Guaranteed Maximum Price Contracts

Serial Number	Project Contracting Type	Main Aspects	Advantages	Disadvantages
1	Cost plus guaranteed maximum price contract	With this type of contract, the owner and the contractor agree to a project cost guaranteed by the contractor as maximum. In this method, the contractor is compensated for actual costs incurred, in connection with design and construction of the project, plus a fixed fee-all subject, however, to a ceiling above which the client is not obligated to pay. This type of contract need adequate cost pricing information for establishing a reasonable firm target price at the time of initial contract negotiation.	• Maximum contract price is certain. • Construction can start at early stage.	• Contractor share the maximum risk. • High administration cost from owner side to monitor what the contractor is actually spending to get the benefit of under-spending. • Contractor's tendering/bidding department should have adequate information to provide necessary data to support negotiation of final cost and incentive price revision.

2.2.3   SELECTION OF PROJECT DELIVERY SYSTEM

Each of the project delivery systems discussed above has advantages and disadvantages. It is a strategic decision which the owner has to take considering the suitability of appropriate system to achieve successful completion of the project. Therefore, while selecting an appropriate project delivery system the owner has to consider the following:

- Size and complexity of the project
- Type of project
- Location of project
- Owner's level of construction expertise (human resources available with owner and owner's knowledge of construction management practices)
- Owner's interest to exert influence/control over the design
- Owner's interest to exert influence/control over the management of planning
- Owner's interest to exert influence/control over the management of construction
- Owner's interest to exert influence/control over the management of project and the end user(s)
- Design
- Schedule
- Budget, funding mechanism
- Quality
- Risk owner can tolerate (risk allocation)

2.2.3.1 Selection of Project Delivery Teams (Designer/Consultant, Contractor)

Construction projects have the involvement of three main parties. These are

1. Owner
2. Designer/Consultant
3. Contractor

Participation involvement of all three parties at different construction phases is required to develop a successful facility/project. These parties are involved at different levels, and their relationship and responsibilities depend on the type of project delivery system and contracting system. Following are the common procurement methods for the selection of project teams:

1. Low bid
   - Selection is based solely on the price.
2. Best value
   a) Total cost
      - Selection is based on total construction cost and other factors.
   b) Fees
      - Selection is based on weighted combination of fees and qualification.
3. Qualification-based selection (QBS)
   - Selection is based solely on qualification.

Table 2.16 illustrates the main aspects of selection methods and selection criteria of the project delivery system team.

TABLE 2.16
Procurement Selection Types and Selection Criteria

Serial Number	Selection Type	Description	Selection Criteria			
			Design–Bid–Build	Design–Build	CM-at-Risk	Integrated Project Delivery
1	Low Bid	Total construction cost including the cost of work is the sole criteria for final selection.	Yes	Yes	Not Typical	Not Typical
2	Best Value					
	2.1 Total cost	Both total cost and other factors are criteria in the final selection.	Yes	Yes	Not Typical	Not Typical
	2.2 Fees	Both fees and qualifications are factors in the final selection.	Not Applicable	Yes	Yes	Yes
3	Qualification-based Selection (QBS)	Total costs of the work are **not** a factor in the final selection. Qualification is the sole factor used in the final selection.	Not Applicable	Yes	Yes	Yes

2.3 QUALITY INFORMATION SYSTEM

Quality Information System (QIS) is an organized method of collecting, analyzing, storing, and reporting the quality-related information for appropriate decision-making.

Figure 2.12 illustrates the QIS process.

QIS is developed based on the organizational requirements to align with goals and objectives to achieve project quality.

QIS can be manual or computerized. The following section discusses the quality in use of computer-aided design (CAD) software.

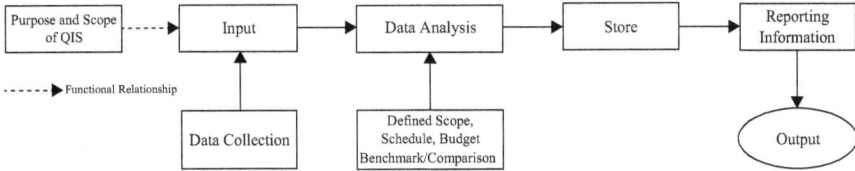

FIGURE 2.12 Quality information system process. (Abdul Razzak Rumane (2017), *Quality Management in Construction Projects*, Second Edition. Reprinted with permission from Taylor & Francis Group.)

2.3.1 QUALITY IN USE OF CAD SOFTWARE

With the advent of the computer age, information automation has become a reality. Computers are now being used in design, presentation of documents, estimation, presentation, analysis, planning, and many other applications.

Engineering design is the partial realization of the designer's concept. Engineering drawing is an abstract universal language used to represent the designer's concept to others. The conventional method of representing an engineering drawing is drafting on paper with a pen or pencil. Manual drafting is tedious and requires a tremendous amount of patience and time. The invention of CAD has had a tremendous effect on the design process, and the computer-aided drafting system has improved drafting efficiency. CAD is a process by which a computer-based program assists in the creation or modification of design. It allows not only passing information more quickly and accurately but also opens the door to integration and automation. A product of the computer era, CAD originated from early computer graphics systems and their development into interactive computer graphics. The 1970s marked the beginning of a new era in CAD—the invention of three-dimensional solid modeling. Since the invention of computer-aided drafting, there has been tremendous progress in CAD, and it is being regularly used in design drafting. In construction projects, CAD is used by all trades such as architectural, structural, mechanical, electrical, infrastructure, and landscape engineers, and others. Design software has helped engineers to produce and modify drawings in the related field of applications and has made it possible to integrate and visualize their overall effects and put them to use to construct the facility. Its application has helped in producing fully coordinated design drawings and avoiding any conflict during construction.

A CAD system consists of three major parts:

1. Hardware: Computer and input/output (I/O) devices
2. Operating system: Software
3. Application software: CAD package

Hardware is used to support the software function. Operating system software is the interface between the CAD application software and the hardware. It is important not only for CAD software but for non-CAD software as well. Application software is the heart of the CAD system. CAD consists of programs that do drafting, engineering analysis, 2-D and 3-D drawings, 3-D modeling, and many other engineering-related functions.

AutoCAD is the most commonly used software in the construction project/ industry. It is the most commonly used software by the construction project professional. It began as a PC-based drafting package running under the MS-DOS environment and gradually evolved into a full-blown CAD system. Its editions are upgraded to meet the current demands of industries. AutoCAD 2010 is the latest version of AutoCAD.

There are numerous off-the-shelf application packages available; it is beyond the scope of this book to list them all, but they are being used in designing systems in construction projects. The following are the most common application software packages used by all the trades:

- General purpose by all trades
  - AutoCAD
  - Microsoft Office
- Planning, scheduling, and controlling
  - Primavera
  - Microsoft Project
- Structural design
  - STADD
  - ETABS
- Mechanical design
  - ELITE
  - Storm CAD, Water CAD, Sewer CAD
- HVAC design
  - HAP
  - APEC
  - HEVACOMP
- Electrical design
  - DIALUX
  - ECODIAL
- Landscape
  - DynaScape
- Infrastructure/roads
- Bently MX
- Eagle Point

Quality in use aims at defining the quality attributes that are important to the end user. In 1991, the International Organization for Standardization (ISO) introduced ISO/IEC 9126 (1991): Software evaluation quality characteristics and guidelines for their use. ISO/IEC 9126 is a four-part model:

Part 1: Quality model
Part 2: External metrics
Part 3: Internal metrics
Part 4: Quality in use metrics

There are six characteristics for both external and internal metrics: functionality, reliability, usability, efficiency, maintainability, and portability. They can be further subdivided into sub-characteristics. The quality in use metrics has four characteristics: effectiveness, satisfaction, productivity, and safety.

ISO/IEC 9126 was superseded in 2005 by ISO/IEC 25000: Software Engineering. All the requirements of ISO/IEC 9126 have been taken care of in this standard.

Table 2.17 lists software quality factors to be considered while selecting a software package.

TABLE 2.17
Software Quality Factors

Sl. No.	Factor	Description
1	Suitability	Whether the application software is suitable and satisfies the intended use and has all the required parameters?
2	Compatibility with the operating system	Application software is usually operating system dependent; therefore, attention must be given to the compatibility of operating system
3	Hardware capability	Whether the RAM and storage memory is sufficient to store the created/modified design data?
4	Reliability	Whether the application software shall perform its intended use with required precision?
5	Acceptability	Whether the designs produced with application software are accepted by the client?
6	Credibility	How much credible the product is?
7	Usability	How easy it is to learn, operate, and transfer input information by translation or interpreting the same to the specific data format?
8	Integrability	Whether the program can be integrated with other application software and how much efforts required for interoperating with other systems?
9	Flexibility	Efforts required to modify an operational system
10	Efficiency	Whether the program can perform the amount of computing resources and code required by a program?
11	Maintainability	How much efforts required for troubleshooting and fix an error?
12	Testability	How much efforts required to test and ensure that the program is performing its intended function?
13	Portability	Efforts required to transfer the program from one hardware to another and to configure the same with new environment
14	Reusability	Whether the program can be used in other applications related to the packaging and scope of the functions that program perform?
15	Safety	How safe is the program to use?

Source: Abdul Razzak Rumane. (2017). *Quality Management in Construction Projects*, Second Edition. Reprinted with permission of Taylor & Francis Group.

2.4 BUILDING INFORMATION MODELING

Building Information Modeling (BIM) is an innovative process of generating a digital database for collaboration and managing building data during its life cycle and for preserving the information for reuse and additional industry-specific applications. It is Autodesk's strategy for the application of information technology to the building industry. It helps in better visualization and clash detection and is an excellent tool to develop project staging plans, study phasing, and coordination issues during the construction project life cycle, preparation of As-Built, and also during maintenance of the project.

BIM uses software and processes to digitally develop building data in a manner that is collaborative and integrative. Using BIM as a robust process and not just a tool has transformed the design and construction industry. Different from computer-aided drafting (CAD), BIM leverages digital information about the building (including materials, furnishings, and equipment requirements) in a searchable database that is additive and scalable, enabling design and construction team members to participate in the virtual co-creation and co-development of a project design and operational requirements. Accurately described as virtual design and construction, BIM facilitates three-dimensional digital development including coordination, estimating, scheduling, and material selection. BIM also creates a rich data source onto which rules-based project studies such as occupancy simulations, fire and life-safety code reviews, and energy use calculations can be applied. The future of BIM will be leveraged in physical facility operations, where model information will be utilized in continuous facility commissioning for operating and maintaining a building.

The advent of BIM and sharing the parametric geometry and information contained in these files during all phases of design and construction have disrupted the traditional incumbent roles within the construction industry. Currently deployed on large and complicated projects, the use of BIM has become the tool of choice for designers and builders for coordination and scheduling.

2.5 SIX SIGMA IN CONSTRUCTION PROJECTS

Six Sigma is, basically, a process quality goal. It is a process quality technique that focuses on reducing variation in the process and preventing deficiencies in product. In a process that has achieved Six Sigma capability, the variation is small compared to the specification limits.

Sigma is a Greek letter—σ—that stands for standard deviation. Standard deviation is a statistical way to describe how much variation exists in a set of data, a group of items, or a process. Standard deviation is the most useful measure of dispersion. Six Sigma means that for a process to be capable at the Six Sigma level, the specification limits should be at least 6 σ from the average point. So, the total spread between the upper specification (control) limit and the lower specification (control) limit should be 12 σ. With Motorola's Six Sigma program, no more than

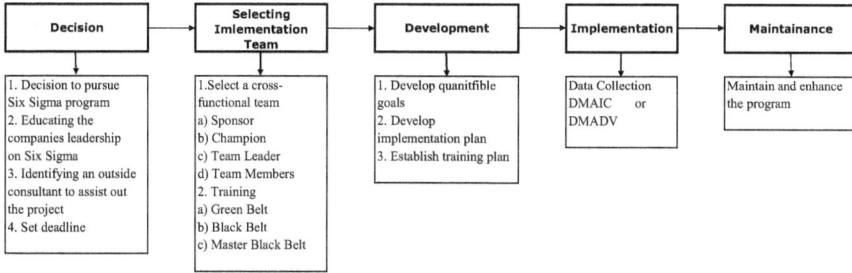

FIGURE 2.13 Six Sigma roadmap. (Abdul Razzak Rumane (2017), *Quality Management in Construction Projects*, Second Edition. Reprinted with permission from Taylor & Francis Group.)

3.4 defects per million fall outside specification limits with a process shift of not more than 1.5 σ from the average or mean. In a process that has achieved Six Sigma capability, the variation is very small. Six Sigma started as a defect reduction effort in manufacturing and was then applied to other business processes for the same purpose.

Six Sigma is a measurement of "goodness" using a universal measurement scale. Sigma provides a relative way to measure improvement. Universal means Sigma can measure anything from coffee mug defects to missed chances to close a sales deal. It simply measures how many times a customer's requirements were not met (a defect), given a million opportunities. Sigma is measured in defects per million opportunities (DPMO). For example, a level of Sigma can indicate how many defective coffee mugs were produced when one million were manufactured. Levels of Sigma are associated with improved levels of goodness. To reach a level of Three Sigma, you can only have 66,811 defects, given a million opportunities. A level of Five Sigma only allows 233 defects. Minimizing variation is a key focus of Six Sigma. Variation leads to defects, and defects lead to unhappy customers. To keep customers satisfied, loyal, and coming back, you have to eliminate the sources of variation. Whenever a product is created or a service performed, it needs to be done the same way every time, no matter who is involved. Only then will you truly satisfy the customer. Figure 2.13 illustrates the Six Sigma roadmap.

2.5.1 SIX SIGMA METHODOLOGY

Six Sigma is an overall business improvement methodology that focuses an organization on

- Understanding and managing customer requirements
- Aligning key business processes to achieve these requirements
- Utilizing rigorous data analysis to minimize variation in these processes

- Driving rapid and sustainable improvement in business processes by reducing defects, cycle time, impact on the environment, and other undesirable variations
- Timely execution

As a management system, Six Sigma is a high-performance system for executing business strategy. It uses the concept of fact and data to drive better solutions. Six Sigma is a top-down solution to help organizations:

- Align their business strategy to critical improvement efforts
- Mobilize teams to attack high-impact projects
- Accelerate improved business results
- Govern efforts to ensure improvements are sustained

Six Sigma methodology also focuses on:

- Leadership Principles
- Integrated Approach to Improvement
- Engaged Teams
- Analytic Tool
- Hard Coded Improvements

Six Sigma methodology has four leadership principles. These are

1. Align
2. Mobilize
3. Accelerate
4. Govern

Teamwork is absolutely vital for complex Six Sigma projects. For teams to be effective, they must be engaged, involved, focused, and committed to meet their goals. Engaged teams must have leadership support.
There are four types of teams. These are

1. Black Belts
2. Green Belts
3. Breakthrough
4. Blitz

For a typical Six Sigma project, four critical roles exist:

1. Sponsor
2. Champion
3. Team Leader
4. Team member

2.5.2 Six Sigma Analytic Tool Sets

Following are the analytic tools used in Six Sigma projects:

2.5.2.1 Ford Global 8D Tool

Ford Global 8D tool is primarily used to bring performance back to a previous level.

2.5.2.2 DMADV Tool Set Phases

DMADV tool is used primarily for the invention and innovation of modified or new products, services, or processes. Using this tool set Black Belts optimize performance before production begins. DMADV is proactive, solving problems before they start. This tool is also called DFSS (Design for Six Sigma). Table 2.18 lists the fundamental objectives of DMADV.

2.5.2.3 DMAIC Tool

DMAIC tool refers to a data-driven quality strategy and is used primarily for improvement of an existing product, service, or process. Table 2.19 lists the fundamental objectives of DMAIC.

2.5.2.4 DMADDD Tool

DMADDD tool is primarily used to drive the cost out of a process by incorporating digitization improvements. These improvements can drive efficiency by identifying non-value-added tasks and using simple web-enabled tools to automate certain tasks and improve efficiency. In doing so, employees can be freed up to work on more value-added tasks. Table 2.20 lists the fundamental objectives of DMADDD.

TABLE 2.18
Fundamental Objectives of Six Sigma DMADV Tool

DMADV	Phase	Fundamental Objective
1	**Define**—What is important?	Define the project goals and customer deliverables (internal and external)
2	**Measure**—What is needed?	Measure and determine customer needs and specifications
3	**Analyze**—How we fulfill?	Analyze process options and prioritize based on capabilities to satisfy customer requirements
4	**Design**—How we build it?	Design detailed process(es) capable of satisfying customer requirements
5	**Verify**—How do we know it will work?	Verify design performance capability

TABLE 2.19
Fundamental Objectives of Six Sigma DMAIC Tool

DMAIC	Phase	Fundamental Objective
1	**Define**—What is important?	Define the project goals and customer deliverables (internal and external)
2	**Measure**—How are we doing?	Measure the process to determine current performance
3	**Analyze**—What is wrong?	Analyze and determine the root cause(es) of the defects
4	**Improve**—What needs to be done?	Improve the process by permanently removing the defects
5	**Control**—How do we guarantee performance?	Control the improved process's performance to ensure sustainable results

TABLE 2.20
Fundamental Objectives of Six Sigma DMADDD Tool

DMADDD	Phase	Fundamental Objective
1	**Define**—Where must we be learner?	Identify potential improvements
2	**Measure**—What's our baseline?	Analog touch points
3	**Analyze**—Where can we free capacity and improve yields?	Task elimination and consolidated operations Value-added/non-value-added tasks Free capacity and yield
4	**Design**—How should we implement?	Future state vision Define specific projects Define draw down timing Define commercialization plans
5	**Digitize**—How do we execute?	Execute Project
6	**Draw Down**—How do we eliminate parallel paths?	Commercialize new process Eliminate parallel path

DMADV PROCESS

DEFINE PHASE: What is important?

(Define the project goals and customer deliverables)
Key deliverables of this phase are

- Establish the goal
- Identify the benefits
- Select project team
- Develop project plan
- Project charter

MEASURE PHASE: What is needed?
(Measure and determine customer needs and specifications)
Key deliverable in this phase is

• Identify specification requirements

ANALYZE PHASE: How do we fulfill?
(Analyze process options and prioritize based on capability to satisfy customer requirements)
Key deliverables in this phase are

• Design generation (data collection)
• Design analysis
• Risk analysis
• Design model (prioritization of data under major variables)

DESIGN PHASE: How do we build it?
(Design detailed process(es) capable of satisfying customer requirements)
Key deliverables in this phase are

• Constructing a detailed design
• Converting CTQs (Critical to Quality) into CTPs (Critical to Process elements)
• Estimate the capabilities of the CTPs in the design
• Preparing a verification plan

VERIFY PHASE: How do we know it works?
(Verify design performance capability)
Key deliverable in this phase is

• Designing a control and transition plan

THE DMAIC PROCESS
The majority of the time, Black and Green Belts approach their projects with the DMAIC analytic tool set, driving process performance to never-before-seen levels.
DMAIC has the following fundamental objectives:

1. Define Phase: Define the project and customer deliverables
2. Measure Phase: Measure the process performance and determine current performance
3. Analyze: Collect, analyze, and determine the root cause(s) of variation and process performance
4. Improve: Improve the process by diminishing defects with alternative remedial
5. Control: Control improved process performance

The DMAIC process contains five distinct steps that provide a disciplined approach to improving existing processes and products through the effective integration of project management, problem solving, and statistical tools. Each step has fundamental objectives and a set of key deliverables, so the team member will always know what is expected of him/her and his/her team.

DMAIC stands for the following:

- Define opportunities
- Measure performance
- Analyze opportunity
- Improve performance
- Control performance

DEFINE OPPORTUNITIES (What is important?)

The Objective of this phase is

To identify and/or validate the improvement opportunities that will achieve the organization's goals and provide the largest payoff, develop the business process, define critical customer requirements, and prepare to function as an effective project team.

Key deliverables in this phase include

- Team charter
- Action plan
- Process map
- Quick win opportunities
- Critical customer requirements
- Prepared team

MEASURE PERFORMANCE (How are we doing?)

The objectives of this phase are

- To identify critical measures that are necessary to evaluate the success or failure, meet critical customer requirements, and begin developing a methodology to effectively collect data to measure process performance.
- To understand the elements of the Six Sigma calculation and establish baseline Sigma for the processes the team is analyzing.

Key deliverables in this phase include

- Input, process, and output indicators
- Operational definitions
- Data collection format and plans
- Baseline performance
- Productive team atmosphere

ANALYZE OPPORTUNITY (What is wrong?)
The objectives of this phase are

- To stratify and analyze the opportunity to identify a specific problem and define an easily understood problem statement.
- To identify and validate the root causes and thus the problem the team is focused on.
- To determine true sources of variation and potential failure modes that lead to customer dissatisfaction.

Key deliverables in this phase include:

- Data analysis
- Validated root causes
- Sources of variation
- Failure modes and effects analysis (FMEA)
- Problem statement
- Potential solutions

IMPROVE PERFORMANCE (What needs to be done?)
The objectives of this phase are

- To identify, evaluate, and select the right improvement solutions.
- To develop a change management approach to assist the organization in adapting to the changes introduced through solution implementation.

Key deliverables in this phase include

- Solutions
- Process maps and documentation
- Pilot results
- Implementation milestones
- Improvement impacts and benefits
- Storyboard
- Change plans

CONTROL PERFORMANCE (How do we guarantee performance?)
The objectives of this phase are

- To understand the importance of planning and executing against the plan and determine the approach to be taken to ensure achievement of the targeted results.
- To understand how to disseminate lessons learned, identify replication and standardization opportunities/processes, and develop related plans.

Key deliverables in this phase include

- Process control systems
- Standards and procedures

- Training
- Team evaluation
- Change implementation plans
- Potential problem analysis
- Solution results
- Success stories
- Trained associates
- Replication opportunities
- Standardization opportunities

Six Sigma methodology is not so commonly used in construction projects; however, DMAIC tool can be applied at various stages in construction projects. These are

1. Detailed Design Stage—To enhance coordination method in order to reduce repetitive work
2. Construction Stage—Preparation of builder's workshop drawings and composite drawings, as it needs a lot of coordination among different trades
3. Construction Stage—Preparation of contractor's construction schedule
4. Execution of works

Impact of Six Sigma Strategy
The Six Sigma strategy affects five fundamental areas of business. These are

1. Process improvement
2. Product and service improvement
3. Customer satisfaction
4. Design methodology
5. Supplier improvement

2.5.3 APPLICATION OF SIX SIGMA DMADV TOOL FOR DESIGN OF A BUILDING

Six Sigma DMADV tool can be used to develop design for a building construction project. Following is an example procedure to develop design for a typical building construction project using the Six Sigma DMADV analytic tool set.
DMADV stands for

1. D → Define
2. M → Measure
3. A → Analyze
4. D → Design
5. V → Verify

DMADV tool

1. Define → What is important?
2. Measure → What is needed?

3. Analyze → How will we fulfill?
4. Design → How do we build it?
5. Verify → How do we know it will work?

DMADV tool has the following usages:

1. DMADV tool is used primarily for the invention and innovation of modified or new products, services, or processes.
2. DMADV tool is used when a product, service, or process is required but does not exist.
3. DMADV is proactive, solving problems before they start. This tool is also called DFSS.
4. Using this tool set Black Belts optimize performance before production begins.

Table 4.3 discussed earlier under Section 2.5.2.2 illustrates the fundamental objectives of DMADV tool.

2.5.3.1 Define Phase

What is important? (Define the project goals and customer deliverables)
Key deliverables of this phase are

- Establish the goal
- Identify the benefits
- Select project team
- Develop project plan
- Develop project charter

2.5.3.1.1 Goal

Develop construction project design using Six Sigma tool to meet owner's needs and satisfy the project quality requirements.

2.5.3.1.2 Benefits

The measurable benefits of adopting this process will

- Minimize design errors
- Minimize omissions
- Reduce design rework
- Reduce risk and liabilities
- Construction project (building) design that will meet owner's needs without rejection of documents by the owner's review team (project manager, construction manager) at the first submission itself and without affecting the construction process thus reducing external failure cost
- It will not affect the construction process thus reducing external failure cost

- Reduce construction overruns
- Increase customer satisfaction
- Improve reputation
- Increase profits

2.5.3.1.3   Selection of Team

The team shall consist of

1. Sponsor
    - Project manager
2. Champion
    - Design manager (civil/structural)
    - Quality manager
3. Team leader
    - Principal engineer (civil/structural)
4. Team members
    - Structural design engineer
    - CAD technician
    - Quality control engineer
    - Quantity surveyor
    - Cost engineer
    - Planner
    - Owner's representative
    - End user

Similar teams can be organized for other trades such as architectural, HVAC, mechanical (P&FF), electrical including low voltage systems, landscape, and external works.

Figure 2.14 illustrates the design management team and their major responsibilities.

2.5.3.1.4   Project Plan

Time frame in the form of a Gantt chart shall be prepared to meet the target dates for

- Data collection and analysis
- Development of concept design
- Submission and approval of concept design for client review and comments
- Development of schematic design
- Submission and approval of schematic design for client review and comments
- Development of detailed design (design development)
- Regulatory/authority approvals
- Preparation of contract documents
- Submission of detailed design and documents for client review and comments
- Submission of the project design and documents (final)
- Review and coordination meetings (both internal and with client)

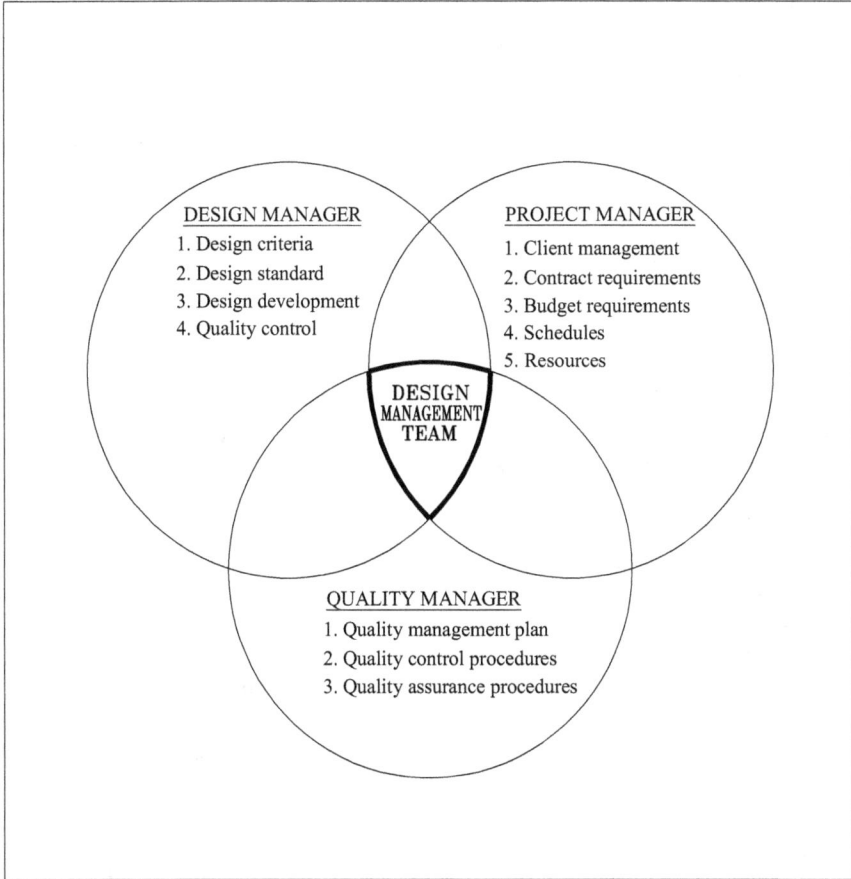

FIGURE 2.14 Design management team. (Abdul Razzak Rumane (2013). *Quality Tools for Managing Construction Project*, CRC Press, Boca Raton FL Reprinted with permission of Taylor and Francis Group.)

- Monitoring of design schedule
- Monitoring and controlling resources

2.5.3.1.5 Project Charter

In this example, the project objective is to create a construction project (building) design within stipulated schedule that

- Will meet owner's need
- Will meet relevant standards, codes, regulatory requirements
- Will be approved by the owner/owner's representative at the first submission without any major comments on the design and specification documents
- The documents shall be accurate and shall not affect construction quality

2.5.3.2   Measure Phase

What is needed? (Measure and determine customer needs and specifications)
Key deliverables in this phase are

1. Matrix of owner's requirements (Terms of Reference—TOR), scope of work
2. Codes and standards to be followed
3. Regulatory requirements
4. Sustainability requirements (economical, environmental, and social)
5. Implementation of LEED requirements
6. Energy conservation requirements
7. Waste management
8. Maintainability
9. Constructability
10. HSE requirements
11. Fire protection requirements
12. Project schedule
13. Project budget
14. Design review procedure by the owner/client
15. Other disciplines' requirements for coordination purposes
16. Procedure to incorporate changes/revisions requested by the client/owner
17. Number of drawings to be produced
18. Duration of completion of project design and submission of documents at different stages/phases of the project
19. Quality system requirements within the organization or portion of organization
20. Any special condition by the client

2.5.3.3   Analyze Phase

How do we fulfill? (Analyze process options and prioritize based on capability to satisfy customer requirements.)
Key deliverables in this phase are

- Data collection
- Prioritization of data under major variables
- Compliance to organization's procedures and guidance

2.5.3.3.1   Data Collection

Data shall be collected at different phases of the construction project life cycle, for design considerations, to meet requirements for the development of following phases:

1. Concept design
2. Schematic/preliminary design
3. Detail design/design development

The following data shall be collected to develop construction project design at different phases:

- Identification of need by the owner
- Identify number of floors
- Identify building usage
- Identify technical and functional capability requirements
- Soil profile and laboratory test of soil
- Identify topography of the project site
- Identify wind load, seismic load, dead load, and live load
- Identify existing services passing through the project site
- Identify existing roads and structure surrounding the project site
- Identify environmental compatibility requirements
- Identify energy conservation requirements
- Identify sustainability requirements
- Identify regulatory/authority requirements
- Identify codes and standards to be followed
- Identify social responsibility requirements
- Identify health and safety features
- Identify fire protection requirements
- Identify aesthetic requirements
- Identify zoning requirements
- Identify project constraints
- Identify ease of constructability
- Identify critical activities during construction
- Identify method statement requirements
- Identify requirements of other disciplines for coordination purpose
- Identification of team members
- Project time schedule
- Financial implications and resources for the project
- Identify cost effectiveness of the project
- Identify 3D information areas
- Identify suitable software program
- Identify number of drawings to be produced
- Identify milestones for the development of each phase of design

2.5.3.3.2   *Arrangement of Data*

The generated data shall be prioritized in an orderly arrangement under the following major variables during designing of concept, schematic, and design development phases:

- Owner's need
- Regulatory compliance
- Sustainability
- Safety
- Constructability

2.5.3.4   Design Phase

How do we build it? (Design detailed process(es) capable of satisfying customer requirements.)

Key deliverables in this phase are

- Development of concept design
- Development of schematic design
- Development of detailed design

2.5.3.4.1   Development of Concept Design

While developing the concept design, designer shall consider the following:

- Project goals/owner needs
- Number. of floors
- Usage of building
- Technical and functional capability
- Regulatory requirements
- Authorities requirements
- Environmental compatibility
- Sustainability requirements
- Constructability
- Health and safety
- Cost effectiveness over the entire life cycle of project

2.5.3.4.2   Development of Schematic Design

While developing the schematic design, designer shall consider the following:

- Concept design deliverables
- Site location in relation to the existing environment
- Building structure
- Floor grade and system
- Tentative size of columns, beams
- Stairs
- Roof
- Authorities' requirements
- Energy conservation issues
- Available resources
- Environmental issues
- Sustainability requirements
- Requirements of all stakeholders
- Optimized life cycle cost (value engineering)
- Constructability
- Functional/aesthetic aspect
- Services requirements
- Project schedule

- Project budget
- Preliminary contract documents (outline specifications)

2.5.3.4.3 Development of Detail Design

While developing the detail design, designer shall consider the following:

- Schematic design deliverables
- Authorities' requirements
- Energy conservation issues
- Environmental issues
- Sustainability requirements
- Requirements of all stakeholders
- Environmental compatibility
- Available resources
- Number of floors
- Property limits/surrounding areas
- Excavation
- Dewatering
- Shoring
- Backfilling
- Substructure
- Design of foundation based on field and laboratory tests of soil investigation
- Subsurface profiles and subsurface conditions, and subsurface drainage
- Co-efficient of sliding on foundation
- Degree of difficulty in excavation
- Method of protection below grade concrete members against the impact of soil and groundwater
- Geotechnical design parameters
- Design load such as dead load, live load, and seismic load
- Grade and type of concrete
- Type of footings
- Type of foundation
- Energy efficient foundation
- Size of bars for reinforcement and the characteristic strength of bars
- Clear cover for reinforcement
- Reinforcement bar schedule, stirrup spacing
- Superstructure
- Columns
- Walls
- Stairs
- Beams
- Slab
- Parapet wall
- Height of each floor
- Beam size and height of beam

- Location of columns in coordination with architectural requirements
- Openings for services
- Deflection which may cause fatigue of structural elements, crack or failure of fixtures, fittings or partitions or discomfort of occupants
- Movement and forces due to temperature
- Equipment vibration criteria
- Expansion joints
- Insulation
- Concrete tanks (water storage)
- Services requirements (shafts, pits)
- Shafts and pits for conveying system
- Building services to fit in the building
- Coordination with other trades and conflict resolution
- Calculations required as per contract requirements

2.5.3.4.5 Preparation of Contract Documents

Contract documents shall be prepared as per MasterFormat ® (2016 edition).

2.5.3.5 Verify Phase

How do we know it works? (Verify design performance capability.)
Key deliverables in this phase are

- Review and check the design for quality assurance using thorough itemized review checklists to ensure that design drawings fully meet the owner's objectives/goal
- Review and check contract documents (design drawings, specifications, and contract documents)
- Check for accuracy
- Check calculations
- Review studies and reports
- Review discipline requirements
- Review inter-discipline requirements and conflict
- Review constructability
- Review for sustainability compliance
- Management review

After verification, the documents can be released for submission.

In case of any comments from the client/client's representative, the design shall be reviewed and modified accordingly.

2.5.4 APPLICATION OF SIX SIGMA DMADV TOOL FOR CONSTRUCTION SCHEDULE

The contractor's construction schedule (CCS) is an important document used during the construction phase. It is used to plan, monitor, and control project activities and resources. The document is voluminous and important. It has to be prepared with accuracy in order to follow the work progress without deviation from the milestones set up in the contract documents. Generally, the project interim payment to the contractor is linked to the approval of the CCS. The contractor is not paid unless the CCS is approved by the construction manager/project manager/consultant.

In most cases, contractors experience problems with getting the CCS approved, at the very first submission, from the construction manager/project manager/consultant. It could be rejected if it does not meet the specifications. Therefore, the contractor has to put all effort into collecting relevant data to be fed to develop the CCS.

2.5.4.1 The DMADV Process

The following is an example procedure to develop the CCS using the Six Sigma DMADV analytic tool set. The DMADV method is used primarily for the invention of modified or new products, services, or processes. DMADV stands for

1. Define →What is important?
2. Measure →What is needed?
3. Analyze →How will we fulfill?
4. Design →How do we build?
5. Verify →How do we know it will work?

2.5.4.1.1 Define Phase (What Is Important?)

The objective of this phase is to define the project goals and customer deliverables.

The key deliverables of this phase are

- Establish the goal
- Identify the benefits
- Select project team
- Develop project plan

Goal: Develop CCS using Six Sigma tools.

Benefits: The measurable benefits in adopting this process will result in CCS that will meet all the requirements of the specifications and shall be approved by the construction manager/project manager/consultant at the first submission itself. This will reduce the repetitive work and help implement the schedule right from the early stage of the project.

Selection of team: The team shall consist of

a. Sponsor—Project manager
b. Champion—Construction manager
c. Team leader—Planning and control manager
d. Team members—Planning engineer, cost engineer, and one representative from each subcontractor

Project plan: Time frame in the form of a Gantt chart shall be prepared to meet the target dates for submitting the CCS.

2.5.4.1.2　Measure Phase (What Is Needed?)

The objective of this phase is to measure and determine customer needs and specifications

The key deliverable in this phase is

- Identify specification requirements

The following are the requirements listed in most contract documents.

The contractor has to submit the construction schedule in a bar chart time-scaled format to show the sequence and interdependence of activities required for the complete performance of all items of work under the contract. The contractor shall use a computerized precedence diagram critical path method (CPM) in preparation of CCS. The schedule shall include, but not be limited to, the following:

1. Project site layout
2. Concise description of the work
3. Milestones (contractual milestones or constraints)
4. Number of working days
5. Work breakdown structure (WBS) activities shall consist of all those activities that take time to carry out execution/installation and on which resources are expended
6. Construction network of project phases (if any), including various subphases
7. Construction network of the project arrangements (activities) and sequence
8. Time schedules for various activities in a bar chart format
9. The minimum work activities to be included in the program shall include items stated in the bill of quantity (BOQ)
10. WBS activities shall consist of all those activities that take time to carry out execution/installation and on which resources are expended
11. Early and late finish dates
12. Time schedule for critical path
13. Schedule text report showing activity, start and finish dates, total float, and relationship with other activities
14. Summary schedule report showing the number of activities, project start, project finish, number of relations, open ends, constraints, and milestones
15. Total float of each activity
16. Cost loading
17. Expected progress cash flow S-curve
18. Resource-loaded S-curve
19. Manpower loading
20. Labor and crew movement and distribution
21. Resource productivity schedule
22. The number of hours per shift
23. Average weekly usage of manpower for each trade
24. Resource histogram showing the manpower required for different trades per time period for each trade (weekly or monthly)

25. Equipment and machinery loading
26. Schedule of mobilization and general requirements
27. Schedule of subcontractors and suppliers' submittal and approval
28. Schedule of materials submittals and approvals
29. Schedule of long lead materials
30. Schedule of procurement
31. Schedule of shop drawings submittals and approvals
32. Regulatory/authorities' requirements
33. Schedule of testing, commissioning, and handover
34. Expected cash flow for executed work (during progress of work)

2.5.4.1.3 Analyze Phase (How Will We Fulfill?)

The objective of this phase is to analyze process options and prioritize based on the capability to satisfy customer requirements.

The key deliverables in this phase are

- Data collection
- Prioritization of data under major variables

Data collection: The objectives of this process are to

1. Identify milestone dates and constraints
2. Identify project calendar
3. Identify resource calendar
4. Review contract conditions and technical specifications
5. Identify mobilization requirements
6. Identify project method statement
7. Identify subcontractors/suppliers
8. Identify materials requirements
9. Identify long lead items
10. Identify procurement schedule
11. Identify shop drawing requirements
12. Identify regulatory/authorities' requirements
13. Identify WBS activities using BOQ
14. Relate WBS activities with BOQ and contract drawings
15. Identify zoning/phasing
16. Identify codes for all activities per contract document divisions/sections per the Construction Specifications Institute (CSI) format
17. Identify volume of work for each activity
18. Identify duration/time schedule of each activity
19. Identify early and late finish dates
20. Identify critical activities and its effect on critical path
21. Identify logical relationship
22. Identify sequencing of activities
23. Identify project progress cash flow (work in place)

24. Identify manpower resources with productivity rate
25. Identify equipment and machinery
26. Identify project constraints such as access, logistics, delivery, seasonal, national, safety, existing workflow discontinuity, and proximity of adjacent concurrent work
27. Identify testing, commissioning, and handover requirements
28. Identify special inspection requirements
29. Identify closeout requirements
30. Identify and include items not listed in the specifications but are important for project scheduling
31. Identify suitable software program
32. Identify submittal requirements

Arrangement of data: The generated data can be prioritized in an orderly arrangement under the following major variables:

1. Milestones
2. WBS activities
3. Time schedule
4. General requirements
5. Resources
6. Engineering
7. Cost loading

Figure 2.15 illustrates these variables along with related sub-variables arranged in the form of the Ishikawa diagram.

2.5.4.1.4   Design Phase (How Do We Build It?)

The objective of this phase is to design detailed processes capable of satisfying customer requirements.

The key deliverable in this phase is

• Preparation of program using suitable (specified) software program

The project and control manager can prepare the CCS based on the collected data and sequence of activities.

2.5.4.1.5   Verify Phase (How Do We Know It Will Work?)

The objective of this phase is to verify design performance capability.

The key deliverables in this phase are

• Review the schedule by the team members to ascertain that all the required elements are included for compliance with specification requirements.
• Submit CCS to construction manager/project manager/consultant.
• Update the schedule as and when required.

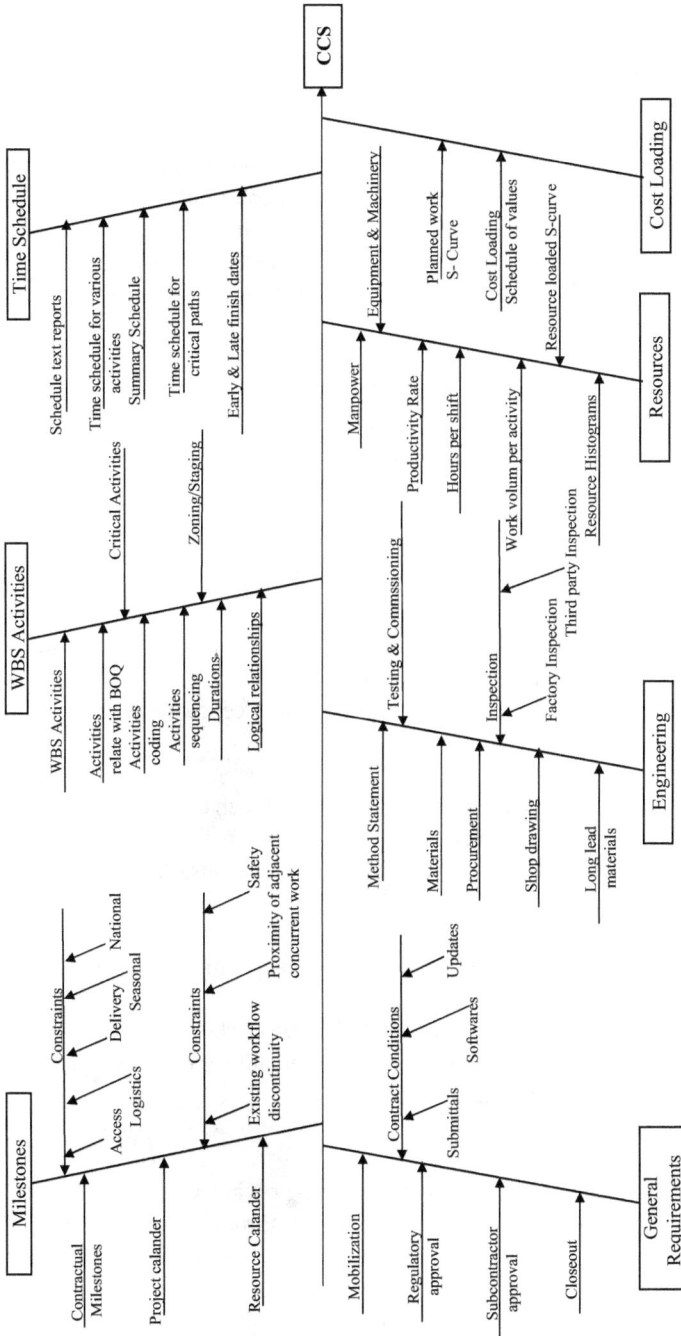

FIGURE 2.15 Ishikawa diagram for CCS data. (Abdul Razzak Rumane (2017), *Quality Management in Construction Projects*, Second Edition. Reprinted with permission from Taylor & Francis Group.)

2.5.5 Application of Six Sigma DMAIC Tool for Concrete Works

Construction projects are unique and non-repetitive in nature and need specific attention to maintain quality. Construction activities mainly consist of construction of concrete foundations, footings, beams, walls, slabs, roofing, finishes, furnishings, conveying systems, electro-mechanical services, low voltage systems, landscape works, and external works.

In construction, an activity may be repeated at various stages, but it is done only one time for a specific work. Therefore, it has to be done right from the onset.

In most cases, contractors experience problems with the approval of executed works, at the very first submission of checklist. In order to achieve "Zero-Defect" policy during construction process, and to avoid rejection of executed works, the contractor has to take the following measures:

- Execution of works as per approved shop drawings
- Use approved material
- Follow approved method of statement
- Conduct continuous inspection
- Employ trained workforce
- Coordinate requirements of other trades

DMAIC tool is primarily used

- For improvement of an existing product, service, or process
- When a product, service, or process is failing to meet customer requirements or is not performing adequately

Using this toolset Black Belts and Green Belts approach their projects driving process performance to never-before-seen levels.

The Six Sigma DMAIC tool can be applied at various stages in construction projects. These are

- Detailed Design Stage—To enhance coordination method in order to reduce repetitive work
- Construction Stage—Preparation of builder's workshop drawings and composite drawings, as it needs lot of coordination among different trades
- Construction Stage—Preparation of CCS
- Execution of works

DMAIC process contains five distinct steps that provide a disciplined approach to improve existing process and processes and products through the effective integration of project management, problem solving, and statistical tools. Each step is designed to ensure that

- Companies apply the breakthrough strategy in a systematic way
- Six Sigma projects are systematically designed and executed
- Incorporating the results of these projects into running the day-to-day business

DMAIC stands for

1. D → Define
2. M → Measure
3. A → Analyze
4. I → Improve
5. C → Control

DMAIC Tool

1. Define → What is important?
2. Measure → How we are doing?
3. Analyze → What is wrong?
4. Improve → What needs to be done?
5. Control → How do we guarantee performance?

Table 2.19 discussed earlier under Section 2.5.2.3 illustrates the fundamental objectives of DMAIC tool.

The following is an example procedure to develop quality management system for the execution of concrete structural works in construction projects using Six Sigma DMAIC tool.

2.5.5.1 Define Opportunities: What Is Important?

The objectives of this phase are

- To identify and/or validate the improvement opportunities that will achieve the organization's goals and provide largest payoff, develop the business process.
- To define critical customer requirements and prepare to function as an effective project team.

Key deliverables of this phase are

- Team charter
- Action plan
- Process map
- Quick win opportunities
- Critical customer requirements
- Develop project plan
- Prepared team

2.5.5.1.1 Team Charter

The scope of team charter is to

- Ensure structural concrete works are executed without any defects and concrete strength is as specified
- Reduce rejection of executed works by the supervisor/independent testing agency

2.5.5.1.2 Action Plan

The objective is to ensure that each activity during the execution of structural concrete works is carried out as specified and as per standards and codes by ensuring that

- Work follows approved shop drawings
- Concrete mix as per approved sample
- Proper transportation of concrete ready mix from batching plant
- Follow casting procedures and method statement

2.5.5.1.3 Process Map (Sequence of Execution of Works)

The structural concrete works mainly consist of the following activities:

- Form work
- Reinforcement work
- Concrete
- Concrete pouring/casting
- Curing

Figures 2.16 and 2.17 illustrate Process Maps.

2.5.5.1.4 Critical Customer Requirements

- Concrete quality as specified
- Control of the design mix as per approved sample
- Concrete casting to comply with specified strength

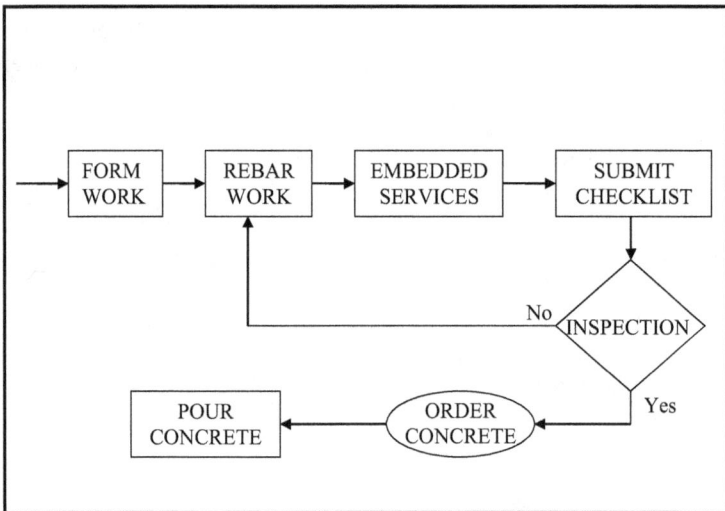

FIGURE 2.16 Flowchart for concrete casting process. (Abdul Razzak Rumane. (2017). *Quality Management in Construction Projects*, Second Edition. Reprinted with permission of Taylor & Francis Group.)

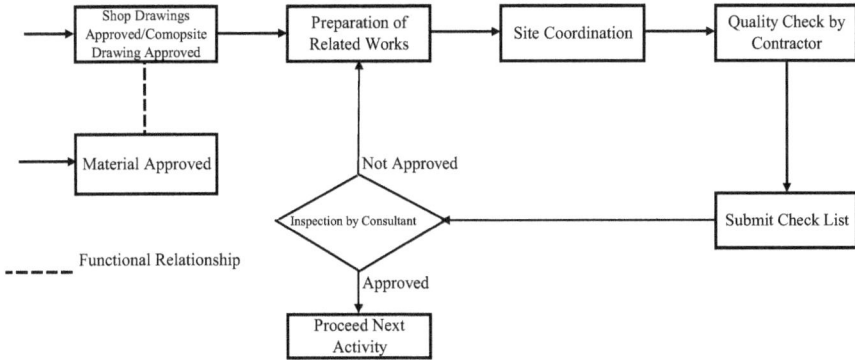

FIGURE 2.17 Sequence of execution of works. (Abdul Razzak Rumane (2017), *Quality Management in Construction Projects*, Second Edition. Reprinted with permission from Taylor & Francis Group.)

- Concrete thickness and level as specified
- Concrete finishes as specified
- Temperature during casting
- Concrete curing
- Laboratory testing of sample cubes during casting
- Construction joints

2.5.5.1.5   Develop Project Plan

Project plan in the bar chart format (time frame) shall be prepared to complete the process within three to five concrete castings to ensure executed works are not rejected and rework is not required for the remaining works.

2.5.5.1.6   Prepared Team

The team shall be selected taking into consideration the qualification of team members, background, and experience and shall consist of

1. Sponsor—Contractor's project manager
2. Champion—Site structural engineer
3. Team leader—Quality control engineer
4. Team members—Coordination engineer, MEP co-coordinator, civil/structural foreman (supervisor), lab technician, batching plant technician

Each member shall be notified about his role, responsibilities, and authority to perform the job.

2.5.5.2   Measure Performance: How Are We Doing?

The objectives of this phase are

- To identify critical measures that are necessary to evaluate success or failure, meet customer requirements

- To establish baseline for the process the team is analyzing
- To measure the process to determine current performance, how exactly the process operates
- To quantify the problem (s)

Key deliverables in this phase are

- Input, process, and output indicators
- Operational definitions
- Data collection format and plans
- Baseline performance
- Productive atmosphere

2.5.5.2.1 Input, Process, and Output Indicators

There are three main areas which may have variations in the structural concrete works. These are

1. Form work
2. Reinforcement
3. Concrete casting

Create process map in detail exactly as it exists for these areas and identify

- What are key Input variables?
- What are key Process variables?
- What are key Output variables?

The process map would include time, people, and material elements to ensure that the current state is clearly understood.

2.5.5.2.2 Operational Definitions

An operational definition is a specific description of the defect, process, product, and/or service to be measured.

Define a specific process or set of validation tests to be used to identify the characteristics you would like to measure.

It is a very clear and very precise explanation of items being measured.

There are four criteria used in validating and operational definition. These are

1. The requirements to be measured must be agreed upon
2. The method of statement must be agreed upon
3. There must be agreement on what the definition will not include
4. Customer must agree with the team that operational definition is appropriate

2.5.5.2.3 Data Collection Format and Plans

In the definition phase, Critical Customer Requirements are developed.

Data collection should relate both to the problem statement and what the customer considers to be CTQ.

The data will then be graphed or charted to obtain a visual representation of the data.

If the team was collecting error data, then a Pareto Chart would be a likely graphical choice.

If a trend chart is needed to show how the process reacts over time, Histograms are another way to observe the process data.

Another widely utilized tool is the Control Chart.

Other quality tools can be used depending on the need.

This data shall be used both as baseline data for improvement efforts and to estimate the current state of process Sigma.

This will be a relative indicator of how close the current process is to delivering zero defects.

Following are the main sources of variations:

- Material
- Method
- Manpower
- Machine
- Measurement
- Environment

The causes of variations are

1. Chance/inherent causes
2. Special/assignable causes

2.5.5.2.4 Baseline Performance

Measure baseline process performance (process capability, yield, Sigma level).

2.5.5.3 Analyze Opportunity: What Is Wrong?

The objectives of this phase are

- Stratify and analyze the opportunity to identify a specific problem and define an easily understood problem statement
- Determine true sources of variation and potential failure modes that lead to customer dissatisfaction

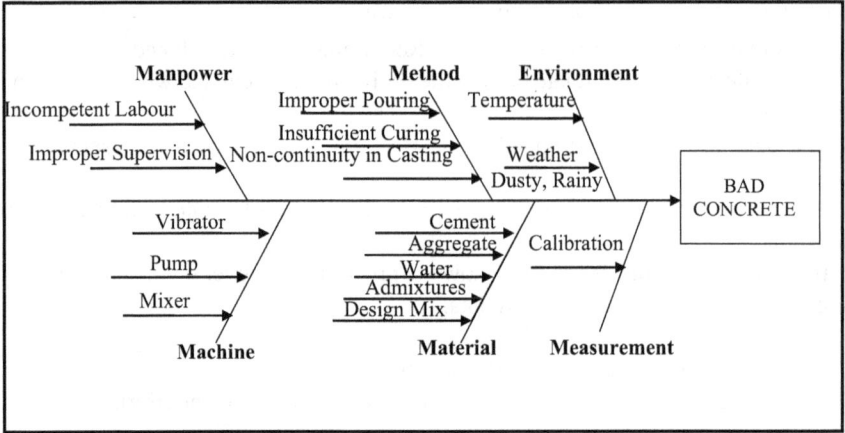

FIGURE 2.18 Cause-and-effect diagram for bad concrete.

Key deliverables in this phase are

- Data analysis
- Validated root causes
- Source of variation
- Failure modes and effects analysis (FMEA)
- Problem statement
- Potential solutions

Figure 2.18 illustrates the root cause analysis diagram.

2.5.5.3.1 Validate Root Causes

Identify the source of variation.

2.5.5.3.2 Problem Statement

To achieve concrete strength as specified without any rework and rejection by the testing agency/consultant.

2.5.5.3.3 Potential Solutions

Identify how problems shall be eliminated.

2.5.5.4 Improve Performance: What Needs to Be Done?

The objectives of this phase are

- To identify, evaluate, and select the right improvement solutions
- To develop a change management approach to assist the organization in adapting to the changes introduced through solution implementation

Key deliverables in this phase are

- Solutions
- Process maps and documentation
- Pilot results
- Implementation milestone
- Improvement impacts and benefits
- Storyboard
- Change plans

2.5.5.4.1   Solutions
- List out solutions

2.5.5.4.2   Process Maps and Documentation
- Establish and revise process map

Figure 2.19 is the revised process for casting.

2.5.5.4.3   Pilot Results
- Run the process as per revised process map and conduct pilot study and note the results.

2.5.5.4.4   Improvement Impacts and Benefits
- Identify the impact on the process due to the application of revised process.

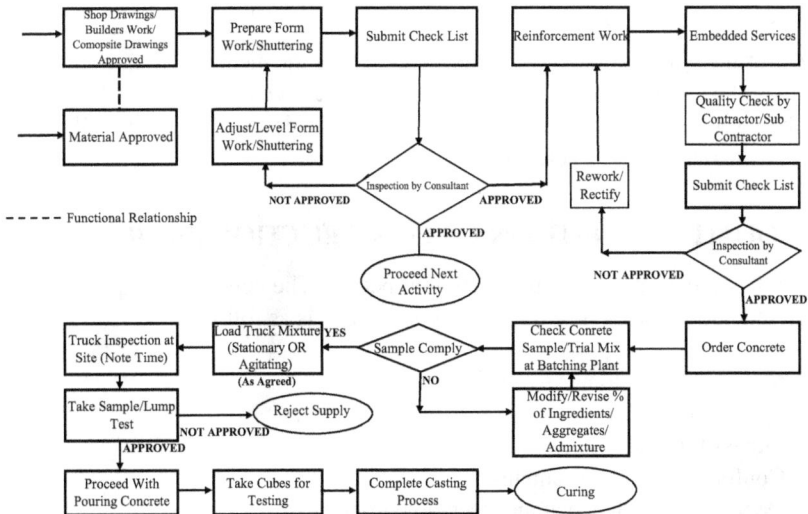

FIGURE 2.19   Sequence of execution of concrete works. (Abdul Razzak Rumane. (2013). *Quality Tools for Managing Construction Projects*. Reprinted with permission from Taylor & Francis Group.)

2.5.5.5   Control Performance: How Do We Guarantee Performance?

The objectives of this phase are

- Understand the importance of planning and executing against the plan
- Understand how to disseminate lessons learned

Key deliverables in this phase are

- Process control systems
- Standards and procedure
- Training
- Team evaluation
- Change implementation plan
- Potential problem analysis
- Solution results
- Success stories
- Trained associates
- Replication opportunities
- Standardization opportunities

2.5.5.5.1   *Process Control Systems*
Follow the revise concrete casting process.

2.5.5.5.2   *Training*
Train team members with the new process.

2.5.5.5.3   *Standardization*
Check results with the new process and standardize the method.

Note: The system is an example system to improve concrete quality. It may not be possible to collect large amount of data; however, the described system will definitely improve the quality and reduce rework.

2.6   QUALITY DEFINITION OF CONSTRUCTION PROJECTS

Quality has different meanings to different people. The definition of quality relating to manufacturing, processes, and service industries is as follows:

- Meeting the customer's need
- Customer satisfaction
- Fitness for use
- Conforming to requirements
- Degree of excellence at an acceptable price

The International Organization for Standardization (ISO) defines quality as "the totality of characteristics of an entity that bears on its ability to satisfy stated or implied needs."

However, the definition of quality for construction projects is different from that of manufacturing or services industries as the product is not repetitive, but unique piece of work with specific requirements. In the case of mass production and batch-oriented production systems, quality can be achieved by getting feedback from the process by observing the actual performance and regulating the process to meet the established standards. Whereas, because of the non-repetitive nature of construction projects, it is not possible to compare actual performance of the project as past experience may be of limited value. The quality management of manufactured products is performed by the manufacturer's own team and has control over all the activities of the product life cycle, whereas construction projects have diversity of interaction and relationship between owners, architects/engineers, and contractors. Construction projects differ from manufacturing or production. Construction projects are custom-oriented and custom designed having specific requirements set by the customer/owner to be completed within a finite duration and assigned budget. Every project has elements that are unique. No two projects are alike. It is always the owner's desire that his project should be unique and better. To a great extent, each project has to be designed and built to serve a specified need. Construction project is custom than a routine and repetitive business.

Quality in construction projects encompasses not only the quality of products and equipment used in the construction but also the total management approach to completing the facility per the scope of works to customer/owner satisfaction within the budget and in accordance with the specified schedule to meet the owner's defined purpose. The nature of the contracts between the parties plays a dominant part in the quality system required from the project, and the responsibility for fulfilling them must therefore be specified in the project documents. The documents include plans, specifications, schedules, BoQ, and so on. Quality control in construction typically involves ensuring compliance with minimum standards of material and workmanship in order to ensure the performance of the facility according to the design. These minimum standards are contained in the specification documents. For the purpose of ensuring compliance, random samples and statistical methods are commonly used as the basis for accepting or rejecting work completed and batches of materials. Rejection of a batch is based on nonconformance or violation of the relevant design specifications.

Quality in construction is achieved through the complex interaction of many participants in the facilities development process. In the case of construction projects, an organizational framework is established and implemented mainly by three parties: owner, designer/consultant, and contractor. Project quality is the result of aggressive and systematic application of quality control and quality assurance.

The quality plan for construction projects is part of the overall project documentation consisting of the following:

1. Well-defined specification for all the materials, products, components, and equipment to be used to construct the facility
2. Detailed construction drawings
3. Detailed work procedure
4. Details of the quality standards and codes to be compiled

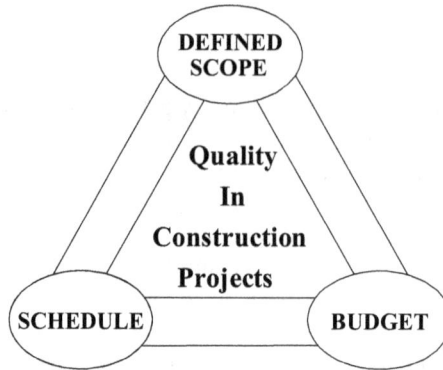

FIGURE 2.20   Construction project quality trilogy.

5. Project completion schedule
6. Cost of the project
7. Manpower and other resources to be used for the project

These definitions when applied to construction projects relate to the contract specifications or owner/end user requirements to be formulated in such a way that construction of the facility is suitable for the owner's use or meets the owner's/user's requirements.

Based on the above, quality of construction projects can be defined as: Construction project quality is the fulfillment of owner's needs as per defined scope of works within a budget and specified schedule to satisfy owner's/user's requirements. The phenomenon of these three components can be called as "Construction Project Trilogy" and is illustrated in Figure 2.20.

2.6.1   QUALITY PRINCIPLES IN CONSTRUCTION PROJECTS

Construction projects are mainly capital investment projects. They are customized and non-repetitive in nature. Construction projects have become more complex and technical, and the relationships and the contractual grouping of those who are involved are also more complex and contractually varied. Quality in construction is achieved through a complex interaction of many participants in the facilities development process.

Construction projects comprised of a cross-section of many different participants. These participants are both influenced and depend on each other in addition to "other players" involved in the construction process. Traditional construction projects have involvement of three main groups. These are as follows:

1. Owner
2. Designer (Consultant)
3. Contractor

The participation and involvement of all three parties at different levels of construction phases of an oil and gas project are required to develop a quality system and application of quality tools and techniques. In oil and gas projects, the project owner engages PMC (project management consultant). PMC plays an important role in the development of oil and gas projects. Construction project quality has three main components. These are

1. Scope
2. Schedule
3. Cost (Budget)

In order to achieve a successful project as per owner/end user satisfaction, project documents are to be formulated in such a way that the construction of project is suitable for owner's use/end user's use or meet the owner's requirements. An ISO document has listed eight quality management principles (CLIPSCFM) on which the quality management system standards of the revised ISO 9000:2000 series are based. These are as follows:

Principle 1—Customer focus
Principle 2—Leadership
Principle 3—Involvement of people
Principle 4—Process approach
Principle 5—System approach to management
Principle 6—Continual improvement
Principle 7—Factual approach to design making
Principle 8—Mutual beneficial supplier relationship

With the application of various quality principles, tools, and methods by all the participants at different stages of construction project, rework can be reduced resulting in savings in the project cost and making the project qualitative and economical. This will ensure completion of construction and making the project most qualitative, competitive, and economical. In order to improve construction project quality and to eliminate/reduce unsatisfactory relations between project owner, designer, and contractor, Table 2.21 summarizes quality principles that are applicable to construction projects. These principles are based on eight quality principles as per ISO 9000:2000.

2.7   QUALITY MANAGEMENT IN CONSTRUCTION PROJECTS

Quality management in construction addresses the management of project and the product of the project and all the components of the product. It also involves incorporation of changes or improvements, if needed. Construction project quality is the fulfillment of owner's needs as per the defined scope of works and as per the specified schedule and within the budget to satisfy owner's/user's requirements.

Quality management system in construction projects mainly consists of

• Quality management planning (plan quality)
• Quality assurance (perform quality assurance)
• Quality control (perform quality control)

TABLE 2.21
Principles of Quality in Construction Projects

Principle	Construction Projects' Quality Principle
Principle 1 (Customer Focus)	1.1 Designer (Consultant) is responsible for providing owner's requirements explicitly and clearly defining the standards of the end products and their compliance in the contract documents.
	1.2 Engineering design should include the process, process equipment, engineering systems requirements clearly and without any ambiguity for ease of operation.
	1.3 The project and end products should satisfy owner's need, requirements, and suitable for intended usage.
Principle 2 (Leadership)	2.1 Owner, designer, consultant, contractor are fully responsible for application of quality management system to meet customer requirements and strive to exceed customer expectations by complying with defined scope of work in the contract documents.
	2.2 Every member of the project team should exert collaborative efforts in all the functional areas to improve the quality of project.
Principle 3 (Involvement/ Engagement of People)	3.1 Each member of project team should participate and fully involve as per their abilities in all the functional areas by adhering to team approach and coordination to continuously improve quality of the project.
Principle 4 (Process Approach)	4.1 Contractor to build the facility as stipulated in the contract documents, plan, specifications as per the approved schedule and within agreed upon budget to meet owner's objectives.
	4.2 Contractor should study all the documents during tendering/ bidding stage and submit his proposal taking into consideration all the requirements specified in the contract documents and identifying, understanding, and managing interrelated processes as a system in achieving the specified product output.
	4.3 Contractor is responsible to provide all the resources, manpower, material, equipment to build the facility as per specifications to produce the specified products.
	4.4 Contractor to check executed/installed works to confirm that works have been performed/executed as specified, using specified/approved materials, approved shop drawings, installation methods and specified references, codes, standards to meet intended use.
Principle 5 (System Approach to Management)	5.1 Contractor to prepare contractor's quality control plan (CQCP) and follow the same to insure meeting the performance standards specified in the contract documents.
	5.2 Method of payments (work progress, material, equipment, etc.) to be clearly defined in the contract documents. Rate analysis of Bill of Quantities (BOQ) or Bill of Materials (BOM) item to be agreed before signing of contract.
	5.3 Contract documents should include a clause to settle the dispute arising during construction stage.

TABLE 2.21 (Continued)
Principles of Quality in Construction Projects

Principle	Construction Projects' Quality Principle
Principle 6 (Continual Improvement)	6.1 Contractor shall follow the submittal procedure specified in the contract documents for detailed design, procurements, checklists, inspection and testing procedures as per the communication matrix. Review the contents of transmittals and executed works prior to submission for approval.
Principle 7 (Factual Approach to Evidence-Based Design Making)	7.1 Contractor shall follow an agreed upon quality assurance and quality control plan. Consultant, PMC shall be responsible to oversee the compliance with contract documents and specified standards and codes.
	7.2 Contractor is responsible to construct the facility to produce the products as specified and use the material, products, systems, equipment, and methods which satisfy the specified requirements (Factual Approach to design making).
Principle 8 (Mutual beneficial relationship) Relationship Management	8.1 Contractor/All team members should participate and put collective efforts to perform the works as per agreed upon construction program and handover the project as per contracted schedule to meet the owner's requirements.
	8.2 All team members should focus on participative management and strong operational accountability at the individual contributory level to follow principles of Total Quality Management.

Source: Abdul Razzak Rumane. (2021). *Quality Management in Oil and Gas Projects*. Reprinted with permission of Taylor & Francis Group.

These activities are to be performed during each stage and phase of the project. Quality management in each stage and phase helps to

- Enhance quality conformance to requirements
- Produce desired results to satisfy customer needs
- Achieve quality objectives in the project deliverables
- Reduce omission/errors in the project activity, element
- Team approach and coordination
- Manage and control quality at every stage/phase
- Engineering design is properly developed to achieve economical and competitive results
- Construction works are properly executed taking into consideration all the activities so as to fully meet all the project requirements
- Timely completion of project with economical use of resources
- Cost saving in the project

- Ease of communication
- Identify risk and take action
- Understand quality issues and take corrective action for improvement
- Auditing at each phase

2.7.1 TOTAL QUALITY MANAGEMENT IN CONSTRUCTION PROJECTS

Total quality management (TQM) is an organization-wide effort centered on quality to improve performance that involves everyone and permeates every aspect of an organization to make quality a primary strategic objective. It is a way of managing an organization to ensure satisfaction at every stage of the needs and expectations of both internal and external customers.

Construction projects are mainly capital investment projects. They are customized and non-repetitive in nature. Construction projects have become more complex and technical, and the relationships and the contractual grouping of those who are involved are also more complex and contractually varied.

Construction projects have the involvement of many participants comprising of owner, designer, contractor, and many other professionals from construction-related industries. Each of these participants is involved in implementing quality in construction projects. These participants are both influenced by and depend on each other in addition to "other players" involved in the construction process. Therefore, the construction projects have become more complex and technical, and extensive efforts are required to reduce rework and costs associated with time, materials, and engineering, and so forth.

Quality in construction projects is different from that of manufacturing. Quality in construction projects is not only the quality of products and equipment used in the construction, it is the total management approach to complete the facility as per the scope of works to customer/owner satisfaction to be completed within specified schedule and within the budget to meet owner's defined purpose. This is achieved through a complex interaction of many participants in the facilities development process.

TQM is an organization-wide effort centered on quality to improve performance that involves everyone and permeates every aspect of an organization to make quality a primary strategic objective. It is a way of managing an organization to ensure satisfaction at every stage of the needs and expectations of both internal and external customers.

TQM focuses on participative management and strong operational accountability at the individual contributor level. TQM is the implementation of quality system by organization to satisfy the customer needs/requirements by developing a system that will make the project qualitative, competitive, and economical. A collaborative effort in the construction project is required among all the participants in order to improve alignment and misunderstanding during the execution of project. This is achieved through complex interaction of many participants in the facilities development process by application of TQM concept in the construction project.

TQM in construction projects typically involves ensuring compliance with minimum standards of material and workmanship in order to ensure the performance of the facility according to the design. TQM in a construction project is a cooperative form of doing business that relies on the talents and capabilities of both labor and management to continually improve quality. The important factor in construction projects is to complete the facility per the scope of works to customer/owner satisfaction within the specified schedule and approved budget and to complete the work to meet the owner's defined purpose. TQM will help reduce the overrun of schedule and cost.

TQM is a management approach centered on quality, based on organization-wide participation, and aimed at long-term success through customer satisfaction.

TQM focuses on customers, both internal (within the organization, the next party in the work process) and external (end users, stakeholders, regulatory agencies).

Given the fluctuating nature of customer satisfaction, continuous improvement is critical to an organization's survival. The concept applies to processes and the people who operate them as well as products.

The plan–do–check–act (PDCA) cycle is a well-known model for continuous process improvement. The four-step process is also referred to as the Shewhart cycle (for Walter A. Shewhart), the Deming cycle (for W. Edwards Deming), and the PDSA cycle (with the S standing for "Study"). First, a plan to effect improvement is developed. Next, the plan is carried out, preferably on a small scale. The effects of the plan are observed. Last results are studied to determine what was learned and what can be predicted.

TQM's emphasis on participation recognizes that every activity contributes to or detracts from quality and productivity. Leadership from management and employee involvement are crucial for success.

Management's role in TQM is to develop a quality strategy aligned with organizational business objectives and based on customer and stakeholder needs. After that strategy is defined, managers must participate in its deployment regularly and at every level.

Employee involvement can take several forms. Typically, quality improvement requires teams involving employees across functional boundaries. When employees are involved in quality, their organizations are more likely to make well-informed quality decisions and feel responsible for those decisions. Organizations empower employees by allowing them to make decisions that improve the work process within defined boundaries.

Figure 2.21 illustrates the elements of TQM in construction projects.

These elements are developed on the TQM concept as follows:

1. Total (all the participants/stakeholders involved in construction project)
2. Quality (customer/owner/end user satisfaction, requirements of regulatory authorities)
3. Management (application of tools, techniques, and systems to complete the project)

T **O** **T** **A** **L**	**Organization-wide**	1-Organizational involvement 2-Corporate long term vision 3-Systems approach to management 4-Management commitment 5-Management plans 6.Team approach, Coordination 7- Involvement of project stakeholders/team members 8-Steering committee 9- Clearly defined roles and responsibilities 10- Employee participation/involvement 11- Strategic Quality Plan

Q
U
A
L
I
T
Y
	Customer Satisfaction	1- Customer driven quality
		2- Identification of Customer, owner and their needs
		3- Project as per properly defined scope
		4- Project as per construction quality
		   (Defined Scope, Schedule, Budget)
		5-Avoid overrun of schedule, cost
		6- Project/Product specifications as per
		   customer needs/requirements
		7- Perceived need for change

M **A** **N** **A** **G** **E** **M** **E** **N** **T**	**Systems of Managing**	1-Strategic quality deployment process 2-Integrated management/leadership system 3-Identification of products and processes 4-Critical processes and measures 5- Factual approach to design making 6-Scientific approach to problem solving 7- Performance/process improvement 8- Application of all categories of quality tools 9- Six Sigma Methodology, Tools 10- Innovation and New technology 11- Focuss on processes and method statement 12- Quality features of critical activity/element 13- Quality Information System (Digitized system) 14- Process Automation 15-Internet of Things (IoT) 16-Codes and Standards (Quality Management System) 17- Defect prevention not Correction 18-Quick response 19-Evidence-based decision making 20- Benficial supplier, vendor, subcontractor relationship 21- Education, traning, and development 22-Better utilization of talents of project team 23- Performance monitoring/Audit 24-Continuos improvement 25- Reward System

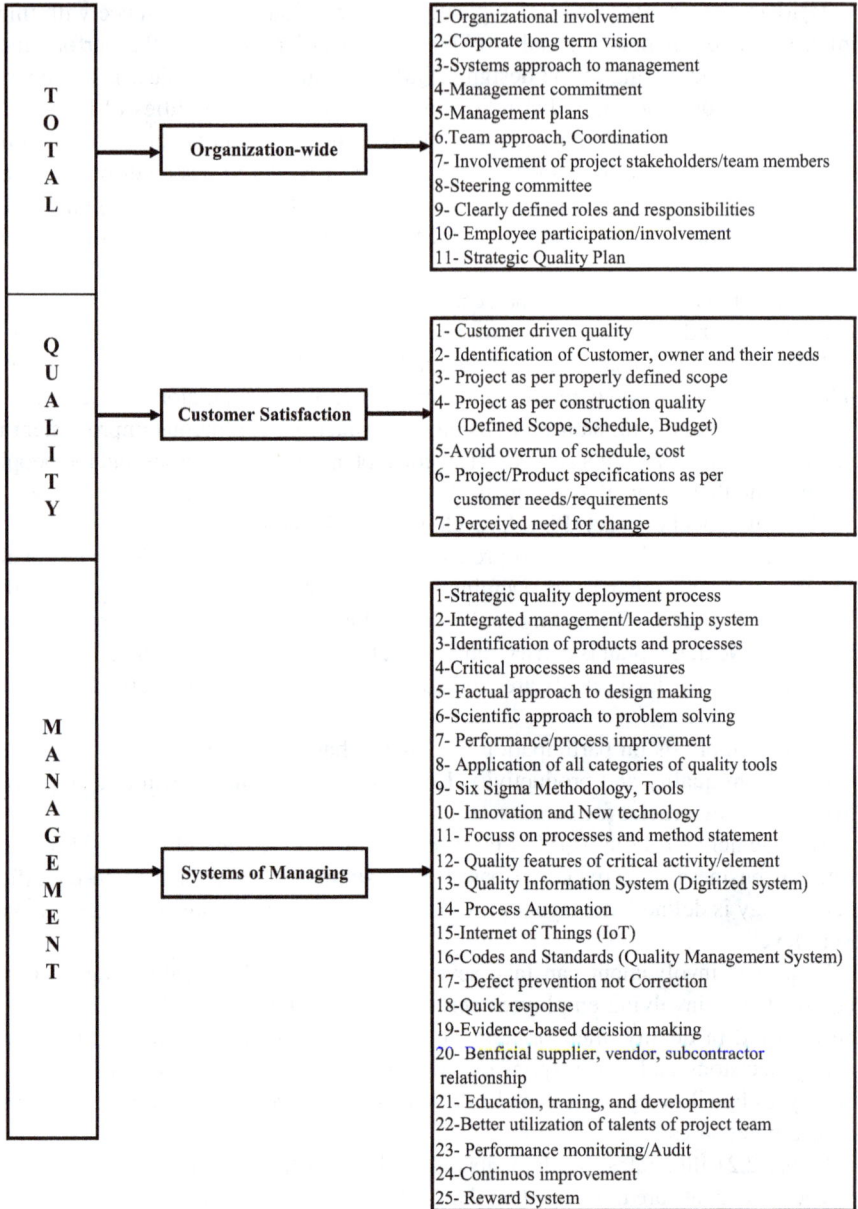

FIGURE 2.21   Elements of total quality management in construction projects.

These elements, in different formats, can be applied in construction projects at different phases/stages of the lifecycle of construction project to achieve customer satisfaction and results in most qualitative, competitive, and economical construction project.

Implementation of TQM in construction projects has tangible results in the following:

- Owner satisfaction
- Project execution time and cost
- Owner/contractor relationship
- Simplification in work process
- Supplier/vendor/subcontractor performance
- Employee morale
- Savings in engineering cost
- Savings in material procurement cost

Quality management in construction projects is different from that of manufacturing.

Quality in construction project is not only the quality of product and equipment used in the construction of facility but also the total management approach to complete the facility.

Quality in manufacturing passes through a series of processes. Material and labor are input through a series of processes out of which a product is obtained. The output is monitored by inspection and testing at various stages of production. Any nonconforming product identified is either repaired, reworked, or scrapped, and proper steps are taken to eliminate problem causes. Statistical process control methods are used to reduce the variability and to increase the efficiency of process. In construction projects, the scenario is not the same. If anything goes wrong, the nonconforming work is very difficult to rectify, and remedial actions are sometimes not possible.

TQM is an organization-wide effort centered on quality to improve performance that involves everyone and permeates every aspect of an organization to make quality a primary strategic objective. It is a way of managing an organization to ensure satisfaction at every stage of the needs and expectations of both internal and external customers.

Construction projects being unique and non-repetitive in nature need specified attention to maintain quality. Each project has to be designed and built to serve a specific need.

In order to conveniently manage and control the construction project, the construction project lifecycle is divided into a number of phases. TQM principles have to be applied in each phase to achieve qualitative, competitive, and economical project to meet owner's requirements/needs.

Figure 2.22 illustrates the TQM concept in construction project lifecycle phases of a traditional contracting system (Design–Bid–Build).

Product Lifecycle (Manufacturing) Phases

NEED → Conceptual Design → Preliminary Design → Detail Design and Development → Production and/or Construction → Product Use, Phaseout, and Disposal

Activities performed, controlled within manufacturing unit (INHOUSE) — INHOUSE or Vendor Location — User/Buyer/Customer

Total Quality Management

Construction Project Lifecycle (Design–Bid–Build) Phases

Conceptual Design → Preliminary Design → Detail Design → Construction Documents → Bidding and Tendering → Construction → Testing, Commissioning, and Handover

	Conceptual Design	Preliminary Design	Detail Design	Construction Documents	Bidding and Tendering	Construction	Testing, Commissioning, and Handover
	Total Quality Management Application	Total Quality Management Application	Total Quality Management Application	Total Quality Management Application	Total Quality Management Application	Total Quality Management Application	Total Quality Management Application
	Major Activities/Elements of Conceptual Phase	Major Activities/Elements of Preliminary Design Phase	Major Activities/Elements of Detail Design Phase	Major Activities/Elements of Construction Documents Phase	Major Activities/Elements of Bidding and Tendering Phase	Major Activities/Elements of Construction Phase	Major Activities/Elements of Testing, Commissioning, and Handover Phase
Total (Organization-Wide)	Stakeholders/Team Members for Conceptual Design Phase	Stakeholders/Team Members for Preliminary Design Phase	Stakeholders/Team Members for Detail Design Phase	Stakeholders/Team Members for Construction Documents Phase	Stakeholders/Team Members for Bidding and Tendering Phase, Bidders/Contractors	Stakeholders/Team Members/Contractor/Subcontractor/Vendors for Construction Phase	Stakeholders/Team Members for Testing, Commissioning, and Handover Phase
Owner is involved from inception of project. Participation/Involvement of other stakeholders, project members differ as per the activities to be performed in various phases							
Quality (Customer Satisfaction)	Owner's requirements, TOR, Regulatory requirements	Owner's requirements, TOR, Regulatory requirements	Owner's requirements, TOR, Regulatory requirements	Owner's requirements, TOR, Regulatory requirements	Owner's requirements, TOR, Regulator requirements, Tender Documents	Contract Documents, Project quality, Regulatory requirements	Contract Documents, Project quality, Regulatory requirements
Management (Systems of Managing)	Quality Tools and Techniques, Principles, Procedures, Methods	Quality Tools and Techniques, Principles, Procedures, Methods	Quality Tools and Techniques, Principles, Procedures, Methods	Quality Tools and Techniques, Principles, Procedures, Methods	Quality Tools and Techniques, Principles, Procedures, Methods	Quality Tools and Techniques, Principles, Procedures, Methods, Sequence of activities, Project schedule	Quality Tools and Techniques, Principles, Procedures, Methods

FIGURE 2.22   TQM concept in construction project life cycle (design–bid–build) phases.

3 Quality Standards

3.1 QUALITY STANDARDS

A quality system is a framework for quality management. It embraces the organizational structure, procedure, and processes needed to implement a quality management system (QMS). The quality system should address everything in the organization related to the quality of products, services, processes, projects, and operations. The adequacy of the quality system and the quality of products, services, processes, projects, operations, and customer satisfaction are judged by their compliance to specified/relevant standards. Standards have important economic and social repercussions. They are useful to industrial and business organizations of all types, to government, and to other regulatory bodies, to conformity assessment professionals, to suppliers and customers of products, projects, and services in both the public and private sectors, and to people in general in their role as customers and users. Standards provide governments with a technical base for health, safety, and environmental legislation.

Standards are documents used to define acceptable conditions or behaviors and to provide a baseline for assuring that conditions or behaviors meet the acceptable criteria. In most cases standards define minimum criteria; world-class quality is, by definition, beyond the standard level of performance. Standards can be written or unwritten, voluntary or mandatory. Unwritten quality standards are generally not acceptable.

3.1.1 IMPORTANCE OF STANDARDS

The ISO has given the importance of standards as follows:

Standards make an enormous contribution to most aspects of our lives. Standards ensure desirable characteristics of products and services such as quality, environmental fitness, safety, reliability, efficiency, and interchangeability at an economical cost.

When products and services meet our expectations, we tend to take this for granted and be unaware of the role of standards. However, when standards are absent, we soon notice. We soon care when products turn out to be of poor quality, do not fit, are incompatible with equipment that we already have, are unreliable or dangerous. When products, systems, machinery, and devices work well and safely, it is often because they meet standards.

DOI: 10.1201/9781003451464-3

Standard setting is one of the first issues in developing a quality assurance system, and increasingly organizations are relying on readily available standards rather than developing their own. Each standard should be

- Clearly written in simple language that is unambiguous
- Convenient in understanding
- Specific in setting out precisely what is expected
- Measurable so that the organization can know whether the standard is being met
- Achievable, that is, the organization must have the resources available to meet the standard
- Constructible

Standards are published documents that establish specifications, guidelines, and procedures designed to ensure the reliability of materials, products, systems, processes, and services people use every day to satisfy their requirements. In the construction sector, compliance with minimum standards in the implementation of workmanship, materials, products, systems, and processes provides a competitive advantage in terms of customer satisfaction and achieves qualitative, competitive, and economical project. Standards have important economic, social, and environmental repercussions.

Standards are used to ensure that a product, system, or service measures up to its specifications and is safe for use. Standards are the key to any conformity assessment activity.

3.2 STANDARDS ORGANIZATIONS

There are many organizations that produce standards; some of the best-known organizations in the quality field of oil and gas industry are

- American Concrete Institute (ACI)
- American Institute of Steel Construction (AISC)
- American National Standards Institute (ANSI)
- American Petroleum Institute (API)
- American Society of Mechanical Engineers (ASME)
- American Society for Heating, Refrigerating, and Air-Conditioning Engineers (ASHRAE)
- American Society for Quality (ASQ)
- American Society for Testing and Materials (ASTM)
- American Welding Society (AWS)
- British Standards Institution (BSI)
- European Committee for Electrotechnical Standardization (CENELEC)
- International Association of Plumbing and Mechanical Officials (IAPMO)
- International Electrotechnical Commission (IEC)
- International Organization for Standardization (ISO)
- Engineering Equipment and Material Users Association (EEMUA)
- Expansion Joint Manufacturers Association (EJMA)

- National Association of Corrosion Engineers (NACE)
- National Electrical Code (NEC)
- National Fire Protection Association (NFPA)
- Tubular Exchange Manufacturers Association (TEMA)

Standards produced by these organizations/institutes are recognized worldwide. These standards are referred to in the contract documents by the designers to specify products or systems or services to be used in a project. They are also used to specify the installation method to be followed or the fabrication works to be performed during the construction process.

Apart from these, there have been many other national and international quality system standards. These various standards have commonalities and historical linkage. However, in order to facilitate international trade, delegates from 25 countries met in London in 1946 to create a new international organization. The objective of this organization was to facilitate international coordination and unification of industrial standards. The new organization, International Organization for Standardization, ISO, officially began operation on February 23, 1947.

3.3   INTERNATIONAL ORGANIZATION FOR STANDARDIZATION (ISO)

ISO is an independent nongovernmental organization with membership of 168 (as of December 2023) national standards bodies/institutes formed on the basis of one member per country, with a Central Secretariat in Geneva, Switzerland, that coordinates the system.

ISO is the world's largest developer and publisher of international standards. It is a nongovernmental organization that forms a bridge between the public and private sectors. The work of preparing International Standards is normally carried out by technical committees. ISO has more than 21,000 international standards. Of all the standards produced by ISO, the ones that are most widely known are the ISO 9000 and ISO 14000 series. ISO 9000 has become an international reference for quality requirements in business-to-business dealings, and ISO 14000 looks to achieve at least as much, if not more, in helping organizations to meet their environmental changes. ISO 9000 and ISO 14000 families are known as "generic management system standards."

The ISO 9000 is a series, or family of standards, primarily concerned with "quality management." This means what the organization does to fulfill

- The customer's quality requirements
- Applicable regulatory requirements, while aiming to enhance customer satisfaction
- Achieve continual improvement of its performance in pursuit of the objectives

The ISO 9000 family addresses various aspects of quality management and contains some of ISO's best-known standards. The standards provide guidance and tools for companies and organizations who want to ensure that their products and services consistently meet customer's requirements and that quality is consistently improved. ISO 9000 outlines the ways to achieve, as well as benchmark, consistent performance and service. Its application makes the business more competitive and credible. ISO 9000 QMS helps to continually monitor and manage quality across all the operations from inception through completion and enhances customer satisfaction.

The ISO 14000 family is primarily concerned with "environmental management." This means to

- Minimize harmful effects on the environment caused by its activities
- Achieve continual improvement with its environmental performance

ISO standards are updated periodically since they were originally published in 1987. ISO 9000 actually comprises several standards.

ISO 9000:2000 specifies requirements for a QMS for any organization that needs to demonstrate its ability to consistently provide product that meets customer and applicable regulatory requirements and to enhance customer satisfaction.

In keeping with the process of updating the standards, certain clauses of ISO 9001:2000 of the QMS were amended in 2008 in order to improve the QMS, and accordingly the amended standard is known as ISO 9001:2008. ISO 9001:2008 includes Annex B, which outlines the text changes that have been made to specific clauses.

ISO 9001:2008 was revised by various committees, societies, and institutes, and ISO 9001:2015 was published in September 2015. It has the following clauses:

1. Context of Organization/Quality Management System (Clause 4)
2. Leadership (Clause 5)
3. Planning for Quality Management System (Clause 6)
4. Support (Clause 7)
5. Operation (Clause 8)
6. Performance Evaluation (Clause 9)
7. Improvement (Clause 10)

Changes in ISO 9001:2015 are an opportunity to revisit organizational areas that yet need to be improved. An awareness of the upcoming changes in ISO 9001:2015 will enable quality professionals to better prepare for the future. The change is to incorporate risk-based thinking into the management system by considering the context of the organization. In other words, all processes are not equal for all organizations with some being more critical than others, resulting in different levels of risk.

The tremendous impact of ISO 9001 and ISO 14001 on organizational practices and on trade stimulated the development of other ISO standards and deliverables that adapt the generic management system to specific sectors or aspects. These are

1. Food Safety Management Systems ISO 22000
2. Information Security Management Systems ISO 27001
3. Supply Chain Security Management Systems ISO 28000

ISO 22000:2005, published on September 1, 2005, is related to the safe food supply management system to ensure that food is safe at the time of human consumption. ISO 27001:2005 is related to information security system. ISO 28000:2005 is related to the supply management system to help combat threats to safe and smooth flow of international trade.

3.4 ISO 9000 QUALITY MANAGEMENT SYSTEM

ISO 9000 quality system standards are a tested framework for taking a systematic approach to managing the business process so that organizations turn out products or services conforming to customer satisfaction. The typical ISO QMS is structured on four levels, usually portrayed as a pyramid. Figure 3.1 illustrates this.

On top of the pyramid is the quality policy, which sets out what management requires its staff to do in order to ensure QMS. Underneath the policy is the quality manual, which details the work to be done. Beneath the quality manual are work instructions or procedures. The number of manuals containing work instructions or procedures is determined by the size and complexity of the organization. The procedures mainly discuss the following:

- What is to be done?
- How is it done?

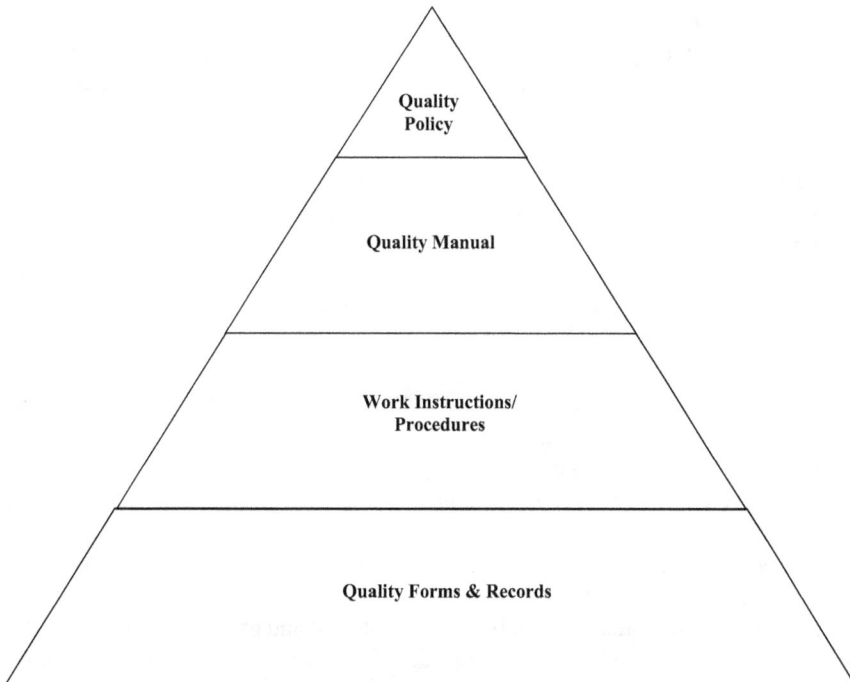

FIGURE 3.1 QMS documentation pyramid.

- How does one know that it has been done properly (e.g., by inspecting, testing, or measuring)?
- What is to be done if there are problems (e.g., failure)?

The bottom level of hierarchy contains forms and records that are used to capture the history of routine events and activities.

The ISO 9000 QMS requires documentation that includes a quality manual and quality procedures, as well as work instructions and quality records. All documentation (including quality records) must be controlled according to a document control procedure. The structure of the QMS depends largely on the management structure in the organization.

ISO 9001:2000 identifies certain minimum requirements that all QMSs must meet to ensure customer satisfaction. ISO 9001:2000 specifies requirements for QMSs when an organization

- Needs to demonstrate its ability to consistently provide a product that meets customer and applicable regulatory requirements
- Aims to enhance customer satisfaction through the effective application of the system, including processes for continual improvement of the system and the assurance of conformity to customer and applicable regulatory requirements

3.4.1 Quality System Documentation

A quality system has to cover all the activities leading to the final product or service. The quality system depends entirely on the scope of operation of the organization and particular circumstances such as number of employees, type of organization, and physical size of the premises of the organization. The quality manual is the document that identifies and describes the QMS.

The adoption of a QMS should be a strategic decision of an organization. The design and implementation of an organization's QMS (Documentation) is influenced by

1) Its organizational environment, changes in that environment, and the risks associated with that environment
2) Its varying needs
3) Its particular objectives
4) The products it provides
5) The processes it employs
6) Its size and organizational structure

The QMS requirements specified in the International Standard are complementary to requirements for products.

This International Standard can be used by internal and external parties, including certification bodies, to assess the organization's ability to meet customer, statutory, and regulatory requirements applicable to the product and the organization's own requirements.

The QMS is based on the guidelines for performance improvement per ISO 9004:2000 and the quality management requirements. The quality management principles stated in ISO 9000 and ISO 9004 have been taken into consideration for the development of this International Standard. ISO 9000:2000 outlines the necessary steps to implement the QMS. These are

1. Identify the processes (activities and necessary elements) needed for QMS.
2. Determine the sequence and interaction of these processes and how they fit together to accomplish quality goals.
3. Determine how these processes are effectively operated and controlled.
4. Measure, monitor, and analyze these processes and implement the action necessary to correct the process and achieve continual requirements.
5. Ensure that all information is available to support the operation and monitoring of the process.
6. Display the most options, thus helping make the right management system.

ISO 9001:2000 requirements fall into the following sections:

1. Quality management system
2. Management responsibility
3. Resource management
4. Product realization
5. Measurement analysis and improvement

These documents are regularly updated and revised to meet the changing industry requirements due to

- Globalization
- Higher levels of expectations by customers
- Higher performance requirements
- Complex nature of environment and works
- Specialized services requirements

ISO 9000:2008 was revised in 2015 and is called ISO 9000:2015 that has ten clauses against eight in the previous version.

In general, QMSs begin with the writing of a quality manual. The quality manual serves as a roadmap for the QMS. It is more practical to build the "road" before preparing the "map." The manual is generic at first and as the QMS develops, the manual is updated.

In the construction industry, a contractor may be working at any time on a number of projects of varied nature. These projects have their own contract documents to implement project quality, which require a contractor to submit a contractor's quality control plan to ensure that specific requirements of the project are considered to meet client's requirements. Therefore, while preparing a QMS at a corporate level, the organization has to take into account tailor-made requirements for the projects and accordingly the manual should be prepared.

3.4.2   QUALITY MANAGEMENT SYSTEM MANUAL

A QMS is a set of coordinated documentation that includes a quality manual, quality procedures, processes, as well as work instructions (details of works to be done), quality forms, and records that are developed to meet customer and regulatory requirements taking into consideration organization's quality policies and business objectives and improve the effectiveness and efficiency on a continuous basis.

ISO 9001 is an international standard that is most recognized and used as an international reference for the development of an effective QMS.

ISO 9001 contains eight quality management principles (CLIPSCFM). These are

Principle 1—Customer focus
Principle 2—Leadership
Principle 3—Involvement of people
Principle 4—Process approach
Principle 5—System approach to management
Principle 6—Continual improvement
Principle 7—Factual approach to design making
Principle 8—Mutual beneficial supplier relationship
These principles were further revised in ISO 9001:2015. Accordingly, there are seven (7) quality management principles. (Please refer to section 1.5 of chapter 1.)

3.4.2.1   Development of QMS

ISO 9000 family is primarily concerned with "quality management." This means what the organization does to fulfill

* The customers' quality requirements
* Applicable regulatory requirements, while aiming to enhance customer satisfaction
* Achieve continual improvement of its performance in pursuit of the objectives

Although any ISO 9000 QMS should be created to address an organization's business objectives, needs, and customer satisfaction, there are some general elements that are common in all ISO-compliant QMS. These elements have importance as per the levels known as QMS Documentation Pyramid that includes quality policy and objectives, a quality manual, quality procedures, as well as work instructions and quality records. The manual is developed taking into consideration the following:

1. Eight principles (CLIPSCFM) of QMS as defined by ISO Technical Committee, TC 176
2. All the related and applicable documents produced taking into consideration ten sections/clauses listed under ISO 9001:2015 to ensure that the manual is in compliant with ISO 9001:2015. Table 3.1 illustrates these sections/clauses.

TABLE 3.1
ISO 9001:2015 Sections (Clauses)

Section (Clause) No.	Relevant Clause in 9001:2015	Description
1	**Scope**	
2	**Normative References**	
3	**Terms and References**	
4	**Context of the Organization**	
	4.1	Understanding the organization and its context
	4.2	Understanding the needs and expectations of interested parties
	4.3	Determining the scope of quality management system
	4.4	Quality management system and its processes
5	**Leadership**	
	5.1	Leadership and commitment
	5.2	Policy
	5.3	Organizational roles, responsibilities, and authorities
6	**Planning**	
	6.1	Actions to address risks and opportunities
	6.2	Quality objectives and planning to achieve them
	6.3	Planning of changes
7	**Support**	
	7.1	Resources
	7.2	Competence
	7.3	Awareness
	7.4	Communication
	7.5	Documented information
8	**Operation**	
	8.1	Operational planning and control
	8.2	Requirements for product and services
	8.3	Design and development of products and services
	8.4	Control of externally provided processes, products, and services
	8.5	Production and service provision
	8.6	Release of products and services
	8.7	Control of nonconforming outputs
9	**Performance evaluation**	
	9.1	Monitoring, measurement, analysis, and evaluation
	9.2	Internal audit
	9.3	Management review
10	**Improvement**	
	10.1	General
	10.2	Nonconformity and corrective action
	10.3	Continual improvement

**QMS
Documentation
Pyramid**

**ISO 9001:2015
Sections**

Quality Principles

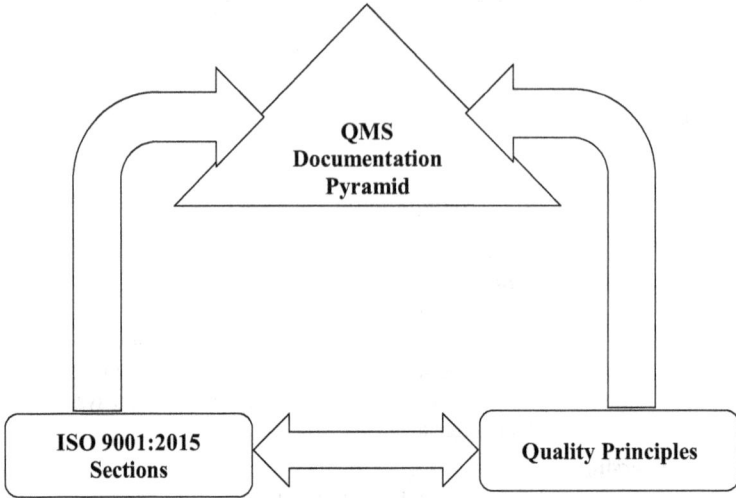

FIGURE 3.2   QMS documentation model.

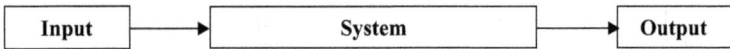

| Input | → | System | → | Output |

FIGURE 3.3   Black box.

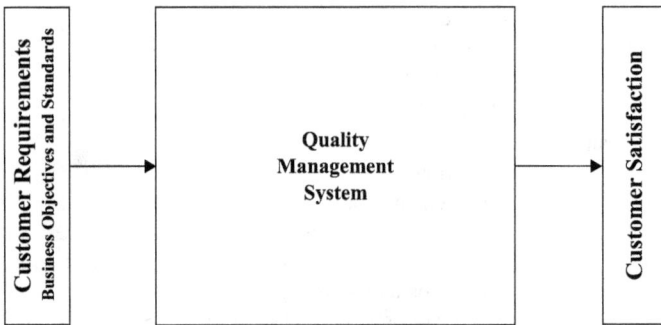

Customer Requirements
Business Objectives and Standards

**Quality
Management
System**

Customer Satisfaction

FIGURE 3.4   Process-based QMS development procedure.

Figure 3.2 illustrates the relationship between QMS principles and ISO sections.

QMS is made up of elements (components) having functional relationship to achieve a common objective (customer satisfaction). These elements need to be coordinated taking a systematic approach to improve and sustain the overall performance of the products and services. Figure 3.3 illustrates a simple behavioral approach to system and is generally known as Black Box.

Systems engineering or process-based approach can be used to develop QMS. Figure 3.4 illustrates the concept of system phenomenon that can be applied to develop QMS.

The application of the process approach in a QMS enhances

- Overall performance of the organization by effectively controlling the interrelationship and interdependencies among the QMS processes
- Consistency in meeting the customer requirements
- Consideration of processes in terms of added value
- Achievement of effective process performance
- Improvement of processes based on evaluation of data and information

The process-based QMS approach incorporates the Plan–Do–Check–Act (PDCA) cycle and risk-based thinking. The PDCA cycle can be applied to all processes and to the QMS as a whole. Figure 3.5 illustrates the organizational structure of ISO 9001:2015 sections (clauses) 4 to 10 and Figure 3.6 describes in brief the related activities in the PDCA cycle.

The PDCA cycle can be applied to all the processes. Figure 3.7 illustrates how clauses 4 to 10 can be grouped in relation to the PDCA cycle.

Development and implementation of an effective QMS is a strategic decision of the organization that helps to improve its overall performance and provides a sound basis for sustainable development initiatives.

Traditional construction projects have involvement of three main groups. These are

1. Owner (client, project, owner)
2. Designer (consultant)
3. Contractor (builder, constructor)

Each of these groups should have their own QMS to meet their business objectives. While developing the QMS, the organizations have to include those documents that are required to perform relevant processes and specific requirements in the organizations that have business interest.

Normally QMS consists of documents produced taking into consideration relevant sections of ISO 9001:2015 and business-related activities.

Figure 3.8 illustrates a flowchart for ISO QMS design and implementation process.

Table 3.2 lists example contents of QMS manual for owner (client, project owner, developer).

Table 3.3 lists example contents of QMS manual for designer (consultant).

Table 3.4 lists example contents of QMS manual for contractor and Annexure 3.4A mentioned in this table lists additional clauses required for Design–Build contractor manual documents.

Table 3.5 lists example contents of QMS manual for manufacturing organization.

Relevant clauses under these tables have to be considered while developing a construction project applying the concept of total quality management.

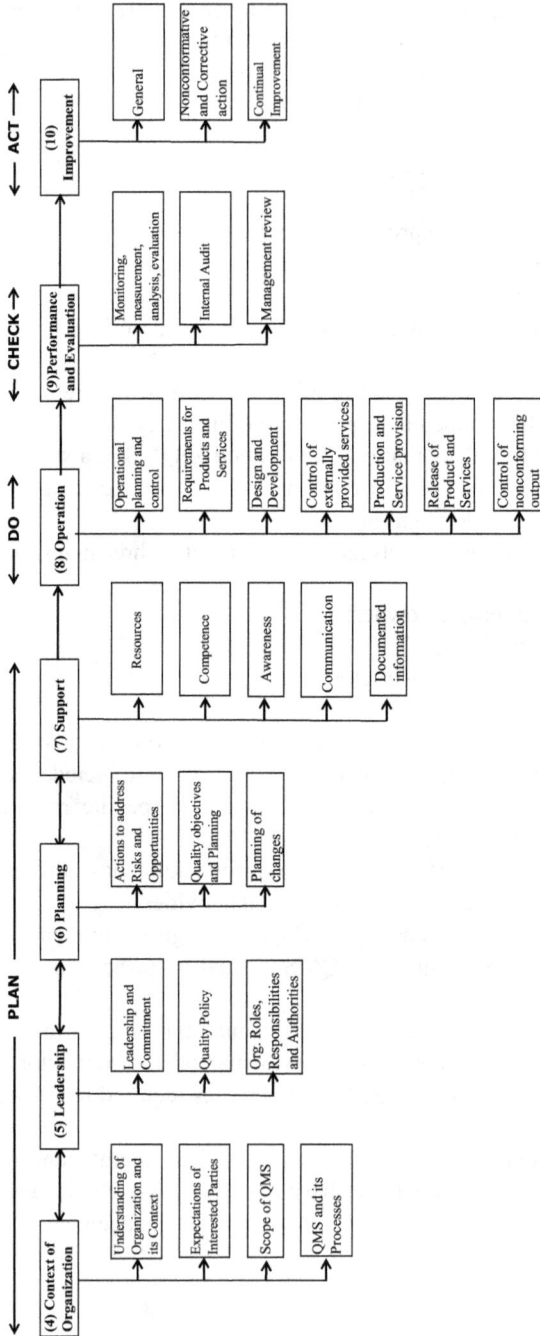

FIGURE 3.5 Organizational structure of ISO 9001:2015 sections.

ACT
- Nonconformity and Corrective action
- Continual improvement

PLAN
- Establish the objectives
  (System and Processes)
- Resources
- Quality Policy
- Roles and Responsibilities
- Risk and Opportunities
- Communication
- Changes
  - Documentation

4. ACT | 1. PLAN

3. CHECK | 2. DO

CHECK
- Monitoring, measurement,
  analysis and evaluation
- Internal Audit
- Management review

DO
- Operation of planning and control
- Design and Development
- Control of processes, products, services
- Control of nonconforming outputs

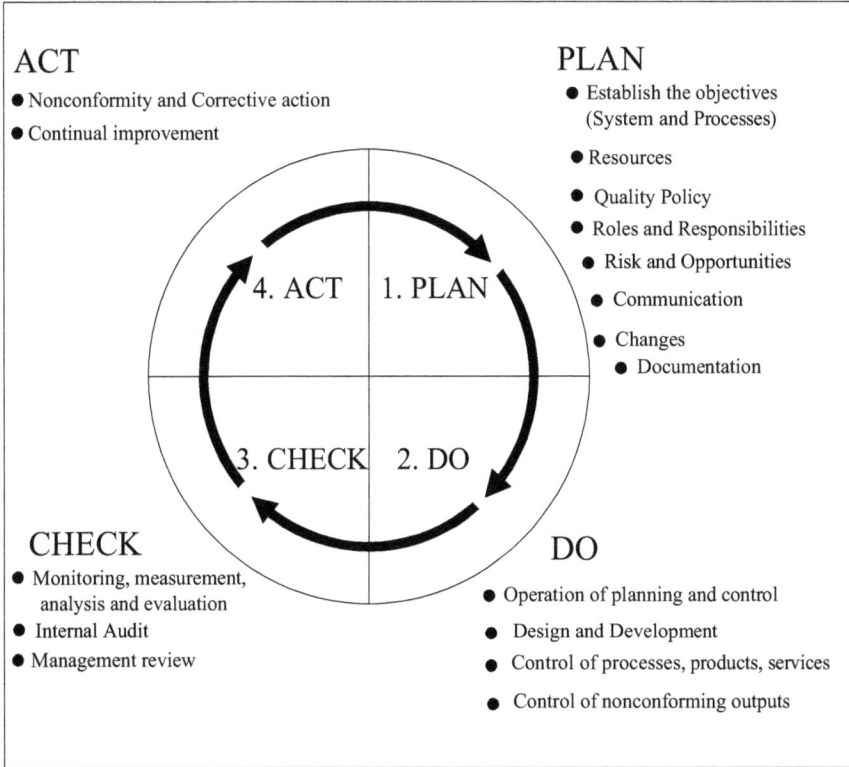

FIGURE 3.6   PDCA cycle for QMS activities.

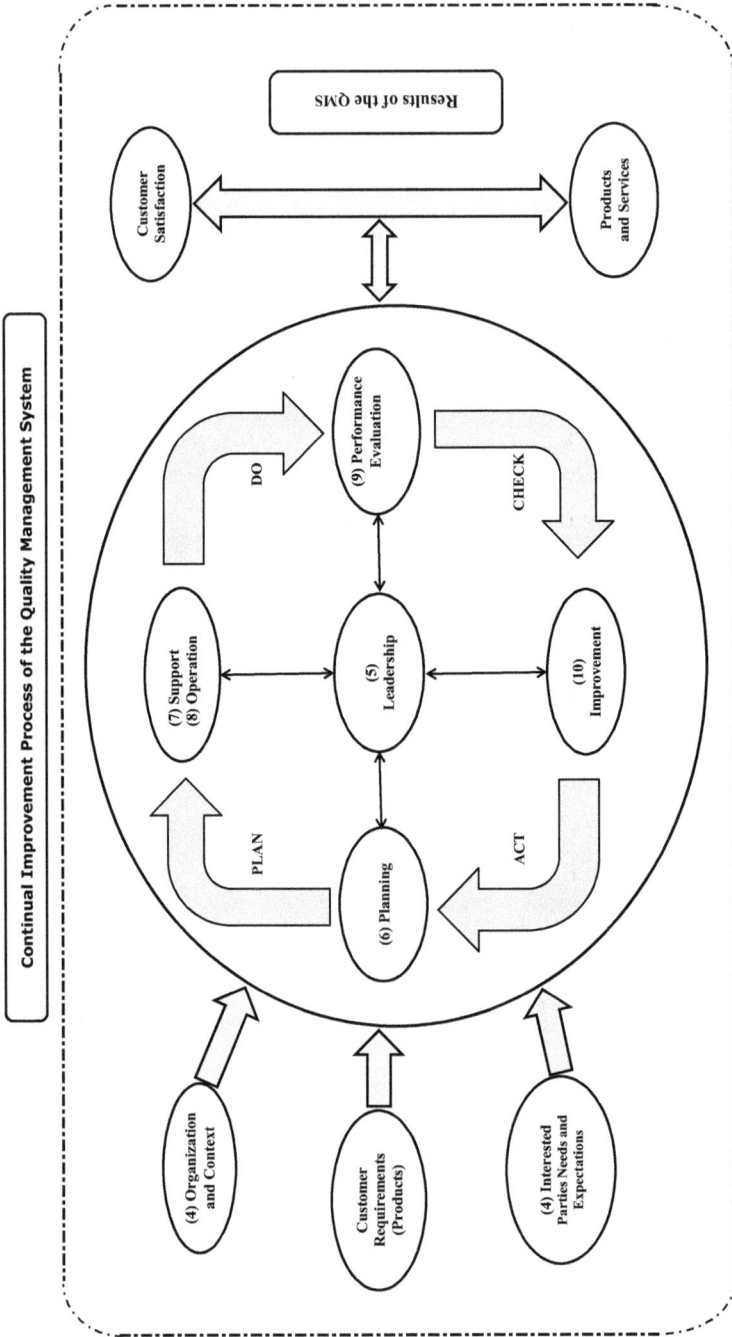

FIGURE 3.7 Quality management process model.

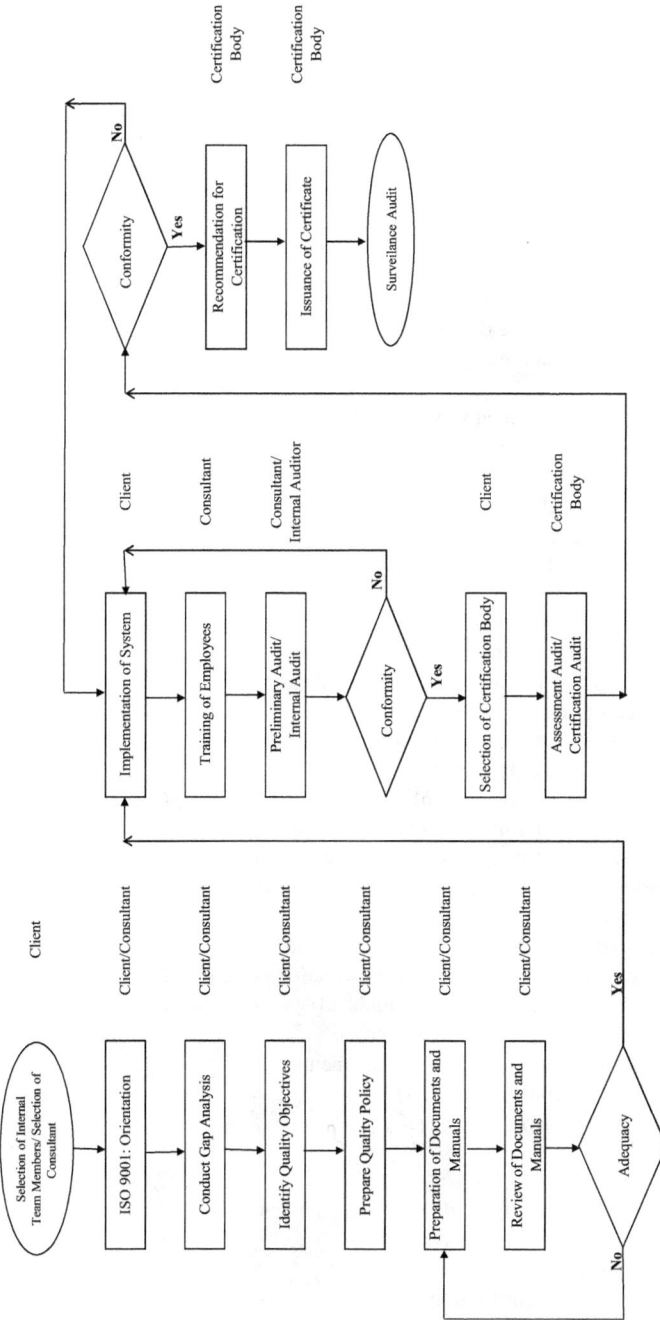

FIGURE 3.8 Flowchart for ISO design and implementation process. (Abdul Razzak Rumane. (2019). *Quality Auditing in Construction Projects*. Reprinted with permission from Taylor & Francis Group.)

TABLE 3.2
List of Quality Manual Documents—Owner (Client, Project Owner, Developer)

Document No.	Document Title	Relevant Clause in 9001:2015	Version/ Revision Date
QC-0	Circulation list		
QC-00	Records of revision		
QC-01	Scope		
QC-02	Normative references		
QC-03	Terms and references		
QC-1.1	Understanding the organization and its context	4.1	
QC-1.2	Monitoring and review of internal and external issues	4.1	
QC-2.1	Relevant requirements of stakeholders	4.2	
QC-2.2	Monitoring and review of stakeholder's information	4.2	
QC-3	Scope of quality management system	4.3/4.4	
QC-4	Project quality management system	4.4	
QC -5	Management responsibilities	5.1	
QC -6	Customer focus	5.1.2	
QC -7.1	Quality policy (organization)	5.2	
QC -7.2	Quality policy (project)	5.2.2	
QC-8	Organizational roles, responsibilities, and authorities (organization chart)	5.3	
QC-9	Preparation and control of project quality plan	6.0	
QC-10.1	Project risk (study stage)	6.1	
QC-10.2	Project risk (during design)	6.1	
QC-10.3	Project risk (during bidding & tendering)	6.1	
QC-10.4	Project risk (during construction)	6.1	
QC-10.5	Project risk (selection of designer, contractor)	6.1	
QC-11	Project quality objective	6.2	
QC-12.1	Change management (during design phase)	6.3	
QC-12.2	Change management (during construction phase)	6.3	
QC-12.3	Change management (owner initiated)	6.3	
QC-13.1	Office resources (office equipment)	7.1	
QC-13-2	Human resources (project management, construction management)	7.1.2	
QC-13.3	Human resources (office staff)	7.1.2	
QC-14.1	Infrastructure	7.1.3	
QC-14.2	Work environment	7.1.4	
QC-15	Monitoring and measuring office equipment	7.1.5	
QC-16	Organizational knowledge	7.1.6	
QC-17.1	Competence	7.2	
QC-17.2	Training in quality system	7.2	
QC-17.3	Training in quality auditing	7.2	
QC-17.4	Training in supervision/technical skills	7.2	

TABLE 3.2 (Continued)
List of Quality Manual Documents—Owner (Client, Project Owner, Developer)

Document No.	Document Title	Relevant Clause in 9001:2015	Version/ Revision Date
QC-17.5	Records of training, skills, experience, and qualifications	7.2	
QC-18	Communication between internal and external	7.4	
QC-19.1	Control of documents for general application	7.5.2/3	
QC-19.2	Control of documents for specific projects	7.5.2/3	
QC-20	Records updates	7.5.2	
QC-21	Control of quality records	7.5.3	
QC-22.1	Planning of project/facility and quality plan	8.1/8.3/ 8.3.1/2/3	
QC-22.2	Project study and feasibility	8.1/8.3/ 8.3.1/2/3	
QC-22.3	Project brief	8.1/8.3/ 8.3.1/2/3	
QC-22.4	Records of feasibility inputs	8.3.3	
QC-22.5	Records of terms of reference	8.3.3	
QC-22.6	Record of changes	8.3.6	
QC-23.1	Selection and evaluation of designer, project manager, construction manager, Contractor	8.4	
QC-23.2	Control of design, contracting services	8.4	
QC-23.3	Communication with designer, contractor	8.4.3	
QC-24.1	Designer selection procedure	8.5	
QC-24.2	Contractor selection procedure	8.5	
QC-25.1	Design review procedure	8.5	
QC-25.2	Construction supervision procedure	8.5	
QC-25.3	Project management procedure	8.5	
QC-25.4	Construction management procedure	8.5	
QC-26.1	Release of project documents for tender	8.6	
QC-26.2	Release of project documents for construction	8.6	
QC-27	Control of nonconforming work (project)	8.7	
QC-28.1	Project document review (during design)	9.1	
QC-28.2	Project document review (during construction)	9.1	
QC-28.3	Project manager/construction manager comments	9.1.2	
QC-29	Internal quality audits	9.2	
QC-30	Management review	9.3	
QC-31	Nonconformity and corrective action	10.2.2	
QC-32	Control of client complaints	10.2.2	
QC-33	Continual improvement	10.3	

TABLE 3.3
List of Quality Manual Documents—Consultant (Design and Supervision)

Document No.	Document Title	Relevant Clause in 9001:2015	Version/ Revision Date
QC-0	Circulation list		
QC-00	Records of revision		
QC-01	Scope		
QC-02	Normative references		
QC-03	Terms and references		
QC-1.1	Understanding the organization and its context	4.1	
QC-1.2	Monitoring and review of internal and external issues	4.1	
QC-2.1	Relevant requirements of stakeholders	4.2	
QC-2.2	Monitoring and review of stakeholder's information	4.2	
QC-3	Scope of quality management system	4.3/4.4	
QC-4	Project quality management system	4.4	
QC -5	Management responsibilities	5.1	
QC -6	Customer focus	5.1.2	
QC -7.1	Quality policy (organization)	5.2	
QC -7.2	Quality policy (project)	5.2.2	
QC-8	Organizational roles, responsibilities, and authorities (organization chart)	5.3	
QC-9	Preparation and control of project quality plan	6.0	
QC-10.1	Project risk (design proposal)	6.1	
QC-10.2	Project risk (supervision proposal)	6.1	
QC-10.3	Project risk (during design)	6.1	
QC-10.4	Project risk (during construction)	6.1	
QC-11	Project quality objective	6.2	
QC-12.1	Change management (during design phase)	6.3	
QC-12.2	Change management (during construction phase)	6.3	
QC-13.1	Office resources (office equipment, design software)	7.1	
QC-13-2	Human resources (design team, supervision team)	7.1.2	
QC-13.3	Human resources (office staff)	7.1.2	
QC-14.1	Infrastructure	7.1.3	
QC-14.2	Work environment	7.1.4	
QC-15	Monitoring and measuring office equipment	7.1.5	
QC-16	Organizational knowledge	7.1.6	
QC-17.1	Competence	7.2	
QC-17.2	Training in quality system	7.2	
QC-17.3	Training in quality auditing	7.2	
QC-17.4	Training in operational/technical skills	7.2	
QC-17.5	Record of training, skills, experience, and qualifications	7.2	
QC-18	Communication between internal and external	7.4	
QC-19.1	Control of documents for general application	7.5.2/3	

TABLE 3.3 (Continued)
List of Quality Manual Documents—Consultant (Design and Supervision)

Document No.	Document Title	Relevant Clause in 9001:2015	Version/ Revision Date
QC-19.2	Control of documents for specific projects	7.5.2/3	
QC-20	Records updates	7.5.2	
QC -21	Control of quality records	7.5.3	
QC-22.1	Proposal (design, supervision) documents	8,2	
QC-22.2	Proposal review	8.2	
QC-22.3	Records of proposal changes	8.2	
QC-23.1	Planning of engineering design and quality plan	8.1/8.3/ 8.3.1/2/3	
QC-23.2	Design development (design–bid–build)	8.1/8.3/ 8.3.1/2/3	
QC-23.3	Design development (design–build)	8.1/8.3/ 8.3.1/2/3	
QC-23.4	Record of design development inputs	8.3.3	
QC-23.5	Records of design development control	8.3.4	
QC-23.6	Records of design development outputs	8.3.5	
QC-23.7	Records of design development changes	8.3.6	
QC-24.1	Selection and evaluation of subconsultant	8.4	
QC-24.2	Control of subconsultant services	8.4	
QC-24.3	Communication with subconsultant	8.4.3	
QC-24.4	Evaluation and selection of equipment, provisions	8.4	
QC-25	Engineering design procedure	8.5	
QC-25.1	Design review procedure	8.5	
QC-26.1	Construction supervision procedure	8.5	
QC-26.2	Project management procedure	8.5	
QC-26.3	Construction management procedure	8.5	
QC-27	Release of project documents	8.6	
QC-28	Control of nonconforming work (design errors)	8.7	
QC-29.1	Project document review (management and control)	9.1	
QC-29.2	Client/owner comments	9.1.2	
QC-30	Internal quality audits	9.2	
QC-31	Management review	9.3	
QC-32.1	Nonconformity and corrective action	10.2.2	
QC-32.2	Preventive action	10.2.2	
QC-33	Control of client complaints	10.2.2	
QC-34	Continual improvement	10.3	

TABLE 3.4
List of Quality Manual Documents—Contractor

Document No.	Document Title	Relevant Clause in 9001:2015	Version/ Revision Date
QC-0	Circulation list		
QC-00	Records of revision		
QC-01	Scope		
QC-02	Normative references		
QC-03	Terms and references		
QC-1.1	Understanding the organization and its context	4.1	
QC-1.2	Monitoring and review of internal and external issues	4.1	
QC-2.1	Relevant requirements of stakeholders	4.2	
QC-2.2	Monitoring and review of stakeholder's information	4.2	
QC-3	Scope of quality management system	4.3/4.4	
QC-4	Project quality management system	4.4	
QC -5	Management responsibilities	5.1	
QC -6	Customer focus	5.1.2	
QC-7.1	Quality policy (organization)	5.2	
QC-7.2	Quality policy (project)	5.2.2	
QC-8	Organizational roles, responsibilities, and authorities (organization chart)	5.3	
QC-9	Preparation and control of project quality plan	6.0	
QC-10.1	Project risk management (tendering)	6.1	
QC-10.2	Project risk management (construction)	6.1	
QC-10.3	Project risk management (selection of subcontractor, supplier)	6.1	
QC-11	Project quality objectives	6.2	
QC-12.1	Change management (scope)	6.3	
QC-12.2	Change management (variation orders, site work instructions)	6.3	
QC-13.1	Office resources (office staff, office equipment)	7.1	
QC-13.2	Construction resources (human resources, equipment, and machinery)	7.1	
QC-14.1	Infrastructure	7.1.3	
QC-14.2	Work environment	7.1.4	
QC-15	Control of construction, material, measuring, and test equipment	7.1.5	
QC-16	Control of human resources	7.1.6	
QC-17	Organizational knowledge	7.1.6	
QC-18.1	Competence	7.2	
QC-18.2	Training and development in quality system	7.2	
QC-18.3	Training in quality auditing	7.2	
QC-18.4	Training in operational/technical skills	7.2	
QC-18.5	Records of training, skills, experience, and qualifications	7.2	

TABLE 3.4 (Continued)
List of Quality Manual Documents—Contractor

Document No.	Document Title	Relevant Clause in 9001:2015	Version/ Revision Date
QC-19	Communication between internal and external	7.4	
QC-20.1	Control of documents for general application	7.5.2/3	
QC-20.2	Control of documents for specific projects	7.5.2/3	
QC-21	Records updates	7.5.2	
QC -22	Control of quality records	7.5.3	
QC-23	Documents control (logs)	7.5.3.2	
QC-24	Project planning and control	8.1	
QC-25	Project specific requirements	8.2	
QC-26	Project specific quality control plan (contractor's quality control plan)	8.2	
QC-27-1	Tender documents	8.2	
QC-27.2	Tender review	8.2	
QC-27.3	Contract review	8.2.1	
QC-28	Construction processes	8.2.2	
QC-29	Variation review	8.2.3	
QC-30.1	Engineering and shop drawings	8.3	
QC-30.2	Records of engineering and shop drawing input	8.3.3	
QC-31	Design developments for design–build projects	8.3	Refer below Annexure 3.4A
QC-32	Selection and evaluation of subcontractors	8.4	
QC-32.1	Selection and evaluation of suppliers	8.4.1	
QC-33	Communication with subcontractors, material suppliers, vendors	8.4.3	
QC-34.1	Inspection of subcontracted work	8.4.2	
QC-34.2	Incoming material inspection and testing	8.4.3	
QC-35	Installation procedures	8.5	
QC-36	Product identification and traceability	8.5.2	
QC-37	Identification of inspection and test status	8.5.2	
QC-38	Control of owner supplied items	8.5.3	
QC-39	Handling and storage	8.5.4	
QC-40	Construction inspection, testing and commissioning	8.6	
QC-41	Project handover	8.6	
QC-42.1	Control of nonconforming work	8.7	
QC-42.2	Control of nonconforming subcontractor	8.7	
QC-43.1	Project performance review	9.1.1	
QC-43.2	Project quality assessment and measurement	9.1.2	
QC-44	Internal quality audits	9.2	
QC-45	Management review	9.3	

(continued)

TABLE 3.4 (Continued)
List of Quality Manual Documents—Contractor

Document No.	Document Title	Relevant Clause in 9001:2015	Version/ Revision Date
QC-46	New technology in construction	10.1	
QC-47.1	Nonconformity and corrective action	10.2.2	
QC-47.2	Preventive action	10.3	
QC-48	Control of client complaints	10.2.2	
QC-49	Continual improvement	10.3	
Annexure 3.4A (Additional Documents for Design–Build Contracting)			
QC-10.4	Project risk management (tendering)	6.1	
QC-10.5	Project risk management (design)	6.1	
QC-10.6	Project risk management (construction)	6.1	
QC-12.3	Change management (design changes)	6.3	
QC-13.3	Design resources (design team, design software)	7.1	
QC-18.4	Competence	7.2	
QC-25.1	Planning of engineering design and quality plan	8.1/8.3	
QC-25.2	Design development	8.1/8.3	
QC-25.3	Records of design development inputs	8.3.3	
QC-41.1	Release of project design	8.6	
QC-43.3	Control of nonconforming work (design errors)	8.7	
QC-44.3	Client/owner/PM/CM comments	9.1	

TABLE 3.5
List of Quality Manual Documents—Manufacturer

Document No.	Document Title	Relevant Clause in 9001:2015	Version/ Revision Date
QC-0	Circulation list		
QC-00	Records of revision		
QC-01	Scope		
QC-02	Normative references		
QC-03	Terms and references		
QC-1.1	Understanding the organization and its context	4.1	
QC-1.2	Monitoring and review of internal and external issues	4.1	
QC-2.1	Relevant requirements of stakeholders	4.2	
QC-2.2	Monitoring and review of stakeholder's information	4.2	

TABLE 3.5 (Continued)
List of Quality Manual Documents—Manufacturer

Document No.	Document Title	Relevant Clause in 9001:2015	Version/ Revision Date
QC-3	Scope of quality management system	4.3/4.4	
QC-4	Project quality management system	4.4	
QC -5	Management responsibilities	5.1	
QC -6	Customer focus	5.1.2	
QC -7.1	Quality policy (organization)	5.2	
QC-7.2	Quality policy (procurement)	5.2.2	
QC-8	Organizational roles, responsibilities, and authorities (organization chart)	5.3	
QC-9	Preparation and control of quality plan	6.0	
QC-10.1	Risk management (quotation/proposal)	6.1	
QC-10.2	Risk management (manufacturing process)	6.1	
QC-10.3	Risk management (selection of vendor, supplier)	6.1	
QC-11	Quality objectives	6.2	
QC-12.1	Change management	6.3	
QC-13.1	Office resources (office staff, office equipment)	7.1	
QC-13.2	Plant resources (human resources, equipment, and machinery)	7.1	
QC-14.1	Infrastructure	7.1.3	
QC-14.2	Work environment	7.1.4	
QC-15	Control of processes, material, measuring and test equipment	7.1.5	
QC-16	Control of human resources	7.1.6	
QC-17	Organizational knowledge	7.1.6	
QC-18.1	Training and development in quality system	7.2	
QC-18.2	Training in quality auditing	7.2	
QC-18.3	Training in operational/technical skills	7.2	
QC-18.4	Records of training, skills, experience, and qualifications	7.2	
QC-19	Communication between internal and external	7.4	
QC-20.1	Control of documents for general application	7.5.2/3	
QC-20.2	Control of documents for specific product/ material/system	7.5.2/3	
QC-21	Records updates	7.5.2	
QC -22	Control of quality records	7.5.3	
QC-23	Process planning and control	8.1	
QC-24	Customer specific requirements	8.2	
QC-25	Customer specific quality control plan	8.2	
QC-26-1	Tender documents	8.2	
QC-26.2	Tender review	8.2	
QC-26.3	Contract (purchase order) review	8.2.1	
QC-27	Manufacturing processes	8.2.2	

(*continued*)

TABLE 3.5 (Continued)
List of Quality Manual Documents—Manufacturer

Document No.	Document Title	Relevant Clause in 9001:2015	Version/ Revision Date
QC-28	Engineering and shop drawings	8.3	
QC-29	Selection and evaluation of vendors, suppliers	8.4	
QC-30	Communication with material suppliers, vendors	8.4.3	
QC-31.1	Inspection of outsources work	8.4.2	
QC-31.2	Incoming material inspection and testing	8.4.3	
QC-32	Fabrication/assembling procedures	8.5	
QC-33	Product identification and traceability	8.5.2	
QC-34	Identification of inspection and test status	8.5.2	
QC-35	Handling and storage	8.5.4	
QC-36	Product inspection and testing	8.6	
QC-37	Product packing and dispatch	8.6	
QC-38.1	Control of nonconforming work	8.7	
QC-38.2	Control of nonconforming work	8.7	
QC-39.1	Production performance review	9.1.1	
QC-39.2	Product quality assessment and measurement	9.1.2	
QC-40	Internal quality audits	9.2	
QC-41	Management review	9.3	
QC-42	New technology in manufacturing	10.1	
QC-43.1	Nonconformity and corrective action	10.2.2	
QC-43.2	Preventive action	10.3	
QC-44	Control of client/customer complaints	10.2.2	
QC-45	Continual improvement	10.3	

4 Applications of Total Quality Management Concepts

4.1 INTRODUCTION

Quality in construction projects is not only the quality of products and equipment used in the construction, it is the total management approach to complete the facility as per the scope of works to customer/owner satisfaction to be completed within specified schedule and within the budget to meet owner's defined purpose.

Construction has unique problems compared to manufacturing. A few of these are listed as follows:

- The construction is a custom rather than a routine, repetitive business and differs from manufacturing.
- Construction is different from both that of mass production and batch (lot) production manufacturing.
- Quality in manufacturing passes through a series of processes. The output is monitored by inspection and testing at various stages of production.
- The location of construction projects varies widely. In a manufacturing plant a given operation is assigned to and carried out in one place. In contrast, specialized construction crews progress from location to location.
- In construction projects, the scenario is not the same as that of manufacturing. If anything goes wrong, the nonconforming work is very difficult to rectify, and remedial action is sometimes not possible. Quality costs play an important role in construction projects.
- The construction is a preliminary step leading to a completed facility; the layout and arrangements may make access for construction difficult and permanent provisions for safety impossible.
- In construction, an activity may be repeated at various stages, but it is done only one time for a specific work. Therefore, it has to be right from the onset.
- In manufacturing, the buyer does not enter the scene until the product comes into being, whereas in construction the buyer (project owner) is involved from beginning to end. Even during the construction phase, it is likely that certain modifications may take place.

DOI: 10.1201/9781003451464-4

- The owner is deeply involved in the construction process, while the purchaser of manufactured goods is not. Buyers of the usual manufactured products seldom have access to the plant where they are made, nor do they deal directly with factory managers.
- Most projects or their individual work phases are of relatively short duration. One consequence is that management teams and possibly the work force must be assembled quickly and cannot be shaken out or restructured before the project or work phase is completed.
- To a great extent, each project has to be designed and built to serve a specific need and therefore it is necessary to make certain modifications in the system process to fit the particular conditions of each construction project and its specific problems.
- Operations are commonly conducted out of doors and are subject to all the interruptions and variations in conditions and the other difficulties that rain, snow, heat, and cold can introduce.
- The final product is usually of unique design and differs from workstation to workstation so that no fixed arrangement of equipment or aids such as jigs and fixtures are possible as is in the case of manufacturing.
- The construction often needs highly skilled craftsmen rather than unskilled workers; individual crews, whether union or nonunion, usually do specialized operations.
- Construction involves installation and integration of various materials, equipment, systems, or other components to complete the facility.
- Construction focuses mainly on overall performance of the project or facility in which a product(s) or a system(s) is a part and assembled/installed to achieve the objectives.
- Construction projects work against defined scope, schedules, and budget to achieve the specified result.
- Performance of construction projects can be evaluated only after it is completed and put into use/operation.

Construction projects comprised of a cross-section of many different participants. Construction projects have involvement of many participants comprising mainly owner, designer (consultant), and contractor. Apart from these, there are many other professionals from construction-related industries. Each of these participants is involved in implementing quality in construction projects. These participants are both influenced by and depend on each other in addition to "other players" involved in the construction process. Therefore, the construction projects have become more complex and technical, and extensive efforts are required to reduce rework and costs associated with time, materials, and engineering, and so forth.

Total quality management (TQM) is an organization-wide effort centered on quality to improve performance that involves everyone and permeates every aspect of an organization to make quality a primary strategic objective. TQM focuses on participative management and strong operational accountability at the individual contributor level.

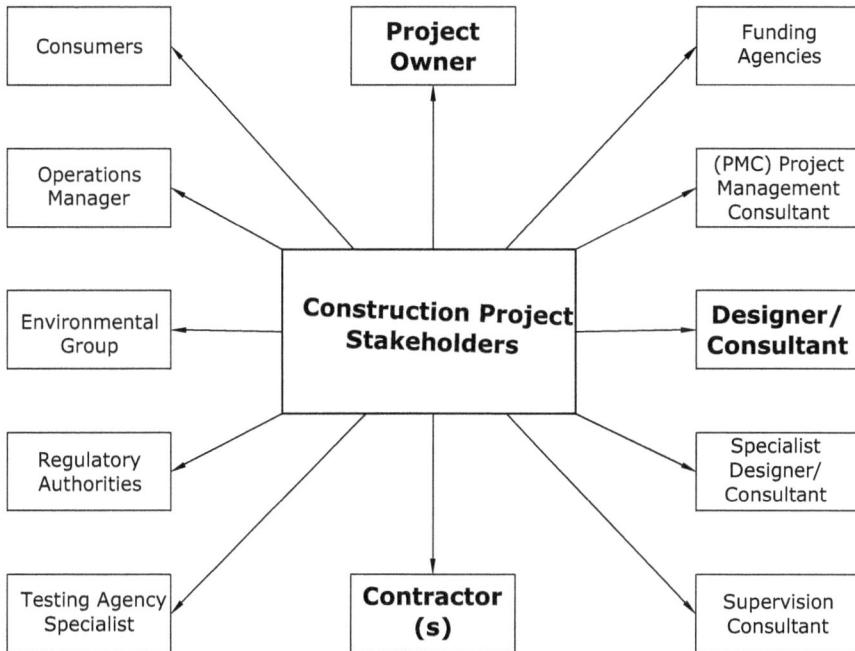

FIGURE 4.1   Construction project stakeholders. (Abdul Razzak Rumane (2016). *Handbook of Construction Management: Scope, Schedule, and Cost Control.* Reprinted with permission from Taylor & Francis Group.)

Construction projects have the involvement of many stakeholders. Their participation involvement is essential in all the related activities for the application of TQM concept. Figure 4.1 illustrates stakeholders having involvement or interest in the construction project.

These stakeholders have significant influence/impact on the outcome of the project. It is important to identify stakeholders who have interest and significant influence in each of the activities/elements of the project phases. In the construction project, the stakeholders include members from within the organization and people, agencies, and authorities outside the organization. It is essential that while developing a construction project, it is important to address the needs of project stakeholders effectively predicting how the project will be affected and how the stakeholders will be affected, and the input from these stakeholders is taken into consideration to develop and to run a successful project,

Table 4.1 is an example stakeholders responsibility matrix for construction project.

In order to conveniently manage and control the project, it is divided into a number of stages/phases. The division of project life cycle helps in application of TQM concept thus satisfying the customer needs and requirements and achieving a successful project.

TABLE 4.1

Stakeholders Responsibility Matrix

Serial Number	Activity	Owner/ client	Construction Manager/ Project Manager	Designer/ Consultant	Contractor	Supervisor	Regulatory Authority	Funding Agency	End User/ Facility Manager	Notes/ Comments
1	Project Initiation	P	-	-	-	-	-	B	B	
2	Selection of Construction Manager	P	-	-	-	-	-	-	-	
3	Selection of Designer	P	B	-	-	-	-	-	-	
4	Preparation of Terms of Reference (TOR)	A	P	-	-	-	-	-	-	
5	Preparation of Design	A	B	P	-	-	R	-	-	
6	Value Engineering	A	R	P	-	-	-	-	-	
7	Preparation of Contract Documents	A	B	P	-	-	-	-	-	
8	Project Schedule	A	B	P	-	-	-	C	C	
9	Project Budget	A	B	P	-	-	-	B	-	
10	Preparation of Tendering Documents	A	P	B	-	-	-	-	-	
11	Submission of Bid	C	C	-	P	-	-	-	-	

12	Evaluation of Bid	C	C	P	-	-	-	-	-
13	Selection of Contractor	A	P	B	-	-	-	C	C
14	Approval of Subcontractor	A	B	B	P	-	-	-	-
15	Approval of Contractor's Staff	A	B	B	P	-	-	-	-
16	Execution of Works	C	C	R	P	R	-	-	-
17	Supervision of Works	C	C	R	P	P	-	-	-
18	Approval of Material	C	A	R	P	B	-	-	-
19	Approval of Shop Drawings	C	C	A	P	B	-	-	-
20	Construction Schedule	C	A	R	P	B	-	-	-
21	Monitoring Progress	C	P	P	P	B			
22	Monitoring Cost	C	P	P	B	B	-	-	-
23	Payments	A	R	R	P	B			
24	Request for Information	C	C	R	P	B	-	-	-
25	Approval of Change	A	B	R	P	B	-	-	-
26	Quality Plan	C	B	R	P	B	-	-	-
27	Project Quality	C	R	R	P	P	-	-	-
28	Meetings	E	E	P	E	E			

(continued)

TABLE 4.1 (Continued)
Stakeholders Responsibility Matrix

Serial Number	Activity	Owner/ client	Construction Manager/ Project Manager	Designer/ Consultant	Contractor	Supervisor	Regulatory Authority	Funding Agency	End User/ Facility Manager	Notes/ Comments
29	Safety Plan	C	B	R	P	B	-	-	-	
30	Site Safety	C	C	B	P	P	-	-	-	
31	Testing and Commissioning	C	C	R	P	D	-	-	C	
32	Authorities Approval	C	C	B	P	B	A	-	-	
33	Snag List	C	C	R	P	P	-	-	C	
34	Substantial Completion Certificate	A	R	P	C	-	-	-	C	

LEGEND: P = Prepare/Initiate/Responsible, R = Review/Comment, B = Advise/Assist, A = Approve, E = Attend, C = Inform.

4.2 PROJECT LIFECYCLE PHASES

Most construction projects are custom-oriented, having a specific need and a customized design. It is always the owner's desire that his project should be unique and better. Further, it is the owner's goal and objective that the facility is completed on time. The expected time schedule is important for both financial and acquisition of the facility by the owner/user.

A systems engineering approach to construction projects helps to understand the entire process of project management and to manage and control its activities at different levels of various phases to ensure timely completion of the project with economical use of resources to make the construction project most qualitative, competitive, and economical.

The system life cycle is fundamental to the application of systems engineering. Detailed presentation of the elaborated technical activities and interaction that must be integrated over the project life cycle is shown in the life cycle phases of the project.

Systems engineering starts from the complexity of the large-scale problem as a whole and moves toward the structural analysis and partitioning process until the questions of interest are answered. This process of decomposition is called a work breakdown structure (WBS). The WBS is a hierarchical representation of system levels. Being a family tree, the WBS consists of a number of levels, starting with the complete system at level 1 at the top and progressing downward through as many levels as necessary to obtain elements that can be conveniently managed.

Benefits of systems engineering applications are

- Reduction in the cost of system design and development, production/construction, system operation and support, system retirement, and material disposal
- Reduction in system acquisition time
- More visibility and reduction in the risks associated with the design decision-making process

Though it is difficult to generalize project life cycle to system life cycle, considering that there are innumerable processes that make up the construction process, the technologies, and processes as applied to systems engineering can also be applied to construction projects. The number of phases shall depend on the complexity of the project.

For a major construction project, it is ideal to divide the construction project life cycle into seven phases for non-process type of construction projects. These are

1. Conceptual Design
2. Schematic Design
3. Detail Design
4. Construction Documents
5. Bidding and Tendering
6. Construction
7. Testing, Commissioning, and Handover

Each phase can further be subdivided into the work breakdown structure (WBS) principle to reach a level of complexity where each element/activity can be treated as a single unit that can be conveniently managed. WBS represents a systematic and logical breakdown of the project phase into its components (activities). It is constructed by dividing the project into major elements with each of these being divided into sub-elements. This is done until a breakdown is done in terms of manageable units of work for which responsibility can be defined. WBS involves envisioning the project as a hierarchy of goals, objectives, activities, sub-activities, and work packages. The hierarchical decomposition of activities continues until the entire project is displayed as a network of separately identified and nonoverlapping activities. Each activity will be of single purpose, of a specific time duration, and manageable; its time and cost estimates easily derived, deliverables clearly understood, and responsibility for its completion clearly assigned. The WBS helps in

- Effective planning by dividing the work into manageable elements, which can be planned, budgeted, and controlled
- Assignment of responsibility for work elements to project personnel and outside agencies
- Development of control and information system

WBS facilitates the planning, budgeting, scheduling, and control activities for the project manager and their team. By application of WBS phenomenon, the construction phases are further divided into various activities. Division of these phases will improve the control and planning of the construction project at every stage before a new phase starts. The components/activities of construction project lifecycle phases divided on WBS principle are listed as follows:

1. Conceptual Design
   - Identification of need
   - Feasibility
   - Project goals and objectives
   - Identification of alternatives/options
   - Selection of preferred activity
   - Project delivery/contracting system
   - Selection of project team (designer A/E)
   - TOR/Project charter
   - Development of concept design
   - Identification of stakeholders
   - Establish concept design scope of work
   - Prepare concept design
   - Prepare time schedule
   - Estimate project cost
   - Quality management
   - Estimate resources
   - Identify/Manage risk

- HSE issues/requirements
- Review concept design
- Finalize concept design

2. Schematic/Preliminary Design
   - Identification of stakeholders/project team
   - Identification of preliminary design requirements
   - Establish preliminary design scope
   - Develop preliminary management plan
   - Develop preliminary design
   - Develop contract terms and conditions
   - Regulatory/authorities' approval
   - Estimate preliminary schedule
   - Estimate project cost
   - Quality management
   - Estimate resources
   - Identification/management of risk
   - HSE issues/requirements
   - Perform value engineering study
   - Review preliminary design
   - Finalize schematic/preliminary design

3. Detail Design
   - Identification of stakeholders/project team
   - Identification of detail design requirements
   - Develop detail design of the works
   - Regulatory/authorities' approval
   - Prepare contract documents and specifications
   - Prepare schedule
   - Estimate cost
   - Quality management
   - Estimate resources
   - Identification/management of risk
   - HSE issues/requirements
   - Finalize detail design

4. Construction Documents
   - Identification of stakeholders/project team members
   - Identification of construction documents requirements
   - Establish construction documents scope
   - Develop construction documents
   - Develop tender documents
   - Estimate project schedule
   - Estimate cost/budget
   - Quality management

- Estimate resources
- Identification/management of risk
- HSE issues/requirements
- Review construction documents (designer)
- Review construction documents (owner)
- Finalize construction documents
- Release for bidding and tendering

5. Bidding and Tendering
   - Identification of stakeholders/project team members
   - Organize tender documents
   - Identification of tendering procedure
   - Identification of bidders
   - Manage tendering process
   - Submit/receive tender documents
   - Manage bidding and tendering quality
   - Identification/management of risk
   - Review bid documents
   - Select contractor
   - Award contract

6. Construction
   - Identification of stakeholders/project teams
   - Mobilization
   - Development of project site facilities
   - Identification of project execution/instruction requirements
   - Identification of sustainability requirements
   - Develop project execution scope
   - Planning and scheduling
   - Construction/execution of works
   - Monitoring and control
   - Inspection of executed/installed works
   - Validate execute works

7. Testing, Commissioning, and Handover
   - Identify stakeholders
   - Identify testing, commissioning, and handover requirements
   - Identify sustainability requirements
   - Develop testing and commissioning scope
   - Develop inspection and testing plan
   - Execute commissioning works/systems
   - Manage testing, commissioning quality
   - Develop documents
   - Train owner's/end user's personnel
   - Handover of facility to owner/end user

TABLE 4.2
Construction Project Life Cycle Phases (Non-process-Type of Projects)

Construction Project Life Cycle (Design-Bid-Build) Phases

Conceptual Design	Schematic Design/ Preliminary Design	Design Development/ Detail Design	Construction Documents	Bidding and Tendering	Construction	Testing, Commissioning, and Handover
• Identification of Need	• Identification of Stakeholders/Project Team Members	• Identification of Stakeholders/Project Team Members	• Identification of Stakeholders/ Project Team Members	• Identification of Stakeholders/Project Team Members	• Identify Stakeholders/ Project Teams	• Identify Stakeholders/ Project Team mambers
• Feasibility	• Identification of Preliminary Design Requirements	• Identification of Detail Design Requirements	• Identification of Construction Documents Requirements	• Organize Tender Documents	• Mobilization	• Identify Testing, Commissioningind Handover Requirements
• Project Goals and Objectives	of	• Establish Detail Design Scope	• Establish Construction Documents Scope	• Identify Tendering Process	• Development of Project Site Facilities	• Identify Sustainability Requirements
• Identification of Alternatives/Options	• Develop Management Plan	• Develop Project Management Plan	• Develop Tender Documents	• Identify Bidders	• Identification of Project Execution/ Installation Requirements	• Develop Testing, Commissioning Scope
• Select Preferred Alternative	• Develop Preliminary Design	• Develop Detail Design	• Develop Tender Documents	Establish Bid Review Process	• Identification Sustainability Requirement	• Develop Inspection and Testing Plan
• Establish Project Delivery and Contracting System	• Prepare Contract Terms and Conditions	• Develop Contract Documents	• Develop Project Schedule	Manage Tendering Process	• Establish Project Execution/ Installation Scope	• Execute Commissioning of Works/Systems
• Selection of Project Team (Designer)	• Regulatory/Authority Approval	• Regulatory Approval	• Estimate Project Cost/Budget	• Submit/Receive Bids	• Project Planning and Scheduling	• Manage Tesing, Commissioning Quality
• Terms of Reference (TOR)/Project Charter	• Estimate Preliminary Schedule	• Prepare Schedule	• Manage Construction Documents Quality	• Manage Bidding and Tendering Quality	• Develop Management Plans	• Develop Documents
• Development of Concept Design	• Estimate Project Cost	• Estimate Cost	• Estimate Resources	• Identification/Management of Tendering Risk	• Construction/ Execution of Works	• Train Owner's/End User's Personnel
• Identify Concept Design Stakeholders	• Quality Management	• Manage Design Quality	• Identification/Management of Risk	• Review Bid Documents	• Monitoring and Control	• Handover Project
• Establish Concept Design Scope of work	• Estimate Resources	• Estimate Resources	• HSE Issues/Requirements	• Select Contractor	• Inspection of Executed Works/Systems	• Issue Substantial Completion Certificate
• Prepare Concept Design	• Identification/Management of Risk	• Identification/Management of Risk	• Review Construction Documents (Designer)	• Award Contract	• Validate Exeduted Works	• Lesson Learned
• Develop Time Schedule	• HSE Requirements	• HSE Issues/Requirements	• Review Construction Documents (Owner)			• Settle Payments
• Estimate Project Cost	• Perform Value Engineering Study	• Review Detail Design	• Finalize Construction Documents			• Settle Claims
• Quality Management	• Review Preliminary Design	• Finalize Detail Design	• Release for Bidding and Tendering			• Close Contract
• Estimate Resources	• Finalize Preliminary Design					
• Management of Risk/HSE Issues/Requirements						
• Review Concept Design						
• Finalize Concept Design						

- Issue substantial completion certificate
- Lesson learned
- Settle payments
- Settle claim
- Close contract

Table 4.2 illustrates the subdivided major activities/components of the construction project life cycle.

These activities may not be strictly sequential; however, the breakdown allows the implementation of project management functions more effectively at different stages.

4.3 TOTAL QUALITY MANAGEMENT IN CONSTRUCTION PHASES

TQM is an organization-wide effort centered on quality to improve performance that involves everyone and permeates every aspect of an organization to make quality a primary strategic objective. It is a way of managing an organization to ensure satisfaction at every stage of the needs and expectations of both internal and external customers.

The TQM concept was born following World War II. It was stimulated by the need to compete in the global market where higher quality, lower cost, and more rapid development are essential to market leadership. Today, TQM is considered a fundamental requirement for any organization to compete, let alone lead, in its market. It is a way of planning, organizing, and understanding each activity of the process and removing all the unnecessary steps routinely followed in the organization. TQM is a philosophy that makes quality values the driving force behind leadership, design, planning, and improvement in activities.

Quality in construction projects is different from that of manufacturing. Quality in construction projects is not only the quality of products and equipment used in the construction, it is the total management approach to complete the facility as per the scope of works to customer/owner satisfaction to be completed within specified schedule and within the budget to meet owner's defined purpose.

Construction projects have involvement of many participants comprising owner, designer, contractor, and many other professionals from construction-related industries. Each of these participants is involved in implementing quality in construction projects. These participants are both influenced by and depend on each other in addition to "other players" involved in the construction process. Therefore, the construction projects have become more complex and technical and extensive efforts are required to reduce rework and costs associated with time, materials, and engineering.

TQM focuses on participative management and strong operational accountability at the individual contributor level. TQM is the implementation of quality system by organization to satisfy the customer needs/requirements by developing a system that will make the project qualitative, competitive, and economical. A collaborative effort in the construction project is required among all the participants in order to improve alignment and misunderstanding during the execution of project. This is achieved through complex interaction of many participants in the facilities development process by application of TQM concept in the construction project.

TQM in construction projects typically involves ensuring compliance with minimum standards of material and workmanship in order to ensure the performance of the facility according to the design. TQM in a construction project is a cooperative form of doing the business that relies on the talents and capabilities of both labor and management to continually improve quality. The important factor in construction projects is to complete the facility per the scope of works to customer/owner satisfaction within the specified schedule and approved budget and to complete the work to meet the owner's defined purpose. TQM will help reduce overrun of schedule and cost.

The quality actually emerged as a dominant thinking since World War II, becoming an integral part of overall business system focused on customer satisfaction and becoming known in the twentieth century as "Total Quality Management" (TQM) with its three constitutive elements:

Total: Organization-wide
Quality: Customer satisfaction
Management: Systems of managing

TQM is a standard management practice wherein each employee within an organization continually analyzes its production processes to improve the manufacturing quality of products and services and enhance customer satisfaction. It involves conducting management training and implementing analytical methods to identify and remove problem areas in business operations.

TQM can be summarized as a management system for a customer-focused organization that involves all employees in effective communications to integrate quality discipline into the culture and activities of the organization. Many management systems are the successor to TQM. It is based on ISO quality principles together with the philosophies of quality gurus (please refer to Sections 1.4.9 and 1.5).

Figure 4.2 illustrates the basic tenets of TQM that are developed considering ISO quality principles and philosophies of quality gurus.

Most construction projects are custom-oriented, having specific need and customized design. Construction projects are constantly increasing in technological complexity, and the relationship and the contractual grouping of those who are involved are also complex and contractually varied. There are innumerable processes that make up the construction project.

Though it is difficult to generalize the number of phases of project life cycle, the number of phases depend on considering the innumerable processes that make up the construction process, the technologies, processes, and complexity of construction projects. Duration of each phase may vary from project to project.

Generally, the major construction projects (non-process type of projects) have seven most common phases. These are

1. Conceptual design
2. Preliminary design
3. Detail design
4. Construction documents
5. Bidding and tendering
6. Construction
7. Testing, commissioning, and handover

Each phase can further be subdivided into the WBS principle to reach a level of complexity where each element/activity can be treated as a single unit that can be conveniently managed. WBS represents a systematic and logical breakdown of the project phase into its components (activities). Each activity is single purpose, of a specific time duration, and manageable; its time and cost estimates are easily derived, deliverables are clearly understood, and responsibility for its completion is clearly assigned.

Figure 4.3 illustrates a typical flow chart for TQM applications in the quality of the activity/element of the construction project phase.

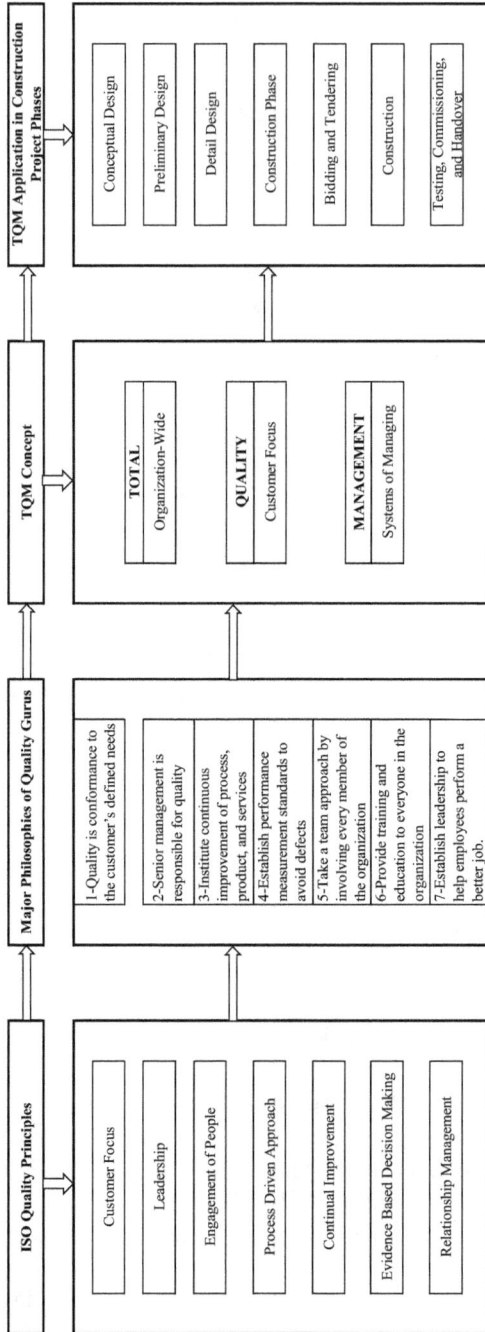

FIGURE 4.2 Basic tenets of TQM.

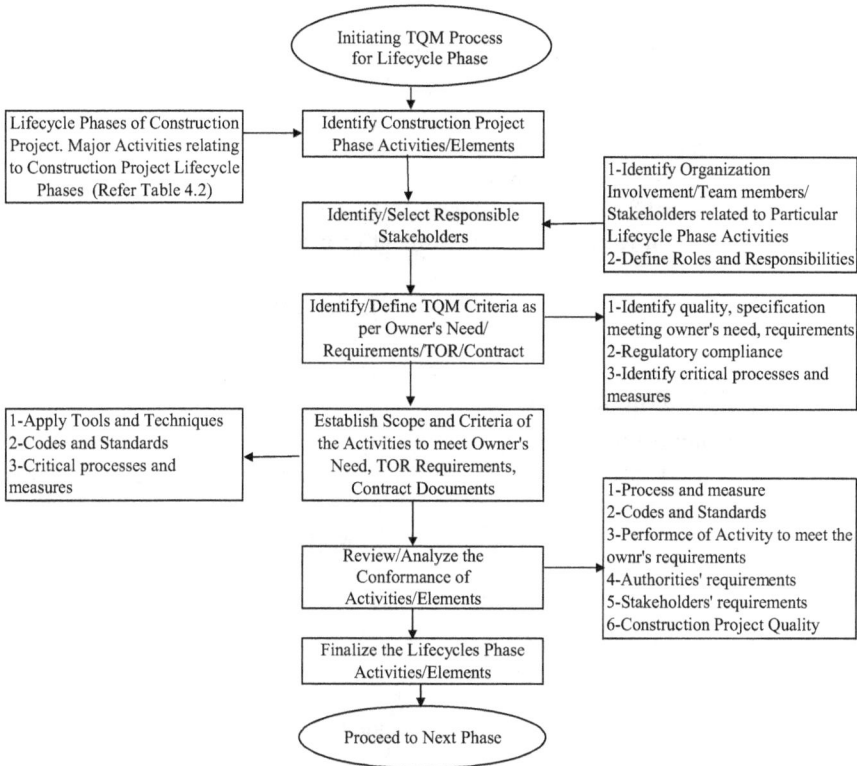

FIGURE 4.3 Typical flowchart for total quality management applications in quality of a construction project phase activity.

4.3.1 TQM IN CONCEPTUAL DESIGN PHASE

Conceptual design is the first phase of the construction project life cycle. The conceptual design is initiated once the need is recognized. In this phase, the idea is conceived and given an initial assessment. Conceptual design, or the design development phase, is often viewed as most critical to achieving outstanding project performance. During the conceptual phase, the environment is examined, forecasts are prepared, objectives and alternatives are evaluated, and the first examination of the technical performance, time, and cost objectives of the project is made. The conceptual phase includes

- Identification of need by the owner, and establishment of main goals
- Feasibility study, which is based on owner's objectives
- Identification of alternatives
- Client brief (Terms of Reference)
- Identification of project team by selecting other members and allocation of responsibilities
- Development of concept design

- Time schedule
- Financial implications and resources based on estimation of life cycle cost of the favorable alternative

The most significant impacts on the quality of a project occur during the conceptual phase. This is the time when specifications, statement of work, contractual agreements, and initial design are developed. Initial planning has the greatest impact on a project because it requires the commitment of processes, resources schedules, and budgets.

Figure 4.4 illustrates a logic flowchart of major activities in the conceptual design phase.

4.3.1.1 Identification of Need

The need of the project is created by the owner. The owner of the facility could be an individual, a public/private sector company, or a governmental agency. The need could be to develop a new facility/project or renovation/refurbishment of existing facility or to produce new products. The need of the project is created by the owner and is linked to the available financial resources to develop the facility/project. The owner's need must be well defined describing the minimum requirements of quality, performance, project completion date, and approved budget for the project.

Owner's needs are quite simple and are based on the following:

- To have the best facility/project for the money, that is, to have maximum profit or services at a reasonable cost
- On-time completion, that is, to meet the owner's/user's schedule
- Completion within budget, that is, to meet the investment plan for the facility.

It is essential to get a clear definition of the identified need or the problem to be solved by the new project. The owner's need must be well defined, indicating the minimum requirements of quality and performance, an approved budget, and required completion date. The need should be based on real (perceived) requirements. The identified need is then assessed and analyzed to develop a need statement. Need assessment is conducted to determine the need. Need assessment is a systematic process for determining and addressing needs or "gaps" between current conditions and desired conditions "want." Need analysis is the process of identifying and evaluating need. The need statement is written based on the need analysis and is used to perform feasibility study to develop project goals and objectives and subsequently to prepare project scope documents.

Table 4.3 illustrates 5W2H analysis of identification of project need.

The owner's need must be well defined indicating the minimum requirements of quality and performance, an approved main budget, and required completion date. Sometimes, the project budget is fixed and therefore the quality of building system, materials, and finishes of the project need to be balanced with the budget. A business case typically addresses the business need for the project and the value the project

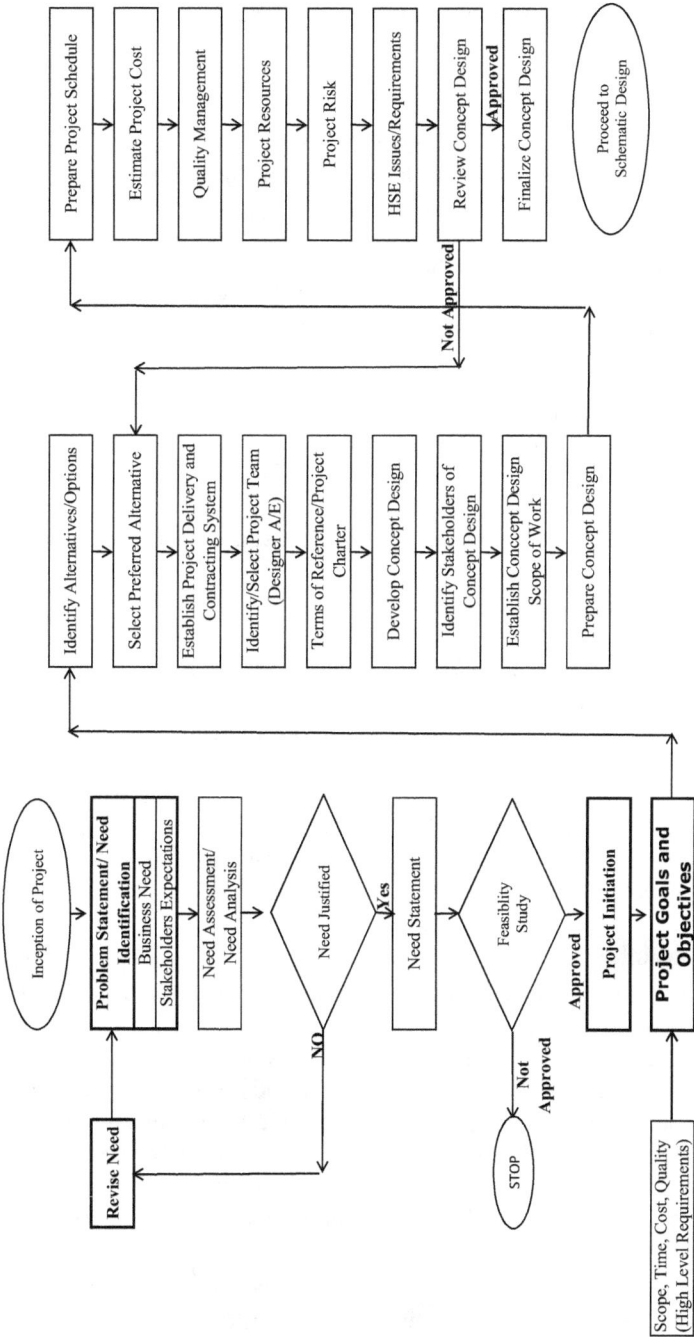

FIGURE 4.4 Logic flowchart of major activities in conceptual design phase.

TABLE 4.3
5 W 2 H Analysis for Identification of Project Need

Serial Number	Why	Related Analyzing Question
1	Why	Why new project?
2	What	What advantage it will have over other similar projects producing similar items?
3	Who	Who will be the customer for the products produced in this plant?
4	Where	Where from the raw material to feed the project requirements to produce desired products will be available?
5	When	When the project will be ready?
6	How many	How many competitors are in the market for the products produced in this project?
7	How much	How much market share we will have for the products/items produced in this project?

brings to the business (project value proposition). A value proposition is a promise of value to be delivered by the project.

Following questions address the value proposition:

a. How the project solves the current problems or improves the current situation?
b. What specific benefits the project will deliver?
c. Why the project is the ideal solution for the problem?

Business need assessment is essential to ensure that the owner's business case has been properly considered before the initial Project Brief (Need Statement) is developed.

Table 4.4 lists major points to be considered for need analysis of a construction project and Table 4.5 illustrates the need statement.

4.3.1.2 Feasibility Study

The feasibility study takes its starting point from the output of the project identification need. Feasibility study is conducted to assist owner/decision makers in making the decision that will be in the best interest of the owner. The main purpose of feasibility study is to evaluate the project need and decide whether to proceed with the project or stop. Depending on the circumstances the feasibility study may be short or lengthy, simple or complex. In any case, it is the principal requirement in project development as it gives the owner an early assessment of the viability of the project and the degree of risk involved. Project feasibility study is usually performed by the owner through his own team or by engaging individuals/organizations involved in the preparation of economical and financial studies. However, the feasibility study can be conducted by specialist consultant in this field.

Table 4.6 lists the qualification requirements of the consultant for feasibility study.

Table 4.7 illustrates major considerations to carry out feasibility study.

TABLE 4.4
Major Considerations for Need Analysis of a Construction Project

Serial Number	Points to be Considered
1	Is the project in line with organization's strategy/strategic plan and mandated by management in support of a specific objective?
2	Is the project a part of mission statement of the organization?
3	Is the project a part of vision statement of the organization?
4	Is the need mandated by regulatory body?
5	Is the need for meeting government regulations?
6	Is the need to fulfill the deficiency/gap of such type of project(s) in the market?
7	Is the need created to meet market demand
8	Is the need to meet the research and development requirements?
9	Is the need for technical advances?
10	Is the need generated to construct a facility/project which is innovative in nature?
11	Is the need aims to improve the existing facility?
12	Is the need a part of mandatory investment?
13	Is the need to develop infrastructure?
14	Is the need will serve the community and fulfill social responsibilities?
15	Is the need created to resolve specific problem?
16	Is the need will have effect on environment?
17	Is the need has any time frame to implement?
18	Is the need has financial constraints?
19	Is the need has major risk?
20	Is the need within capability of the owner/client, either alone or in cooperation with other organizations?
21	Is the need can be managed and implemented?
22	Is the need realistic and genuine?
23	Is the need measurable?
24	Is the need shall be beneficial?
25	Is the need comply with environmental protection agency requirements?
26	Is the need comply with government's health and safety regulations?

The objective of feasibility study is to review technical/financial viability of the project to give sufficient information to enable the client to proceed or abort the project. Feasibility study is undertaken to analyze the ability to complete a project successfully, taking into account various factors such as economic, technological, scheduling, and so on. A feasibility study looks into the positive and negative effects of a project before investing the company resources viz. time and money.

Feasibility study is defined as an evaluation or analysis of the potential impact of the identified need of the proposed project. The feasibility study assists decision makers (investors/owners/clients) in determining whether or not to implement the project. Since the feasibility study is a very crucial stage, in which all kinds of professionals and specialists are required to bring many kinds of knowledge and experience into

TABLE 4.5
Need Statement

Serial Number	Points to be Considered
1	Project purpose and need
	a) Project description
2	What is the purpose of project?
	a) Project justification
3	Why the project is needed now?
4	How did the need of the project determined?
	a) Supporting data
5	Is it important to have the needed project
6	Whether such facility/project is required?
7	What are the factors contributing to the need?
8	What is the impact of the need?
9	Is the need will improve the existing situation and beneficial?
10	What are the hurdles?
11	What is the time line for the project
12	What are funding sources for the project?
13	What are the benefits of the projects?
14	What are the environmental impact?

TABLE 4.6
Consultant's Qualification for Feasibility Study

Serial Number	Description
1	Experience in conducting feasibility study
2	Experience in conducting feasibility study in similar type and nature of projects
3	Fair and neutral with no prior opinion about what decision should be made
4	Experience in strategic and analytical analysis
5	Knowledge of analytical approach and background
6	Ability to collect large number of important and necessary data via work sessions, interviews, surveys, and other methods
7	Market knowledge
8	Ability to review and analysis of market information
9	Knowledge of market trend in similar type of projects/facility
10	Multidisciplinary experienced team having proven record in following fields:
	a) Financial analyst
	b) Engineering/Technical expertise
	c) Policy experts
11	Experience in review of demographic and economic data

TABLE 4.7
Major Considerations for Feasibility Study of a Construction Project

Serial Number	Points to be Considered
1	Technical suitability of facility for intended use by the owner/end user
2	Economical feasibility to ascertain value of benefit that results from the project exceeds the cost that results from the project
3	Financial payback period
4	Market demand
5	Environmental impact
6	Social and cultural assessment
7	Legal and regulatory impacts
8	Political aspects
9	Resources availability
10	Scheduling of the project
11	Operational
12	Risk analysis

broad-ranging evaluation of feasibility, it is required to engage a firm having expertise in the related fields. The feasibility study establishes the broad objectives for the project and so exerts an influence throughout subsequent stages. The successful completion of the feasibility study marks the first of several transition milestones and is therefore most important to determine whether or not to implement a particular project or program. The feasibility study decides the possible design approaches that can be pursued to meet the need.

Following are the contents of feasibility study report:

1. Purpose of feasibility study
2. Project history (project background information)
3. Description of proposed project
   a. Project location
   b. Plot area
   c. Interface with adjacent/neighboring area
   d. Expected project deliverables
   e. Key performance indicators
   f. Constraints
   g. Assumptions
4. Business case
   a. Project need
   b. Stakeholders
   c. Project benefits
   d. Estimated time
   e. Financial benefits
   f. Estimated cost
   g. Justification

5. Feasibility study details
    a. Technical
    b. Economical
    c. Time scale
    d. Financial
    e. Environmental
    f. Ecological
    g. Sustainability
    h. Political, legal
    i. Social
6. Risk
7. Environmental impact(considerations)
8. Social impact (considerations)
9. Final Recommendation

4.3.1.3 Project Goals and Objectives

The outcome of the feasibility study helps in selection of a defined project which meets the stated project objectives, together with a broad plan of implementation. If the feasibility study shows that the objectives of the owner are best met through the ideas generated, then the project is moved to a further stage to deliver the intended objectives of the project passing through different stages of project life cycle.

After completion and approval of the feasibility study, it is possible to establish project goals and objectives. Project goals and objectives are prepared taking into consideration the final recommendations/outcome of the feasibility study. Clear goals and objectives provide the project team with appropriate boundaries to make decisions about the project and ensure the project/facility will satisfy the owner's/end user's requirements fulfilling the owner's needs. Establishing properly defined goals and objectives is the most fundamental element of project planning. Therefore, the project goals and objectives must be

- Specific (Is the goal specific?)
- Measurable (Is the goal measurable?)
- Agreed upon/Achievable (Is the goal achievable?)
- Realistic (Is the goal realistic or result-oriented?)
- Time (cost) limited (Does the goal have a time element?)

The project objective definition usually includes the following information:

1. Project scope and project deliverables
2. Preliminary project schedule
3. Preliminary project budget
4. Specific quality criteria the deliverables must meet
5. Type of contract to be employed
6. Design requirements

7. Regulatory requirements
8. Potential project risks
9. Environmental considerations
10. Logistic requirements

4.3.1.4   Identification of Alternatives/Options

Once the owner/client defines the project objectives, a project team (in-house or out-side agency) is selected to start development of alternatives. In certain cases, the owner assigns the designer (consultant) to develop conceptual alternatives, evaluation of conceptual alternatives, and selection of preferred alternatives in consultation with the owner and is included in the Terms of Reference.

The project goals and objectives serve as a guide for the development of alternatives. The team develops several alternative schemes and solutions. Each alternative is based on the predetermined set of performance measures to meet the owner's requirements. In the case of construction projects, it is mainly the extensive review of development options which are discussed between the owner and the team members. The team provides engineering advice to the owner to enable him assess the feasibility and relative merits of various alternative schemes to meet his requirements.

4.3.1.4.1   Analyze Alternatives

Quantitative comparison and evaluation of identified alternatives is carried out by considering the advantages and disadvantages of each item systematically. Social, economic, and environmental impacts, functional capability, safety, and reliability should be considered while development of alternatives. Each alternative is compared by considering the advantages and disadvantages of each systematically to meet predetermined set of performance measures and owner's requirements. The team makes a brief presentation to the owner, and the project is selected based on preferred conceptual alternatives.

Following elements are considered to analyze and evaluate each of the identified alternatives:

1. Suitability to the purpose and objectives
2. Performance parameters
3. Economy
4. Cost efficiency
5. Life cycle costing
6. Sustainable (environmental, social, economical)
7. Environmental impact
8. Environment-preferred material and products
9. Physical properties, thermal comfort, insulation, fire resistance
10. Utilization of space
11. Accessibility
12. Ventilation
13. Indoor air quality

14. Water efficient
15. Power consumption (energy saving measures—renewable energy—alternate energy)
16. Daylighting
17. High-performance lighting
18. Green building concept
19. Aesthetic
20. Safety and security
21. Statutory/regulatory requirements
22. Codes and standards
23. Any other critical issues

4.3.1.5 Selection of Preferred Alternative

Based on the analysis of identified alternatives, the preferred alternative that satisfies the project goals and objectives is selected.

With the approval of Preferred Alternative by the owner, the project proceeds toward the next stage of project development process.

Figure 4.5 illustrates Evaluation and Analysis Method for selection of Preferred Alternative.

4.3.1.6 Project Delivery System/Contracting System

Construction projects have the involvement of three major groups or parties. These are

1. Owner
2. Designer (consultant)
3. Contractor

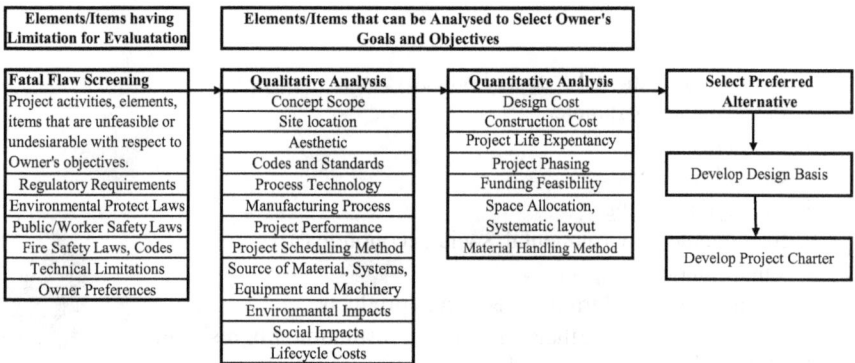

Elements/Items having Limitation for Evaluatation	Elements/Items that can be Analysed to Select Owner's Goals and Objectives		
Fatal Flaw Screening	**Qualitative Analysis**	**Quantitative Analysis**	**Select Preferred Alternative**
Project activities, elements, items that are unfeasible or undesiarable with respect to Owner's objectives.	Concept Scope	Design Cost	
	Site location	Construction Cost	
	Aesthetic	Project Life Expentancy	
Regulatory Requirements	Codes and Standards	Project Phasing	Develop Design Basis
Environmental Protect Laws	Process Technology	Funding Feasibility	
Public/Worker Safety Laws	Manufacturing Process	Space Allocation,	
Fire Safety Laws, Codes	Project Performance	Systematic layout	
Technical Limitations	Project Scheduling Method	Material Handling Method	Develop Project Charter
Owner Preferences	Source of Material, Systems, Equipment and Machinery		
	Environmantal Impacts		
	Social Impacts		
	Lifecycle Costs		

FIGURE 4.5 Evaluation and analysis of alternatives method for selection of preferred alternative.

These parties have different types of organizational arrangements considering the project objectives and procurement strategy of the owner/client and organizational requirement.

Construction project delivery system is an organizational relationship of three elemental parties—owner, designer (consultant), and contractor—the roles and responsibilities of the parties and the general sequence of activities to be performed to deliver the project. The roles and responsibilities of these parties vary considerably under different project delivery systems.

Please refer to Section 2.5 in Chapter 2 that details project delivery systems in construction projects and the selection of project delivery system.

4.3.1.7   Selection of Project Team (Designer A/E)

The owner is the first member of the project team. The relationship and responsibilities with other team members depend upon the type of deliverable system the owner would prefer to go with.

For Design–Bid–Build type of contract system, the first step the owner has to do is to select design professionals/consultants (A/E).

Designer (A/E) consists of architects or engineers or consultants. They are the owner's appointed entity accountable to convert owner's conception and need into specific facility with detailed directions through drawings and specifications within the economic objectives. They are responsible for the design of the project and in certain cases supervision of the construction process.

Table 4.8 lists pre-qualification questionnaires for the selection of designer (A/E) and Table 4.9 lists the contents of Request for Proposal for designer/consultant. Generally, the owner selects a designer/consultant on the basis of qualifications (qualification-based selection—QBS) and prefers to use the one he or she has used before and with which he or she has had a satisfactory result.

The QBS can be considered as meeting one of the 14 points of Deming's principles of transformation which states "End the practice of awarding business on the basis of price alone." The basis of selection is solely based on demonstrated competence, professional qualification, and experience for the type of services required. In QBS the contract price is negotiated after selection of the best-qualified firm. Table 4.10 lists the items to be evaluated for QBS of architect/engineer (consultant), Figure 4.6 illustrates a logic flowchart for selection of designer (A/E), and Table 4.11 illustrates designer's selection criteria on weightage basis.

Upon selection of designer (A/E), an agreement is made between the owner and designer (A/E). The following are typical contents of the contract between owner and designer (A/E):

1. Project definition
2. Project schedule
3. Scope of work
4. Design deliverables
5. Owner responsibilities
6. Fees for the services
7. Variation order

TABLE 4.8
Pre-Qualification Questionnaires (PQQ) for Selecting Designer (A/E)

Serial Number	Question	Answer
1	Name of the Organization and Address	
2	Organization's Registration and License Number	
3	ISO Certification	
4	LEED or Similar Certification	
5	Total Experience (years) in Designing Following Type of Projects	
	5.1 Residential	
	5.2 Commercial (Mix Use)	
	5.3 Institutional (Governmental)	
	5.4 Industrial	
	5.5 Infrastructure	
	5.6 Design-Build (Specify Type)	
6	Size of Project (Maximum Amount Single Project)	
	6.1 Residential	
	6.2 Commercial (Mix Use)	
	6.3 Institutional (Governmental)	
	6.4 Industrial	
	6.5 Infrastructure	
	6.6 Design-Build (Specify Type)	
7	List Successfully Completed Projects	
	7.1 Residential	
	7.2 Commercial (Mix Use)	
	7.3 Institutional (Governmental)	
	7.4 Industrial	
	7.5 Infrastructure	
	7.6 Design-Build	
8	List Similar Type (Type to be Mentioned) of Projects Completed	
	8.1 Project Name and Contracted Amount	
	8.2 Project Name and Contracted Amount	
	8.3 Project Name and Contracted Amount	
	8.4 Project Name and Contracted Amount	
	8.5 Project Name and Contracted Amount	
9	Total Experience in Green Building Design	
10	Joint Venture with any International Organization	
11	Resources	
	11.1 Management	
	11.2 Engineering	
	11.3 Technical	
	11.4 Design Equipment	
	11.5 Latest Software	

TABLE 4.8 (Continued)
Pre-Qualification Questionnaires (PQQ) for Selecting Designer (A/E)

Serial Number	Question	Answer
12	Design Production Capacity	
13	Design Standards	
14	Present Work Load	
15	Experience in Value Engineering (List Projects)	
16	Financial Capability (Turnover for Last 5 years)	
17	Financial Audited Report for Last 3 Years	
18	Insurance and Bonding Capacity	
19	Organization Details	
	19.1 Responsibility Matrix	
	19.2 CV's of Design Team Members	
20	Design Review System (Quality Management during Design)	
21	Experience in Preparation of Contract Documents	
22	Knowledge about Regulatory Procedures and Requirements	
23	Experience in Training of Owner's Personnel	
24	List of professional Awards	
25	Litigation (Dispute, Claims) on Earlier Projects	

8. Penalty for delay
9. Liability toward design errors
10. Insurance
11. Arbitration/dispute resolutions
12. Taxes
13. Appointment of subconsultant
14. Selection of team members
15. Duties (responsibilities)
16. Compliance to authorities' requirements
17. Suspension of contract
18. Termination
19. Glossary

4.3.1.8 Term of Requirements (TOR)/Project Charter

Once the project delivery system is finalized and the designer/consultant is selected and contracted by the owner to proceed with the project design, a client brief or Terms of Reference (TOR) is issued to designer/consultant to prepare design proposal and contract documents. A client brief (TOR) is prepared by the owner/client or by the project manager on behalf of the owner describing the objectives and requirements to develop the project. A well-prepared, accurate, and comprehensive TOR is essential to achieve a qualitative and competitive project. TOR gives the project team (designer) a

TABLE 4.9
Contents of Request for Proposal (RFP) for Designer/Consultant

Serial Number	Content
Project Details (Project Objectives)	
1	Introduction
2	Project description
3	Project delivery system
4	Designer's/Consultant's scope of work
5	Preliminary project schedule
6	Preliminary cost of project
7	Type of project delivery system
Sample Questions (Information for Evaluation)	
1	Consultant name
2	Address
3	Quality management system certification
4	Organization details
5	Type of firm such as partnership or limited company
6	Is the firm listed in stock exchange
7	List of awards, if any
8	Design production capacity
9	Current workload
10	Insurance and bonding
11	Experience and expertise
12	Project control system
13	Design submission procedure
14	Design review system
15	Design management plan
16	Design methodology
17	Submission of alternate concept
18	Quality management during design phase
19	Design firm's organization chart
	a) Responsibility matrix
	b) CVs of design team members
20	Designer's experience with green building standards or highly sustainable projects
21	Conducting value engineering
22	Authorities approval
23	Data collection during design phase
24	Design responsibility/Professional indemnity
25	Designer's relationship during construction
26	Preparation of tender documents/Contract documents
27	Review of tender documents
28	Evaluation process and criteria
29	Any pending litigation
30	Price schedule

Source: Abdul Razzak Rumane. (2013). *Quality Tools for Managing Construction Projects*. Reprinted with permission from Taylor & Francis Group.

TABLE 4.10
Qualification-Based Selection of Architect/Engineer (Consultant)

Serial Number	Items to be Evaluated
1	Organization's registration and license
2	Management plan and technical capability
3	Quality certification and quality management system
4	LEED or similar certification
5	Number of awards
6	Design capacity to perform the work
7	Financial strength and bonding capacity
8	Professional indemnity
9	Current load
10	Experience and past performance in similar type of work
11	Experience and past performance of proposed individuals in similar projects
12	Professional certification of proposed individuals
13	Experience and past performance with desired project delivery system
14	Design of similar value projects in past
15	List of successfully completed projects
16	Proposed design approach in terms of:
	a) Performance
	b) Effectiveness
	c) Maintenance
	d) Logistic support
	e) Environment
	f) Green building
17	Design team composition
18	Record of professionals in timely completion
19	Safety consideration in design
20	Litigation
21	Price schedule

Source: Abdul Razzak Rumane. (2013). *Quality Tools for Managing Construction Projects*. Reprinted with permission from Taylor & Francis Group.

clear understanding for the development of project. Further, TOR is used throughout the project as a reference to ensure that the established objectives are achieved. Client brief or TOR describes information such as

- The need or opportunity which has triggered the project
- Propose the location of the project
- Project/facility to be developed
- Procurement strategy
- Project constraints
- Estimated timescale
- Estimated cost

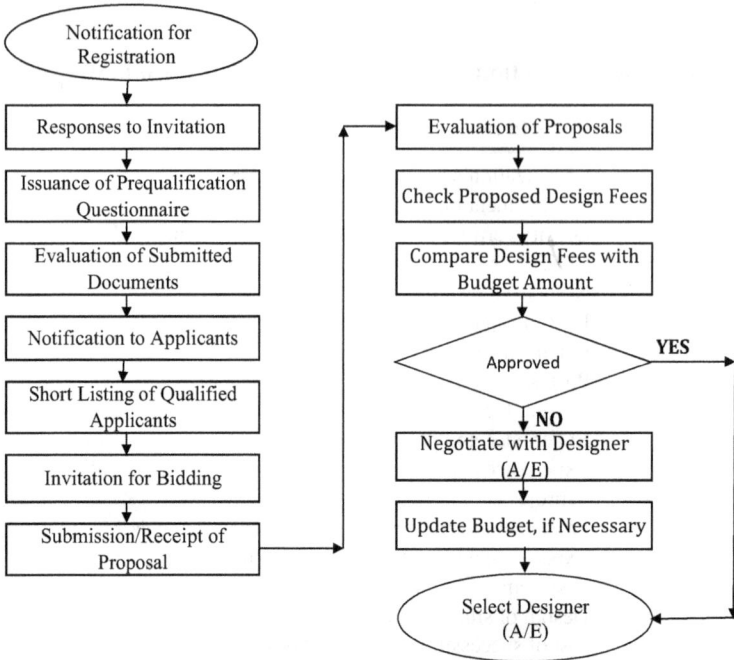

FIGURE 4.6   Logic flowchart for selection of designer (A/E).

Figure 4.7 illustrates logic flowchart for the development of TOR.

TOR generally details the services to be performed by the designer (consultant) which include, but are not limited to, the following:

- Pre-development studies, collection of required data, and analyzing the same to prepare design drawings and documents for the project
- Development of alternatives
- Preparation of concept design
- Preparation of schematic design
- Preparation of detailed design
- Obtaining authorities approvals
- Compliance standards, codes, and practices
- Coordinating and participating in value engineering study
- Preparation of construction schedule
- Preparation of construction budget
- Preparation of contract documents for bidding purpose
- Pre-qualification/selection of contractor
- Evaluation of proposals
- Recommendation of contractor to the owner/client

TABLE 4.11
Designer's (A/E) Selection Criteria

Serial Number	Evaluation Criteria	Weightage	Notes
1	**General Information**		
	a) Company information		
2	**Business**	**10%**	
	a) LEED or similar certification	5%	
	b) ISO Certification	5%	
3	**Financial**	**20%**	
	a) Turnover	5%	
	b) Financial standing	5%	
	c) Insurance and Bonding limit	10%	
4	**Experience**	**30%**	
	a) Design experience	10%	
	b) Similar type of projects	10%	
	c) Current projects	10%	
5	**Design capability**	**10%**	
	a) Design approach	5%	
	b) Design capacity	5%	
6	**Resources**	**20%**	
	a) Design team qualification	10%	
	b) Design team composition	5%	
	c) Professional certification	5%	
7	**Design Quality**	**5%**	
8	**Safety consideration in Design**	**5%**	

Note   The weightage mentioned in the table is indicative only. The % can be determined as per the owner's strategy.

Following are the requirements for a building construction project, normally mentioned in TOR, to be prepared by the designer during the conceptual phase for submission to the owner:

1. Site Plan
   A) Civil
   B) Services
   C) Landscaping
   D) Irrigation
2. Architectural Design
3. Building and Engineering Systems
   A) Structural
   B) Mechanical (HVAC)
   C) Public Health
   D) Fire Suppression Systems

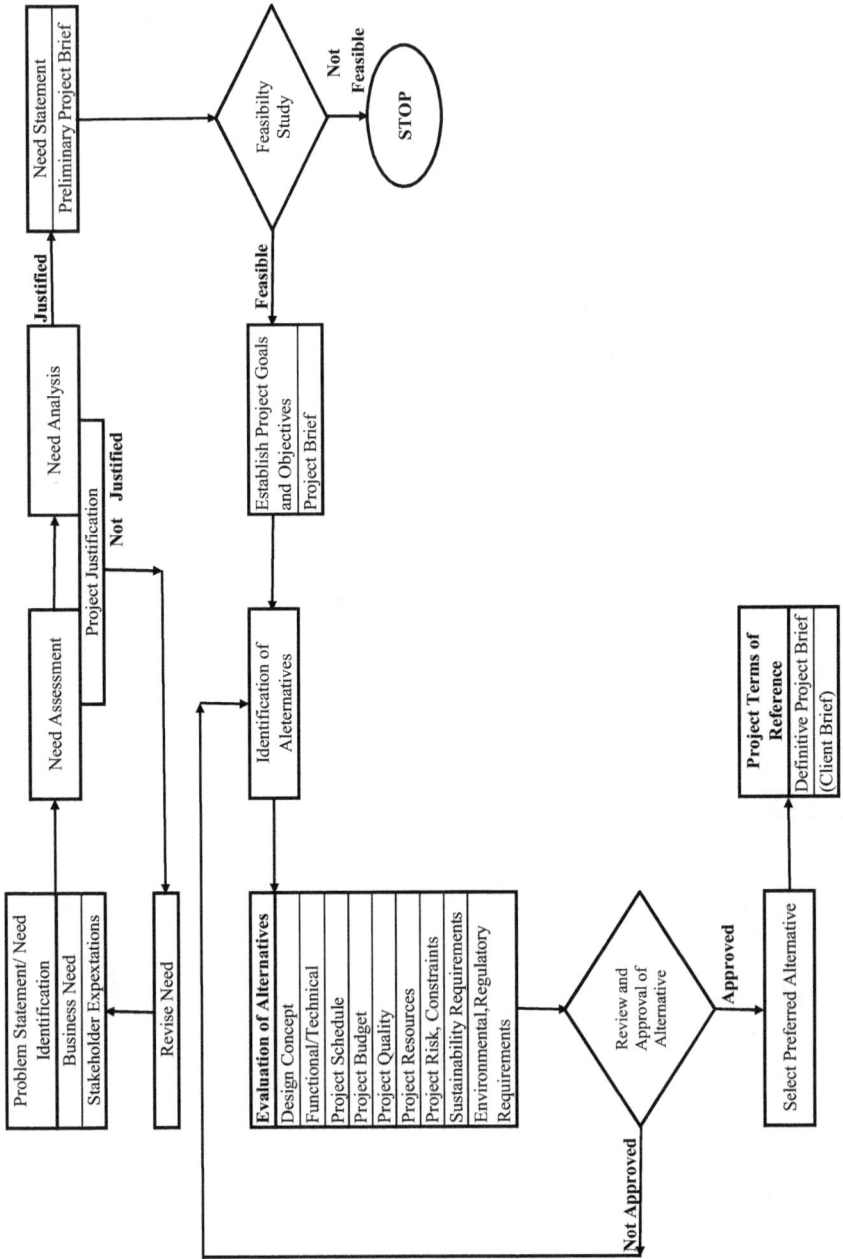

FIGURE 4.7	Logic flowchart for development of terms of reference.

  E) Electrical
  F) Low Voltage Systems
  G) Others
 4. Schedule
 5. Cost Estimates

Table 4.12 lists typical contents of TOR for building construction projects.

TABLE 4.12
Typical Contents of Terms of Reference (TOR) Documents

Serial Number	Topics
1	Project Objectives
	Background
	Project Information
	General Requirements
	Special Considerations
2	Project Requirements
	2.1 Scope of Work
	2.2 Work Program
	2.2.1 Study Phase
	2.2.2 Design Phase
	2.2.3 Tender Stage
	2.2.4 Construction Phase
	2.3 Reports and Presentations
	2.4 Schedule of Requirements
	2.5 Drawings
	2.6 Energy Conservation Considerations
	2.7 Cost Estimates
	2.8 Time Program
	2.9 Interior Finishes
	2.10 Aesthetics
	2.11 Mechanical
	2.12 HVAC
	2.13 Lighting
	2.14 Engineering Systems
3	Opportunities and Constraints
	3.1 Site Location
	3.2 Site Conditions
	3.3 Land Size and Access
	3.4 Climate
	3.5 Time
	3.6 Budget

(*continued*)

TABLE 4.12 (Continued)
Typical Contents of Terms of Reference (TOR) Documents

Serial Number	Topics
4	Performance Target
	4.1 Financial Performance
	4.1.1 Performance Bond
	4.1.2 Insurance
	4.1.3 Delay Penalty
	4.2 Energy Performance Target
	4.2.1 Energy Conservation
	4.3 Work Program Schedule
5	Environmental Considerations
6	Design Approach
	6.1 Procurement Strategy
	6.2 Design Parameters
	6.2.1 Architectural Design
	6.2.2 Structural Design
	6.2.3 Mechanical Design
	6.2.4 HVAC Design
	6.2.5 Electrical Design
	6.2.6 Information and Communication Technology
	6.2.7 Conveying System
	6.2.8 Landscape
	6.2.9 External Works
	6.2.10 Parking
	6.3 Sustainable Architecture
	6.4 Engineering Systems
	6.5 Value Engineering Study
	6.6 Design Review by Client
	6.7 Selection of Products/Systems
7	Specifications and Contract Documents
8	Project Control Guidelines
9	Submittals
	9.1 Reports
	9.2 Drawings
	9.3 Specifications
	9.4 Models
	9.5 Sample Boards
	9.6 Mock up
10	Presentation
11	Project Team Members
	11.1 Number of Project personnel
	11.2 Staff Qualification
	11.3 Selection of Specialists
12	Visits

Source: Abdul Razzak Rumane. (2013). *Quality Tools for Managing Construction Projects*. Reprinted with permission from Taylor & Francis Group.

4.3.1.9 Development of Concept Design

Upon signing of contract with the client to design the project and offer other services, the designer (consultant) assigns the project manager to execute the contract and is responsible to manage the development of design and contract documents to meet the client's need and objectives.

The TOR is a guideline to develop concept design. The selected preferred alternative is the base for development of the concept design. Project charter (TOR) defines the activities to be performed by the designer during this phase and also the deliverables that have to be prepared for further proceedings and action in the next phase of the design development. The most significant impacts on the quality of a project occur during the conceptual phase. This is the time when specifications, statement of work, contractual agreements, and initial design are developed. Initial planning has the greatest impact on a project because it requires the commitment of processes, resources schedules, and budgets. A small error that is allowed to stay in the plan is magnified several times through subsequent documents that are second or third in the hierarchy. Concept design or the basic design is often viewed as most critical to achieving outstanding project performance. Table 4.13 lists the points to be considered while developing the concept design.

TABLE 4.13
Development of Concept Design

Serial Number	Points to be Considered
1	Project goals
2	Usage
3	Incorporate requirements from collected data
4	Technical and functional capability
5	Aesthetics
6	Constructability
7	Sustainability (economical, environmental, and social)
8	Environmental compatibility
9	Conservation of energy
10	Waste management
11	Clean air
12	Pollution control
13	LEED (Equivalent) compliance
14	Comfortable lighting
15	Health and safety
16	Reliability
17	Fire protection measures
18	Process automation requirements
19	Facility management requirements
20	Supportability during maintenance/maintainability
21	Cost-effective over the entire life cycle (economy)
22	Reports, drawings, models

4.3.1.10   Identification of Concept Design Stakeholders

Following stakeholders have direct involvement in the project during the development of concept design;

- Owner
- Consultant
- Designer
- Regulatory authorities
- Project/construction manager (if the owner decided to engage them during this phase)

Table 4.14 illustrates the responsibilities of various participants during development of concept design.

4.3.1.10.1   Project Team (Designer) Organization

The project manager coordinates with other departments and acquires design team members to develop project design. A project team leader along with respective design engineer(s), quality engineer, and AutoCAD technician from each trade is assigned to work on the project. The project team is briefed by the project manager about the project objectives and the roles and responsibilities and authorities of each team member. The quality manager also joins the project team to ensure compliance with the organization's quality management system. Figure 4.8 illustrates the project design organization structure.

TABLE 4.14
Responsibilities of Various Participants during Development of Concept Design

Phase	Responsibilities		
	Owner	**Designer**	**Regulatory**
Development of Concept Design	Approval of team members	Data collection	Approval of project submittals
	Approval of time schedule	Preparation of schedule	
	Approval of budget	Estimation of project cost	
	Approval of concept design	Project quality requirements (concept design quality, project quality)	
		Estimation of project resources	
		Identify project risks	
		Identification of HSE issues and requirements	
		Regulatory approvals	
		Development of concept design	

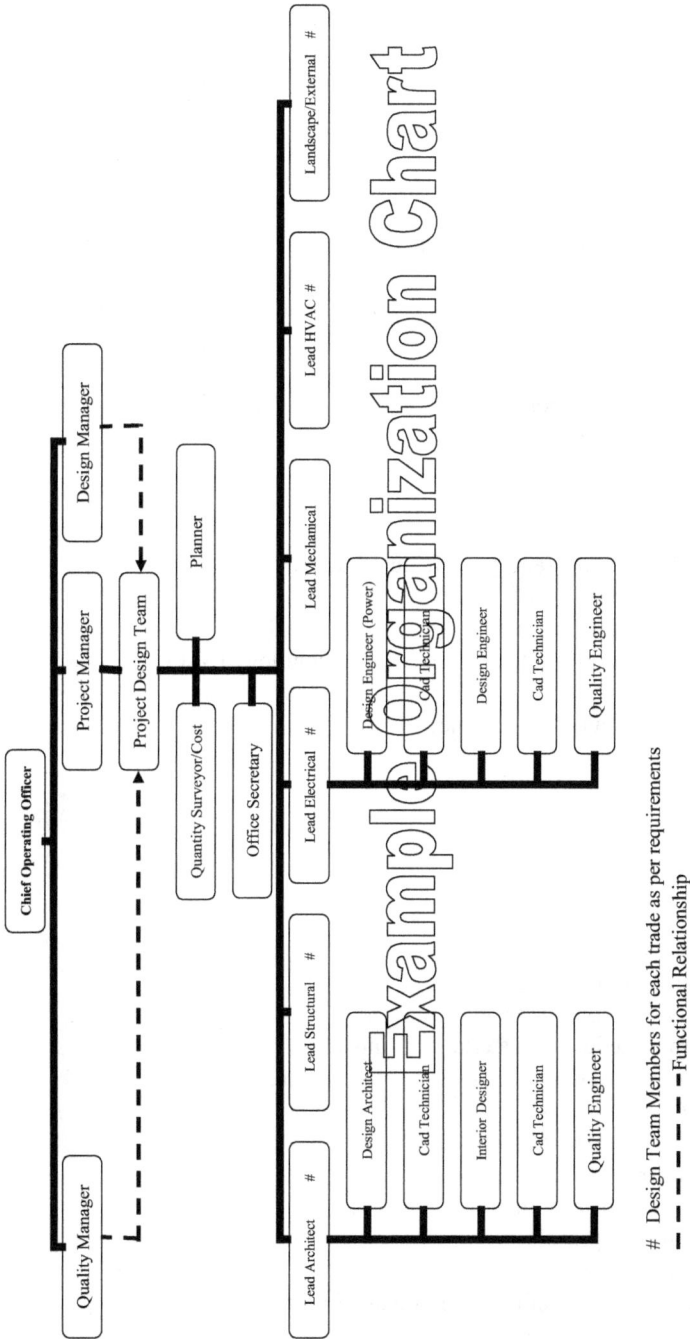

FIGURE 4.8 Project design team organization chart. (Abdul Razzak Rumane (2013), *Quality Tools for Managing Construction Projects*. Reprinted with permission from Taylor & Francis Group Company.

4.3.1.11 Develop Concept Design Scope

In order to establish the concept design scope, the designer has to review all the documents that are part of the TOR and the contract that designer has signed with the owner of the project. It basically includes project charter (TOR). In case of oil and gas project, the TOR generally requires designer (consultant) to perform the following:

- Development of concept design
- Preparation of project design, schedule, budget, and obtaining authority approvals
- Preparation of contract documents for bidding and tendering purposes

Following are the concept design deliverables for a building (non-process-type) project:

1. Concept design report (Narrative/descriptive report)
   a. Space program
   b. Building exterior
   c. Building interior
   d. Structural system
   e. MEP systems (energy-efficient systems)
   f. Fire safety
   g. Conveyance system
   h. Information and Communication Technology (ICT)
   i. Landscape
   j. Traffic plan (if applicable)
   k. HSE issues
2. Drawings
   a. Overall site plan
   b. Floor plan
   c. Elevations
   d. Sketches
   e. Sections (indicative to illustrate overall concept)
3. Data collection, studies reports
4. Existing site conditions
5. Concept schedule of material and finishes
6. Lighting/daylight studies/sensor-controlled lighting
7. LEED requirements
8. Facility management requirements
9. Preliminary project schedule
10. Preliminary cost estimate
11. Regulatory approvals
12. Models
13. Evaluation criteria of the alternatives (if part of TOR)

4.3.1.12   Prepare Concept Design

In order to prepare concept design, the designer has to

1) Collect data/information about the project
2) Collect owner's requirements
3) Collect regulatory requirements

The designer has to consider concept design deliverables developed based on TOR and contract documents.

4.3.1.12.1   Collect Data/Information

The purpose of data collection is to gather all the relevant information on existing conditions, both on project site and surrounding area, that will impact the planning and design of the project. The data related to the following major elements required to be collected by the designer is listed in Table 4.15.

TABLE 4.15
Major Items for Data Collection during Concept Design Phase

Serial Number	Items to be Considered
1	Certificate of title
	a. Site legalization
	b. Historical records
2	Topographical survey
	a. Location plan
	b. Site visits
	c. Site coordinates
	d. photographs
3	Geotechnical investigations
4	Field and laboratory test of soil and soil profile
5	Existing structures in/under the project site
6	Existing utilities/services passing through the project site
7	Existing roads, structure surrounding the project site
8	Shoring and underpinning requirements with respect to adjacent area/structure
9	Requirements to protect neighboring area/facility
10	Environmental studies
11	Daylighting requirements
12	Wind load, seismic load, dead load, and live load
13	Site access/traffic studies
14	Applicable codes, standards, and regulatory requirements
15	Usage and space program
16	Design protocol
17	Scope of work/client requirements

TABLE 4.16
Check List for Owner's Requirements

Owner's Preferred Requirements (Trade Name)				
Sr. No.	Description of Item/Activity	Yes	No	Notes
1				
2				
3				
4				
5				
6				
7				
8				
9				
10				
11				
12				
13				
14				
15				
16				
17				
18				
19				
N...				

Sample Check List

4.3.1.12.2   Collect Owner's Requirements

Based on the scope of work/requirement mentioned in the TOR, a detailed list of requirements is prepared by the designer (consultant). The project design is developed taking into consideration the owner's requirements and all other design criteria. The designer can prepare checklists of various trades by listing the items to obtain owner preferential requirements.

Table 4.16 is a sample checklist to collect owner's preferred requirements.

4.3.1.12.3   Collect Regulatory Requirements

The designer has to collect regulatory requirements to be taken into consideration while preparing concept design.

The designer should also collect sustainability requirements that are essential to comply with authorities' requirements.

The designer can use techniques such as quality function deployment to translate owner's need into technical specifications. Figure 4.9 illustrates a house of quality for an office building project based on certain specific requirements by the customer.

While preparing concept design, the designer has to consider various available options/systems to ensure project economy, performance, and operation. The concept design should be suitable for further development.

Table 4.17 lists the elements that comprise the concept design drawings of various trades (disciplines) in building construction projects.

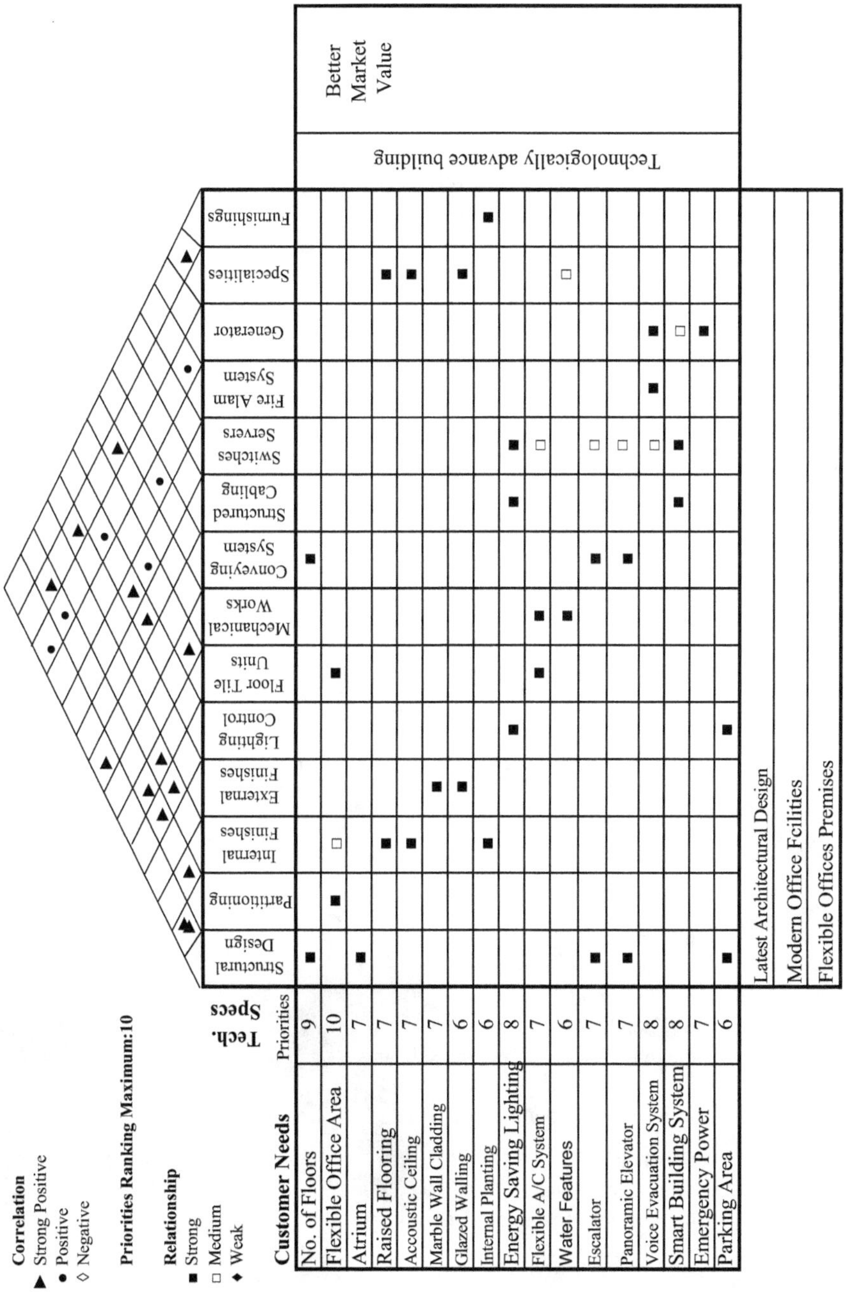

Correlation
▲ Strong Positive
● Positive
◊ Negative

Priorities Ranking Maximum:10

Relationship
■ Strong
□ Medium
◆ Weak

Roof (Correlation matrix): Better Market Value; Technologically advance building

Customer Needs	Tech. Specs Priorities	Structural Design	Partitioning	Internal Finishes	External Finishes	Lighting Control	Floor Tile Units	Mechanical Works	Conveying System	Structured Cabling	Switches Servers	Fire Alarm System	Generator	Specialities	Furnishings
No. of Floors	9	■							■						
Flexible Office Area	10	■	■	□			■								
Atrium	7	■													
Raised Flooring	7			■											
Accoustic Ceiling	7			■											
Marble Wall Cladding	7				■									■	
Glazed Walling	6				■									■	
Internal Planting	6			■										■	■
Energy Saving Lighting	8					■				■	■				
Flexible A/C System	7						■	■						□	
Water Features	6							■	■						
Escalator	7	■							■		□				
Panoramic Elevator	7	■									□				
Voice Evacuation System	8									■	□	■	■		
Smart Building System	8									■	■		□		
Emergency Power	7										■		■		
Parking Area	6	■				■									

Latest Architectural Design
Modern Office Fcilities
Flexible Offices Premises

FIGURE 4.9 House of quality for office building project.

TABLE 4.17
Elements to Be Included in Concept Design Drawings

Serial Number	Elements to Be Included in Drawing
1	Architectural
1.1	Overall site plans
	a. Existing site plan
	b. Location of building, roads, parking, access, and landscape
	c. Project boundary limits
	d. Site utilities
	e. Water supply, drainage, storm water lines
	f. Zoning
	g. Reference grids and axis
	h. Demolition plan, if any
1.2	Floor plans
	a. Floor plans of all floors
	b. Structural grids
	c. Vertical circulation elements
	d. Vertical shafts
	e. Partitions
	f. Doors
	g. Windows
	h. Floor elevations
	i. Designation of rooms
	j. Preliminary finish schedule
	k. Services closets
	l. Raised floor, if required
1.3	Roof plans
	a. Roof layout
	b. Roof material
	c. Roof drains and slopes
2	Structural
2.1	a. Building structure
	b. Floor grade and system
	c. Foundation system
	d. Tentative size of columns, beams
	e. Stairs
	f. Roof and general sections
3	Elevator
3.1	a. Traffic studies
	b. Elevator/escalator location
	c. Equipment room

TABLE 4.17 (Continued)
Elements to Be Included in Concept Design Drawings

Serial Number	Elements to Be Included in Drawing
4	Plumbing and fire suppression
4.1	a. Sprinkler layout plan
	b. Piping layout plan
	c. Water system layout plan
	d. Water storage tank location
	e. Development of preliminary system schematics
	f. Location of mechanical room
5	HVAC
5.1	a. Ducting layout plan
	b. Piping layout plan
	c. Development of preliminary system schematics
	d. Calculations to allow preliminary plant selections
	e. Establishment of primary building services distribution routes
	f. Establishment of preliminary plant location and space requirements
	g. Determine heating and cooling requirements based on heat dissipation of equipment, lighting loads, type of wall, roof, glass, etc.
	h. Estimation of HVAC electrical load
	i. Development of BMS schematics showing interface
	j. Location of plant room, chillers, cooling towers
6	Electrical
6.1	a. Lighting layout plan
	b. Power layout plan
	c. System design schematic without any sizing of cables and breakers
	d. Substation layout and location
	e. Total connected load
	f. Location of electrical rooms and closets
	g. Location of MLTPs, MSBs, SMBs, EMSB, SEMBs, DBs, EDBs, etc.
	h. Location of starter panels, MCC panels, etc.
	i. Location of generator, UPS
	j. Raceway routes
	k. Riser requirements
	l. Information and Communication Technology (ICT)
	a) Information technology (computer network)
	b) IP telephone system (telephone network)
	c) Smart building system
	m. Loss prevention systems
	a) Fire alarm system
	b) Access control security system layout
	c) Intrusion system
	n. Public address, audiovisual system layout
	o. Schematics for F.A and other loss prevention systems
	p. Schematics for ICT system and other low voltage systems
	q. Location of LV equipment

(*continued*)

TABLE 4.17 (Continued)
Elements to Be Included in Concept Design Drawings

Serial Number	Elements to Be Included in Drawing
7	Landscape
7.1	a. Green area layout
	b. Selection of plants
	c. Irrigation system
8	External
8.1	a. Street/road layout
	b. Street lighting
	c. Bridges (if any)
	d. Security system
	e. Location of electrical panels (feeder pillars)
	f. Pedestrian walkways
	g. Existing plans
9	Narrative description

Concept Design Reports should be concise, covering all the related information about concept design considerations. Reports are prepared for each of the trades mentioned in the TOR. Below is the example Table of Contents of concept design report for architectural works and civil/structural works for the building construction project.

    A. Table of Contents for Architectural Works
        1. Introduction
           1.1 Project Goals and Objectives
        2. Owner's Schedule Requirements
        3. Project Directory
        4. Architecture
           4.1 Applicable Codes and Standards
           4.2 International Codes
           4.3 Local Codes
           4.4 Building Height
           4.5 Projections
           4.6 Fire Safety Requirements
           4.7 Usage of Building
           4.8 Stairways
        5. Project Requirements
           5.1 Basement
           5.2 Ground Floor
           5.3 Other Floors
        6. Existing Site Conditions
           6.1 Location Maps
           6.2 Project Site and Surrounding Area Survey Reports

    a. Design Criteria
    b. Sprinkler System
    c. Smoke Ventilation System
    d. Firefighting Pumps
    e. Piping System
    f. Fire Suppression System for Diesel Generator Room
    g. Fire Suppression for Low Voltage Equipment Room
6. Mechanical (Plumbing) Work
    a. Design Criteria
    b. Water Supply System
        i. Cold Water
        ii. Hot Water
    c. Water Distribution Network
    d. Water Storage System
    e. Plumbing Fittings and Fixtures
    f. Plant Room (Pumps, Heaters) Location
    g. Storm Water Drainage System
    h. Sewage System
    i. Irrigation Water System
    j. Piping for Mechanical Works

D. Table of Contents for HVAC Works
1. Introduction
2. Codes and Standards
3. Regulatory Compliance
4. HVAC System Design Criteria
5. Environmental Conditions
6. HVAC Equipment Selection
    a. Chiller
    b. Cooling System
    c. AHU
    d. Energy Recovery Equipment
    e. Direct Digital Control System
    f. Pumps
    g. Fans and Ventilation
    h. Ducting
    i. Duct Insulation
    j. Piping
    k. Risers
    l. Piping Insulation
    m. Acoustic and Vibration
7. Location of Plant Room
8. Building Management System (BMS)

E. Table of Contents for Electrical Works
1. Introduction
2. Codes and Standards
3. Regulatory Compliance

    4. Design Criteria
       a.   Safety and Protection
    5. Electrical Substation
    6. Cables and Wires
    7. Main Electrical Supply
       a.   Main Distribution System
       b.   Secondary Distribution System
       c.   Distribution Equipment (Panels, Boards, Switchgear)
       d.   Circuits
    8. Internal and External Lighting
       a.   Light Fixtures
    9. Power Receptacles
  10. Emergency Power System
  11. Earthing (Grounding) System
  12. Lightning Protection System
  13. Alternate Energy (Solar) System
  14. Fire Alarm and Voice Evacuation System
  15. Audio Visual System
  16. Security System (CCTV, Access Control)
  17. Communication System

The designer has to also prepare the following reports:

  1.   Interior works
  2.   Elevator works
  3.   External works
  4.   Landscape works

Contents of Design Report are illustrated in Table 4.18.

4.3.1.13   Estimate Project Time Schedule

The duration of a construction project is finite and has a definite beginning and a definite end; therefore, during the conceptual phase the expected time schedule for the completion of the project/facility is worked out. The expected time schedule is important for both financial and acquisition of the facility by the owner/end user. It is the owner's goal and objective that the facility is completed in time. Figure 4.10 illustrates a time schedule for a typical construction project.

4.3.1.14   Estimate Project Cost

The next step is to refine cost estimates for the conceptual alternatives as this is required by the owner to determine the capital cost of construction so that he or she can arrange the finances. It is the owner's responsibility to provide an approved maximum finance to complete the facility. It is required that the owner formulates his or her thoughts on project financing, as the financial conditions will affect the

TABLE 4.18
Contents of Concept Design Report (Trade Name)

Section	Topic
1	Introduction
	1.1 Description of project
	1.2 Project goals and objectives
2	Owner's schedule requirement
3	Project directory
4	Trade name (Process, Architecture, MEP, etc.)
	4.1 Applicable codes and standards
	4.2 International codes
	4.3 Local codes
	4.4 Regulatory requirements
	4.5 Applicable design system/Details of the project trade item/element
5	Project requirement
6	Existing site conditions
7	Data collection
8	Design options/strategy/criteria
9	Design software
10	Geographical investigation
11	Risk assessment
12	HSE issues
13	Drawings
14	Models (if applicable)

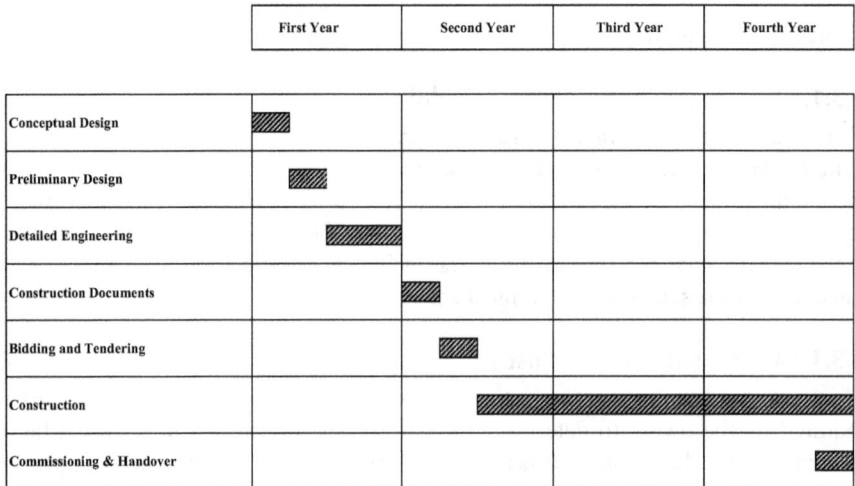

	First Year	Second Year	Third Year	Fourth Year
Conceptual Design	▨			
Preliminary Design	▨			
Detailed Engineering	▨			
Construction Documents		▨		
Bidding and Tendering		▨		
Construction			▨▨▨	▨▨
Commissioning & Handover				▨

FIGURE 4.10 Typical time schedule of construction project.

TABLE 4.19
Cost Estimation Levels for Construction Projects

Project Stage/ Phase	Tools/ Methodology	Accuracy	Purpose
Inception	Analogous	−50% to +100%	Project Initiation (Rough Order of Magnitude)
Feasibility	Analogous	−25% to +75%	Justification to Proceed (Screening Estimate)
Concept Design	Parametric	−10% to +25%	Budgetary (Conceptual Estimate)
Preliminary Design	Elemental Parametric	−10% to +25%	Budgetary (Preliminary Estimate)
Detail Design	Elemental Parametric/ Detailed Costing	−5% to +10%	Detailed Estimate
Bidding and Tendering	Detailed Costing	−5% to +5%	Bid Estimate/Definitive Estimate
Construction	Detailed Costing	Project Cost Baseline (Contracted Value)	Contract Cost (Control Estimate)

possible options from the beginning. Normally the following points should be considered:

1. What are the sources of funding?
2. What criteria or rules apply?
3. How could the project best respond to those rules?

In case any funding agency is involved in financing the project, it may impose certain conditions that affect the project feasibility and implementation. It is likely that such funding agencies may also insist on the adoption of a particular contract strategy.

The cost estimated by the designer (consultant) is based on assumptions and historical data available from experience on similar projects.

Table 4.19 illustrates cost estimation levels for construction projects.

4.3.1.15 Quality Management
The designer has to plan and establish quality criteria for the project. This includes mainly the following:

- Owner's requirements
- Quality standards and codes to be complied
- Regulatory requirements

- Conformance to owner's requirements
- Conformance to requirements listed under project charter (TOR)
- Design review procedure
- Drawings review procedure
- Document review procedure
- Quality management during all the phases of project life cycle

Table 4.20 lists the contents of designer's quality control plan.

The designer has to also manage the quality for the development of concept design and also plan and establish quality criteria for quality management during all the phases of project life cycle.

TABLE 4.20
Contents of Designer's Quality Control Plan

Section	Topic
1	Introduction
2	Description of Project
3	Quality Control Organization
4	Qualification of QC Staff
5	Responsibilities of QC Personnel
6	Procedure for Submittals
7	Quality Control Procedure for Concept Design
	7.1 Establish Concept Design requirements
	7.1.1 Review of Project Charter, Terms of Requirements (TOR)
	7.1.2 Identify Project Goals and Objectives
	7.1.3 Identify Concept Design Scope
	7.1.4 Review Feasibility Reports
	7.1.5 Review Preferred Alternative Selection Report
	7.1.6 Establish Concept Design Deliverables
	7.1.7 Establish Design Procedure
	7.2 Identify Quality Management Requirements
	7.2.1 Plan Quality
	7.2.1.1 Owner's Requirements
	7.2.1.2 TOR Requirements
	7.2.1.3 Numbers of Drawing, Reports, Models
	7.2.1.4 Scope of Design Work
	7.2.1.5 Responsibility Matrix
	7.2.1.6 Codes and Standards
	7.2.1.7 Regulatory Requirements
	7.2.1.8 Submittal Plan
	7.2.1.9 Drawings, Specifications, Documents
	7.2.1.10 Design Review Plan
	7.2.2 Quality Assurance
	7.2.2.1 Information/Data Collection
	7.2.2.2 Site Investigation

TABLE 4.20 (Continued)
Contents of Designer's Quality Control Plan

Section	Topic
	7.2.2.3 Engineering Surveys
	7.2.2.4 Preparation of Drawings
	7.2.2.5 Interdisciplinary Coordination
	7.2.2.6 Project Risks
	7.2.2.7 Environmental Issues
	7.2.2.8 Preparation of Specification, Documents
	7.2.2.9 Functional and Technical Compatibility
	7.2.3 Quality Control
	7.2.3.1 Design Drawing
	7.2.3.2 Quality of Drawings
	7.2.3.3 Project Schedule
	7.2.3.4 Project Cost
	7.2.3.5 Project Resources
	7.2.3.6 Specification, Contract Documents
8	Project-specific Design Requirements
9	Design Software
10	Design Development Procedure
	10.1 Design Criteria
	10.2 Preparation of Drawings
	10.3 Interdisciplinary Coordination
	10.4 Design Review
11	Company's Quality Manual and Procedure
12	Subconsultant's Work
13	Value Engineering
14	Quality Auditing Program
15	Quality Control Record
16	Innovative and Latest Technology
17	Quality Updating Program

Figure 4.11 illustrates quality management procedure for the development of concept design.

4.3.1.16 Estimate Resources

Designer has to estimate the resources required to complete the project. This includes the estimation of manpower required during construction phase, testing, commissioning, and handover phase. The designer has to give preference to available local resources to comply with sustainability requirements.

The designer has to also prepare equipment and machinery data sheets and schedule of material.

Start Concept Design Development

Review Project Charter, Terms of Reference (TOR)

Identify Project Goals and Objectives,Establish Scope of Concept Design, Concept Design Deliverables

Identify Project Team/ Organization Chart

Establish Design Criteria

Establish Responsibility Matrix/ Communication Plan

Establish Submittal/ Owner Review Procedure

Establish Shedule for Submission to Owner

Select Specialist Consultant (if applicable)

Establish Design Procedure

Establish Quality Assurance/ Quality Control Procedure

Estimate Prpject Time Schedule

Estimate Project Cost

Estimate Project Resources

Identify Risk/HSE Issues

Review and Validate Concept Design

Update Concept Design Documents

Finalize Concept Design

Submit Design

Owner's Requirements/Expectations

Codes and Standards to be Followed

Regulatory Requirements

Quality Metrics and Measures

Project Risks

HSE Issues and Requirements

Organizational Policy

Design Basis

Coordination Among All Disciplines

Design Input from Specialist (if applicable)

Compliance with all Requirement

Constructability Review

Design Review

Organizational Quality Management System

Quality Audit/Quality Check Procedure

Corrective and Preventive Action

Remedial Action Plan

Quality Documents and Records

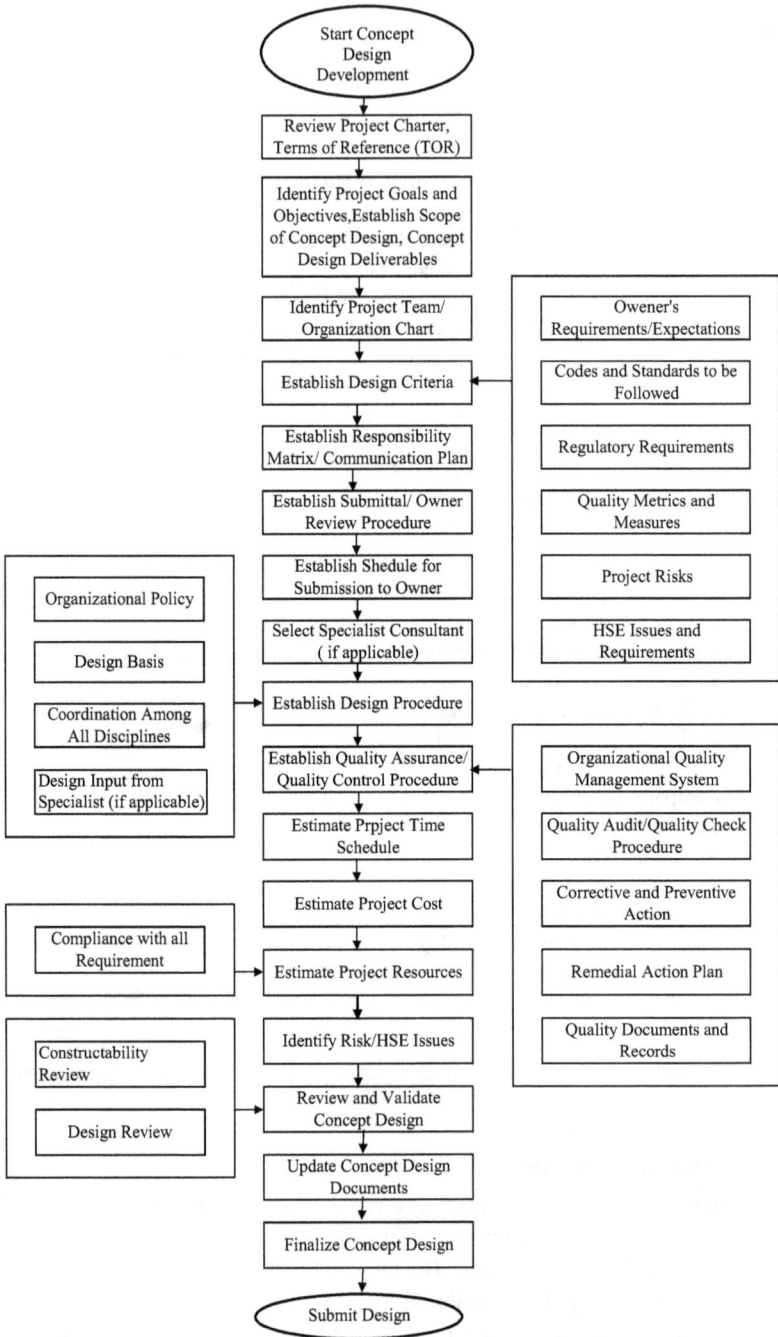

FIGURE 4.11   Quality management procedure for development of concept design.

4.3.1.17   Identify/Manage Risk

The Designer has to identify the risks that will affect the successful completion of project. Following are typical risks that normally occur during conceptual design phase:

- Lack of input from owner about the project goals and objectives
- Project objectives not defined clearly
- Feasibility study not done properly
- Alternative selection is not suitable for further development
- The related project data and information collected is incomplete
- The related project data and information collected likely to be incorrect and wrongly estimated
- Environmental consideration
- Regulatory requirements
- Errors in estimating the project schedule
- Errors in cost estimation

Designer has to take into account above-mentioned risk factors while developing the concept design.

Further, the designer has to consider the following risks while planning the duration for completion of conceptual phase:

- Impractical conceptual design preparation schedule
- Delay to obtain authorities' approval
- Delay in environmental approval
- Delay in data collection
- Delay in deciding project delivery system

4.3.1.18   HSE Issues/Requirements

The designer has to identify HSE issues that could affect the environment due to the project. While developing the design, the designer has to consider the following:

- Hazardous properties of the materials and products used in the project
- Safety in process and operations
- Hazardous emissions
- Pollution and its impact
- Impact on health
- Safety in design (safe design)
- Regulatory and other environmental protection agencies' requirements
- Environmental compatibility

4.3.1.19   Review Concept Design

Table 4.21 illustrates major points for the review of concept design.

TABLE 4.21
Major Points for Review of Concept Design

Serial Number	Description	Yes	No	Notes
A: Concept Design				
A.1	Does the design support owner's project goals and objectives?			
A.2	Does the design meet all the elements specified in project charter (TOR)?			
A.3	Does the design meet all the performance requirements/parameters?			
A.4	Whether constructability has been taken care?			
A.5	Whether usage of space is optimal?			
A.6	Whether technical and functional capability considered?			
A.7	Whether suitability of the project for the defined purpose and objectives considered?			
A.8	Whether ease of operations is considered while selecting the equipment?			
A.9	Whether accessibility is considered?			
A.10	Does the design meet ease of maintenance?			
A.11	Whether the design meets all the specified codes and standards?			
A.12	Whether the designer has considered into design all the data and information collected?			
A.13	Whether design confirm with fire and egress requirements?			
A.14	Whether design risks been identified, analyzed, and responses planned for mitigation?			
A.15	Whether risk assessment report is prepared?			
A.16	Whether health and safety requirements in the design are considered?			
A.17	Whether environmental constraints considered?			
A.18	Whether environmental impact assessment report is prepared and considered?			
A.19	Whether environment-preferred material and products considered?			

TABLE 4.21 (Continued)
Major Points for Review of Concept Design

Serial Number	Description	Yes	No	Notes
A.20	Whether green building concept is considered?			
A.21	Does energy conservation is considered?			
A.22	Does daylighting system is considered?			
A.23	Does sensor-controlled lighting is considered?			
A.24	Does the systems are energy efficient?			
A.25	Does sustainability requirements considered in design?			
A.26	Whether cost-effectiveness over the entire project life cycle is considered?			
A.27	Whether all reasonable design options/ systems are considered for project economy?			
A.28	Does the design have provision for inclusion of facility management requirements?			
A.29	Does the design have provision for interface with all the low voltage and control systems?			
A.30	Whether the design is coordinated with all trades?			
A.31	Whether all the regulatory/statutory requirements taken care?			
A.32	Does the design support proceed to next design development stage?			

B: Schedule

B.1	Project schedule is practically achievable?			

C: Financial

C.1	Project cost is properly estimated?			

D: Resources

D.1	Whether availability of resources during construction phase is considered?			
D.2	Whether schedule of equipment and machinery is prepared?			
D.3	Whether locally available resources are considered?			

(*continued*)

TABLE 4.21 (Continued)
Major Points for Review of Concept Design

Serial Number	Description	Yes	No	Notes
E: Reports				
E.1	Whether the reports are complete and include adequate information about the project?			
E.2	Whether the reports are prepared for all the trades mentioned in TOR?			
E.3	Whether the report is properly formatted and has table of contents for each report?			
F: Drawings, Sketches				
F.1	Whether drawings, sketches for all trades prepared as per TOR?			
F.2	Whether number of drawings is as per TOR requirements?			
G: Models				
G.1	Whether the models meet the design objectives? (as applicable)			
H: Submittals				
H.1	Whether number of sets prepared as per TOR			

4.3.1.20   Finalize Concept Design

Final design is prepared incorporating the comments, if any, found during the analysis and review of the drawings and documents for submission to the owner/client.

Table 4.22 summarizes the TQM concept for conceptual design phase activities.

4.3.2   TQM IN PRELIMINARY DESIGN PHASE

Preliminary design is mainly a refinement of the elements in the conceptual design phase. Preliminary design is also known as Schematic design. It is a design intent documents which quantify functional performance expectations and parameters for each system to be commissioned. It is traditionally labeled as 30% design. Preliminary design adequately describes information about all proposed project elements in sufficient detail for obtaining regulatory approvals, necessary permits, and authorization. The central activity of preliminary design is the architect's design concept of the owner's objective which can help make the detailed engineering and design for the required facility. At this phase, the project is planned to the level where sufficient details are available for initial schedule and cost. This phase also includes the initial preparation of all documents necessary to build the facility/construction project. The primary goal of this phase is to develop a clearly defined design based on the client's

TABLE 4.22

Total Quality Management Concept for Conceptual Design Phase Activities

Serial Number	Phase Activity	Stakeholders/Team Members	Customer Specifications/ Requirements	System for Managing/Tool
1	Identification of Need	1. Project owner 2. Consultant/ Project manager	Owner's need/ requirements	Analytical tool (Please refer Table 4.3 and Table 4.4)
2	Feasibility Study	1. Specialist consultant 2. Owner 3. Regulatory bodies 4. Funding agencies	1. Owner's need/ requirements 2. Financial constraints	Please refer Table 4.6
3	Project Goals and Objectives	Project owner	SMART Need • Specific • Measurable • Achievable • Time bound	Please refer Table 4.5
4	Identification of Alternatives/ Options	1. Project owner 2. Consultant/ Designer	Owner's need/ requirements	
5	Selection of Preferred Alternative	1. Consultant/ Designer 2. Owner 3. Regulatory authorities 4. Funding agencies	1. Owner's need/ requirements 2. Compliance to regulatory requirements 3. Schedule 4. Cost	Please refer Figure 4.5
6	Project Delivery/ Contracting System	Project owner/project manager	Please refer Section 2.2 under Chapter 2	
7	Selection of Designer (A/E)	Project owner/project manager	Competitive having expertise in the similar projects	Please refer Table 4.8, Table 4.9, Table 4.10, Table 4.11, and Figure 4.6
8	TOR/Project Charter	Project owner/project manager	Owner's need/ requirements	Please refer Table 4.12 and Figure 4.7
9	Develop Concept Design	Designer (A/E) and team members	TOR requirements	Please refer Figure 4.8

(continued)

TABLE 4.22 (Continued)
Total Quality Management Concept for Conceptual Design Phase Activities

Serial Number	Phase Activity	Stakeholders/Team Members	Customer Specifications/ Requirements	System for Managing/Tool
10	Identification of Stakeholders	1. Project owner 2. Designer (Design manager) 3. Project manager	Project need and requirements	Please refer Table 4.14
11	Establishment of Scope of Works	1. Designer/Design manager 2. Project manager 3. Quality manager	1. TOR requirements 2. Project data/ information 3. Owner's requirements 4. Sustainability requirements 5. Regulatory requirements	Please refer Table 4.15, Table 4.16
12	Prepare Concept Design	1. Designer/Design manager 2. Project manager 3. Quality manager	Please refer Section 2.5.3 (Application of Six Sigma DMADV Tool for Design)	Please refer Figure 2.11, Figure 4.9, Table 4.17, and Table 4.18
13	Estimate Project Time Schedule	1. Designer/Design manager 2. Planning manager 3. Project manager 4. Quality manager	Owner's requirements	Please refer Figure 4.10
14	Estimate Project Cost	1. Designer/Design manager 2. Cost Engineer 3. Project manager 4. Quality manager	Owner's requirements	Please refer Table 4.19
15	Quality Management	1. Quality manager 2. Designer (Design manager)	1. Project quality 2. Corporate quality management system	Please refer Figure 4.11
16	Project Resources	1. Designer (Design Manager) 2. Planning manager	Resources for construction project	
17	Project Risk	Designer/Design team members	Typical risks to complete construction project	

TABLE 4.22 (Continued)
Total Quality Management Concept for Conceptual Design Phase Activities

Serial Number	Phase Activity	Stakeholders/Team Members	Customer Specifications/ Requirements	System for Managing/Tool
18	HSE Issues/ Requirements	Designer/Safety manager	Environmental and safety management compliance	
19	Review Concept Design	Designer	1. Owner's requirements 2. Regulatory requirements 3. Funding requirements 4. Project quality	Please refer Table 4.21
20	Finalize Concept Design	1. Designer 2. Project owner	1. Owner's requirements 2. Regulatory requirements 3. Funding requirements 4. Project quality	

requirements. Figure 4.12 illustrates a typical flowchart for TQM applications in preliminary design phase.

Preliminary design is a subjective process of transforming ideas and information into plans, drawings, and specifications of the facility to be built. Component/ equipment configurations, material specifications, and functional performance are decided during this stage. At this stage, the owner can alter the scope and consider alternatives. The owner seeks to optimize certain facility features within the constraints of other factors such as cost, schedule, vendor capabilities, and so on.

Design is a complex process. Before design is begun, the scope must adequately define deliverables, that is, what will be furnished. These deliverables are design drawings, contract specifications, type of contracts, construction inspection record drawings, and reimbursable expenses.

Preliminary design is the basic responsibility of the architect (designer/consultant or A/E). In the case of building construction projects, a preliminary design determines

1. General layout of the facility/building/project
2. Required number of buildings/number of floors in each building/area of each floor

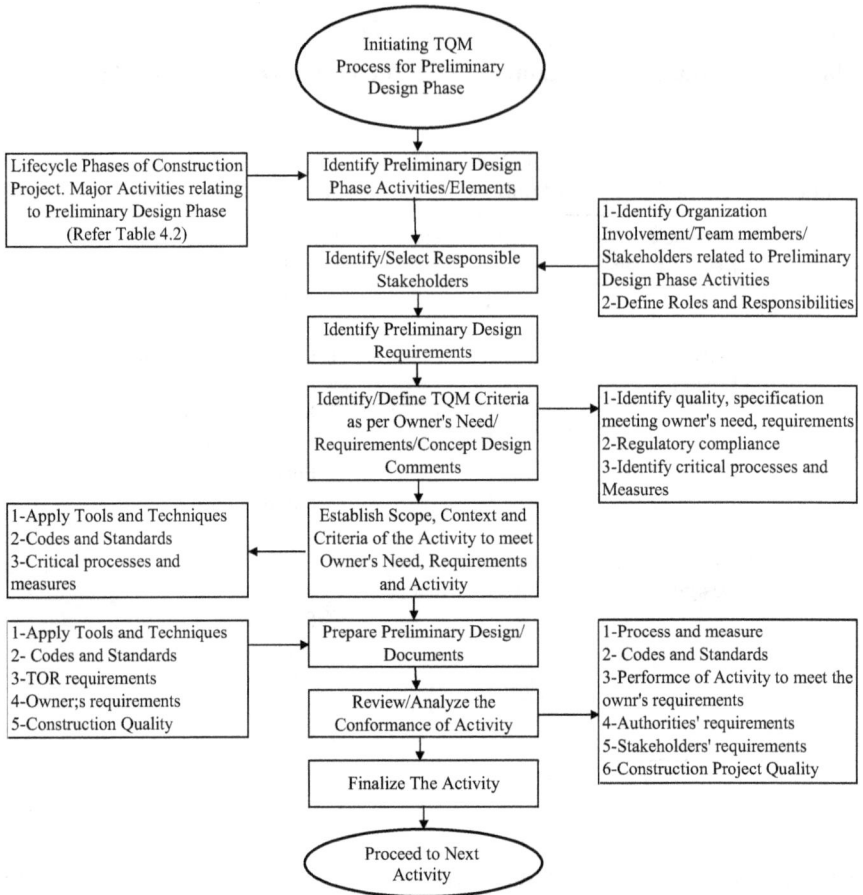

FIGURE 4.12 Typical flowchart for total quality management applications in quality of a preliminary design phase activity.

3. Different types of functional facilities required such as offices, stores, workshops, recreation, training centers, and parking
4. Type of construction such as reinforcement concrete or steel structure, precast, or cast in situ
5. Type of electromechanical services required
6. Type of infrastructure facilities inside the facilities area
7. Type of landscape

The designer has to consider the following points while preparing the preliminary design:

1. Concept design deliverables
2. Calculations to support the design

3. System schematics for electromechanical system
4. Coordination with other members of the project team
5. Authorities' requirements
6. Availability of resources
7. Constructability
8. Health and safety
9. Reliability
10. Energy conservation issues
11. Energy-efficient systems
12. Environmental issues
13. Selection of systems and products that support the functional goals of the entire facility
14. Sustainability requirements
15. Requirements of all stakeholders
16. Optimized life cycle cost (value engineering)

4.3.2.1   Identify Stakeholders

During preliminary design phase, the following stakeholders have direct involvement in the project:

- Owner
- Consultant
- Designer
- Regulatory authorities
- Project/construction manager (if the owner decides to engage them during this phase, depending on the type of project delivery system)

Table 4.23 illustrates the responsibilities of various participants during preliminary design phase.

TABLE 4.23
Responsibilities of Various Participants (Design–Bid–Build Type of Contracts) during Preliminary Design Phase

	Responsibilities		
Phase	**Owner**	**Designer**	**Regulatory Authorities**
Preliminary Design	• Approval of preliminary design	• Develop general layout/scope of facility/project • Regulatory approval • Schedule • Budget • Contract terms and conditions • Value Engineering	• Approval of project submittals

4.3.2.2 1Identify Design Team Members

Design team members are selected based on the organizational structure and suitable skills required to perform the job. Normally the design team consists of

1. Project Manager
2. Design Managers (One for each Trade)
3. Quality Manager
4. Team Leader (Principal Engineer)—Each Trade
5. Team Members (Engineers, CAD Technicians for each discipline)
6. Quantity Surveyor (Cost Engineer)

Figure 4.13 illustrates an example structural/civil engineering design team organization chart.

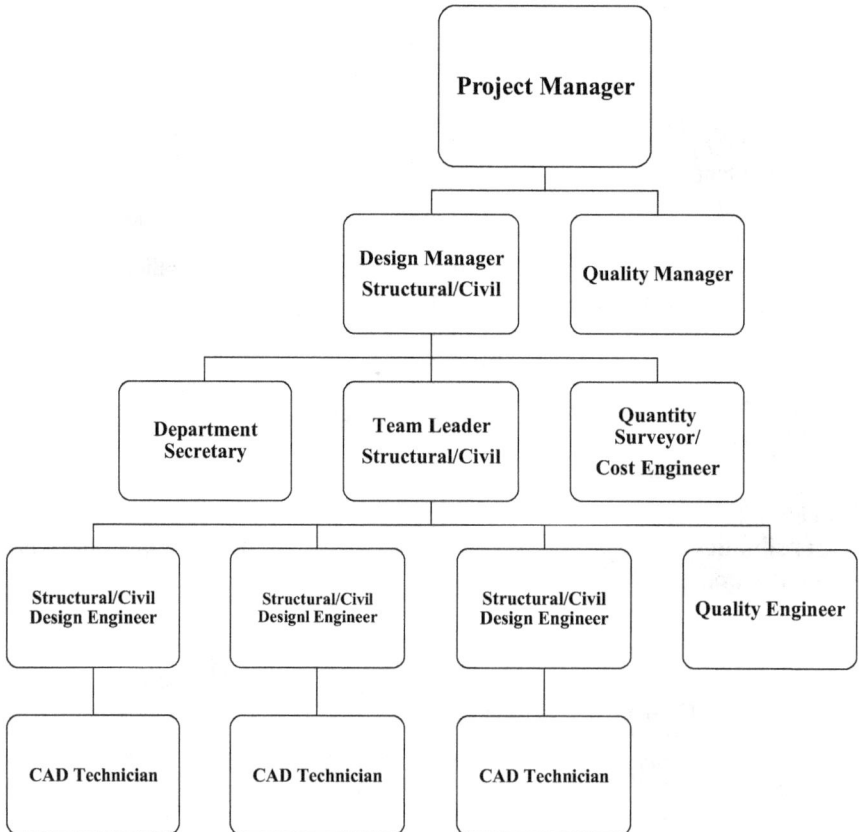

FIGURE 4.13   Structural/civil design team organization chart.

4.3.2.3   Identify Preliminary Design Requirements

In order to identify requirements to develop schematic design, the designer has to gather comments made by the owner/project manager on the submitted concept design, collect TOR requirements, regulatory requirements, and other related data to ensure that the developed design is error-free and with minimum omissions.

The following information is identified/collected to establish preliminary design development scope:

1. Gathering of comments on concept design deliverables
2. Collection of owner's preferred requirements
3. Collection of regulatory requirements
4. TOR requirements
5. Collect LEED requirements
6. Provision for facility management system/equipment

4.3.2.3.1   Identify Concept Design Comments Requirements

The objective of preliminary Design is to refine and develop a clearly defined design based on the client's requirements. While developing the schematic design, the designer reviews the submitted concept design drawings and reports and takes into consideration the comments, if any, made on the concept design. The designer can discuss in detail with the owner/project manager and incorporate all their requirements to develop preliminary design

4.3.2.3.2   Collect Owner's Preferred Requirements

The owner requirements discussed under Section 4.3.1.12.2 are further developed in detail in order to consider in the schematic design.

4.3.2.3.3   Collect Regulatory Requirements

During this phase, the preliminary design drawings are submitted to the regulatory bodies for their review and approval for compliance with the regulations, codes, and licensing procedures. Any comments on the drawings are incorporated into the drawings and are resubmitted, if required.

4.3.2.3.4   TOR Requirements

Normally the TOR lists the Requirements Guidelines to develop schematic design. It mainly consists of the following:

1. Complete Schematic Drawings for the Selected/Approved Concept Design
2. Schematic Design Report
3. Preliminary Models

4.3.2.3.5   LEED Requirements

Leadership in Energy and Environmental Design (LEED) is an internationally recognized green building certification system, providing verification by third party that the building is designed and constructed to the standards and requirements outlined by LEED rating system. The design team and contractor have to integrate all the required features in their design to ensure that building is more durable, healthy, and more energy efficient.

In order to construct a building that merits LEED certification, the designer has to consider the following:

- Optimize site selection
- Orientation of building to make maximum advantages of sunlight
- Water efficiency
- Use of energy as efficiently as possible
- Use of water as efficiently as possible
- Indoor environmental quality
  - Indoor air quality
  - Ventilation
  - Outdoor airflow monitoring
- Maximum use of alternate (renewable) energy
- Minimize wastewater
- Sustainable material selection
- Regional priority
- Innovation in design

4.3.2.3.6 *Provision for Facility Management*
The designer has to make all the related provisions for facility management.

4.3.2.4 Establish Preliminary Design Scope
The purpose of development of preliminary design scope is to provide sufficient information to identify the work to be performed and to allow the detail design to proceed without significant changes that may affect the project schedule and budget. The scope of work during schematic design phase mainly comprises of

- Preparation of schematic drawings
- Outline specifications
- Preliminary schedule
- Preliminary cost estimate
- Authorities approvals
- Narrative reports

The preliminary design scope is established taking into consideration all the requirements discussed under Section 4.3.2.2.

4.3.2.4.1 *Identify Preliminary Design Deliverables*
Based on the scope of preliminary design, following are the deliverables to be developed:

1. Design drawings
   1.1 Architectural
   1.2 Structural
   1.3 Conveying system

    1.4   Mechanical (public health and fire suppression)
    1.5   HVAC
    1.6   Electrical
    1.7   Landscape
    1.8   External
2. Outline specifications
3. Narrative reports
4. Preliminary schedule
5. Preliminary cost estimate
6. Regulatory approvals
7. Value engineering
8. Models

Table 4.24 lists schematic design deliverables to be developed during schematic design phase.

4.3.2.5 Develop Preliminary Management Plan

Preliminary project management plan developed during concept design phase is updated based on additional information collected during schematic design phase. Typical contents of project management plan are as follows:

1. Project Description
2. Project Objectives
3. Project Organization
    3.1   Organization chart
    3.2   Responsibility matrix
    3.3   Project directory
4. Project (Scope) Deliverables
5. Project Schedule
6. Project Budget
7. Project Quality Plan
8. Project Resources
9. Risk Management
10. Communication Matrix
11. Contract Management
12. HSE Management
13. Project Finance Management
14. Claim Settlement

4.3.2.6 Develop Preliminary Design

The purpose of preliminary design is to provide sufficient information to identify the works to be performed and to allow detailed design to proceed without significant changes that may adversely affect the project budget and schedule.

At the preliminary design stage, identified deliverables established in the preliminary design scope must be furnished. It should include a schedule of dates for delivering drawings, specifications, calculations, and other information, forecasts,

TABLE 4.24
Preliminary Design Deliverables

Serial Number	Deliverables
1	General
1.1	a. Preliminary/Outline specifications
	b. Zoning
	c. Permits and regulatory approvals
	d. Energy code requirements
	e. Construction methodology narration
	f. Descriptive report of Environmental, Health and Safety requirements
	g. Estimate construction period (Preliminary schedule)
	h. Estimated cost
	i. Value Engineering suggestions and resolutions
	j. Life safety requirements
	k. Sketches/Perspective
	i. Interior
	ii. Exterior
	m. Graphic presentation
2	Preliminary Design Drawings
2.1	Architectural
2.1.1	Overall site plans
	a. Existing site plans
	b. Location of building, roads, parking, access, and landscape
	c. Project boundary limits
	d. Site utilities
	e. Water supply, drainage, storm water lines
	f. Zoning
	g. Reference grids and axis
2.1.2	Floor plans
	a. Floor plans of all floors
	b. Structural grids
	c. Vertical circulation elements
	d. Vertical shafts
	e. Partitions of various types
	f. Doors
	g. Windows
	h. Floor elevations
	i. Designation of rooms
	j. Preliminary finish schedule
	k. Door schedule
	l. Hardware schedule
	m. Ceiling plans
	n. Services closets
	o. Raised floor, if required

TABLE 4.24 (Continued)
Preliminary Design Deliverables

Serial Number	Deliverables
2.1.3	Roof plans a. Roof layout b. Roof material c. Roof drains and slopes
2.1.4	Elevations a. Wall cladding b. Curtain wall c. Stones cladding d. Exterior Insulation and Finishing System (EIFS) e. Sections at various locations
2.2	Structural a. Building structure b. Floor grade and system c. Foundation system d. Tentative size of columns, beams e. Stairs f. Elevations and sections through various axis g. Roof
2.3	Elevator a. Traffic studies b. Elevator/escalator location c. Equipment room
2.4	Plumbing and fire suppression a. Sprinkler layout plan b. Piping layout plan c. Water system layout plan d. Water storage tank location e. Clean water supply system f. Development of preliminary system schematics g. Location of mechanical room
2.5	HVAC a. Ducting layout plan b. Piping layout plan c. Development of preliminary system schematics d. Calculations to allow preliminary plant selections e. Establishment of primary building services distribution routes f. Establishment of preliminary plant location and space requirements g. Determine heating and cooling requirements based on heat dissipation of equipment, lighting loads, type of wall, roof, glass, etc.

(*continued*)

TABLE 4.24　(Continued)
Preliminary Design Deliverables

Serial Number	Deliverables
	h. Estimation of HVAC electrical load i. Development of BMS schematics showing interface j. Location of plant room, chillers, cooling towers k. Clean air system/Ventilation
2.6	Electrical a.　Lighting layout plan b.　Daylighting c.　Power layout plan d.　System design schematic without any sizing of cables and breakers e.　Substation layout and location f.　Total connected load g.　Location of electrical rooms and closets h.　Location of MLTPs, MSBs, SMBs, EMSB, SEMBs, DBs, EDBs, etc. i.　Location of starter panels, MCC panels, etc. j.　Location of generator, UPS k.　Raceway routes l.　Riser requirements m. Information and Communication Technology (ICT) 　　a) Information technology (computer network) 　　b) IP telephone system (telephone network) 　　c) Smart building system n.　Loss prevention systems 　　a) Fire alarm system 　　b) Access control security system layout 　　c) Intrusion system o.　Public address, audiovisual system layout p.　Schematics for F.A and other loss prevention systems q.　Schematics for ICT system and other low voltage systems r.　Location of LV equipment
2.7	Renewable/Alternate energy system
2.8	Waste management system
2.9	Pollution control system
3.0	Landscape a. Green area layout b. Plantation c. Selection of plants d. Irrigation system

TABLE 4.24 (Continued)
Preliminary Design Deliverables

Serial Number	Deliverables
3.1	External
	a. Street/road layout
	b. Street lighting
	c. Bridges (if any)
	d. Security system
	e. Location of electrical panels (feeder pillars)
	f. Pedestrian walkways

Source: Abdul Razzak Rumane. (2013). *Quality Tools for Managing Construction Projects*. Reprinted
with permission from Taylor & Francis Group.

estimates, contracts, materials, and construction. The designer develops preliminary design with the plan, elevation, and other related information that meet the owner's requirements. The designer also develops a concept of how various systems such as heating and cooling systems, communication systems will fit into the system.

In order to develop preliminary design, the designer has to collect following information:

1. Project-related data/information
   Following data need to be collected to develop preliminary design for a building project:
   - Needs of the owner
   - Building/project usage
   - Space program
   - Technical and functional capability requirements
   - Zoning requirements
   - Aesthetics requirements
   - Fire protection requirements
   - Indoor air quality
   - Lighting/daylighting requirements
   - Conveying system traffic analysis
   - Health and safety features
   - Environmental compatibility requirements
   - Energy conservation requirements
   - Sustainability requirements
   - Facility management requirements
   - Regulatory/authority requirements (permits)
   - Codes and standards to be followed
   - Social responsibility requirements
   - Project constraints
   - Ease of constructability
   - No. of drawings to be produced

- Milestone for development of each phase of design
- Disabled (special needs) access requirements

2. Site investigation
   - Soil profile and laboratory test of soil
   - Topography of the project site
   - Hydrological information
   - Wind load, seismic load, dead load, and live load
   - Existing services passing through the project site
   - Existing roads, structure surrounding the project site
   - Shoring and underpinning requirements with respect to adjacent area/structure
3. Topographical details
4. Geotechnical details
5. Health, safety requirements
6. Environmental requirements

4.3.2.6.1 Prepare Preliminary Design

During preliminary design phase, several alternative schemes are reviewed and one scheme which meets the owner's objectives and TOR is selected.

4.3.2.6.2 Prepare Drawings

Following schematic drawings for building projects are generally prepared during this phase:

1. Architectural
   - Site location in relation to existing environment
   - Overall site plans
   - Floor plans
   - Roof plans
   - Sections
   - Elevations
2. Structural
   - Building structure
   - Foundation system
   - Floor grade and systems
   - Stairs
   - Roof
3. Interior finishes
4. External/interior walls and partitions
5. Conveying system
   - Elevator location
   - Machine room
6. MEP
   - General arrangement of each system
   - Single-line riser diagram
   - Electrical schematic diagram
   - Layout of equipment rooms

- Layout of electrical substation
- Electrical rooms
- Vertical shafts
7. Security system
8. External Works
9. Landscape plans
10. Specialty items
11. Functional/aesthetic aspect of the project
12. Description of critical details
13. Compatibility with surrounding environment
14. Compliance with sustainability requirements

The designer can use Building Information Modeling (BIM) tool for designing the project

4.3.2.6.3   Develop Contract Documents

During this phase following documents are developed in line with the type of contracting system:

1. Preliminary specifications
2. Preliminary contract documents

1. Preliminary specifications (Outline specifications)

Outline specifications indicating project-specific features of major equipment, systems, and material are prepared during this phase. Normally, specifications are prepared per MasterFormat® contract documents produced jointly by the Construction Specification Institute (CSI) and Construction Specifications Canada (CSC) which are widely accepted as standard practice for preparation of contract documents.

2. Preliminary contract documents

The consultant/designer team is responsible for developing a set of contract documents that meet the owner's needs and specifies the required level of quality, schedule, and budget. There are numerous combinations of contract arrangements for handling the construction projects; however, design–bid–build is predominantly used in most construction project contracts. This delivery system has been chosen by owners for many centuries and is called the traditional contracting system. In the traditional contracting system, the detailed design for the project is completed before tenders for construction are invited.

Based on the type of contracting arrangements, the owner would like to handle the project; necessary documents are prepared by establishing a framework for the execution of project. Generally, FIDIC (Federation International des Ingénieurs Counseils) model conditions for international civil engineering contracts are used as a guide to prepare these contract documents.

4.3.2.6.4   Preliminary Design Report

The preliminary design phase reports mainly consist of the following:

- Systems
- Materials

- Finishes
- Design features describing the selected option
- Construction methodology
- Project risk
- HSE requirements

4.3.2.7   Regulatory Approval

Once the preliminary design is approved, the requisite drawings are to be submitted to regulatory bodies for their review and approval for compliance with the regulations, codes, and licensing procedure.

4.3.2.8   Develop Preliminary Schedule

After the preliminary scope of works, the preliminary design and budget (preliminary cost) for the facility/project are finalized; and the logic of the construction program is set. On the basis of logic, a critical path method (CPM) schedule (bar chart) is prepared to determine the critical path and set the contract milestones.

4.3.2.9   Estimate Preliminary Cost

Based on the preliminary design, the budget (preliminary cost) is prepared by estimating the cost of activities and resources. The preparation of the budget is an important activity that results in a timed phased plan summarizing the expected expenses toward the contract and also the income or the generation of funds necessary to achieve the milestone. The budget for a construction project is the maximum amount the owner is willing to spend for design and construction of the facility that meets the owner's needs. The budget is determined by estimating the cost of activities and resources and is related to the schedule of the project. If the cash flow or resulting budget is not acceptable, the project schedule should be modified. It is required that while preparing the budget, the risk assessment of the project is also performed.

4.3.2.10   Manage Design Quality

In order to minimize design errors, minimize design omissions, and reduce rework during schematic design, the designer has to plan quality (planning of design work), perform quality assurance, and control quality for preparing schematic design. This will mainly consist of the following;:

1. Plan Quality
    - Establish owner's requirements
    - Determine number of drawings to be produced
    - Establish scope of work
    - Identify quality standards and codes to be complied
    - Establish design criteria
    - Identify regulatory requirements
    - Identify requirements listed under TOR
    - Establish quality organization with responsibility matrix
    - Develop design (drawings and documents) review procedure

- Establish submittal plan
- Establish design review procedure

2. Quality Assurance:
   - Collect data
   - Investigate site conditions
   - Prepare preliminary drawings
   - Prepare outline specifications
   - Ensure functional and technical compatibility
   - Coordinate with all disciplines
   - Select material to meet owner objectives

3. Control Quality
   - Check design drawings
   - Check specifications/contract documents
   - Check for regulatory compliance
   - Check preliminary schedule
   - Check cost of project (preliminary cost)

BIM tool can be used to control quality of the project.

4.3.2.11 Estimate Resources

Designer has to estimate the resources required to complete the project. At this stage, more detail about the activities and works to be performed during the construction and testing, commissioning phase is available, the designer has to update the earlier estimated resources and prepare manpower histogram. Also, the designer can estimate total number of design team members to develop design and construction documents.

4.3.2.12 Manage Risks

Following are typical risks which normally occur during schematic design phase:

- Concept design deliverables and review comments are not taken into consideration while preparing the preliminary design
- Regulatory authorities' requirements are not taken into consideration
- Preliminary design scope of work is incomplete
- The related project data and information collected are incomplete
- The related project data and information collected are likely to be incorrect and wrongly estimated
- Site investigations for existing conditions not carried out
- Fire and safety considerations
- Environmental consideration
- Incomplete design
- Prediction of possible changes in design during construction phase
- Inadequate and ambiguous specifications
- Wrong selection of materials and systems
- Undersize HVAC equipment selection
- Incorrect water supply requirements

- Estimated total electrical load is much lower than expected actual consumption
- Errors in calculating traffic study for conveying system
- Errors in estimating the project schedule
- Errors in cost estimation
- Number of drawings not as per TOR requirements

Designer has to take into account above-mentioned risk factors while developing the schematic design.

Further, the designer has to consider the following risks while planning the duration for completion of schematic phase:

- Impractical schematic design preparation schedule
- Delay to obtain authorities' approval
- Delay in site investigations
- Delay in data collection

4.3.2.13 HSE Issues and Requirements

The designer has to identify HSE issues and requirements to manage the activities during preliminary design stage. Designer has to do Environmental Impact Assessment study and take into consideration while developing preliminary design.

4.3.2.14 Perform Value Engineering

Value engineering (VE) studies can be conducted at various phases of a construction project; however, the studies conducted in the early stage of a project tend to provide the greatest benefit. In most projects, VE studies are performed during the schematic phase of the project. At this stage, the design professionals have considerable flexibility to implement the recommendations made by the VE team, without significant impacts on the project's schedule or design budget. In certain countries for a project over US$5 million, a VE study must be conducted as part of the schematic design process. The team members who perform the VE study depend on the client's/owner's requirement. It is advisable that a SAVE international registered certified value specialist be assigned to lead this study. Figure 4.14 illustrates Value Engineering Process Activities.

4.3.2.15 Review Preliminary Design

Table 4.25 illustrates major points for the review of preliminary design.

4.3.2.16 Finalize Preliminary Design

Final preliminary design for submission to the owner/client is prepared incorporating the comments, if any, found during the analysis and review of the drawings and documents.

Table 4.26 summarizes the TQM concept for preliminary design phase activities.

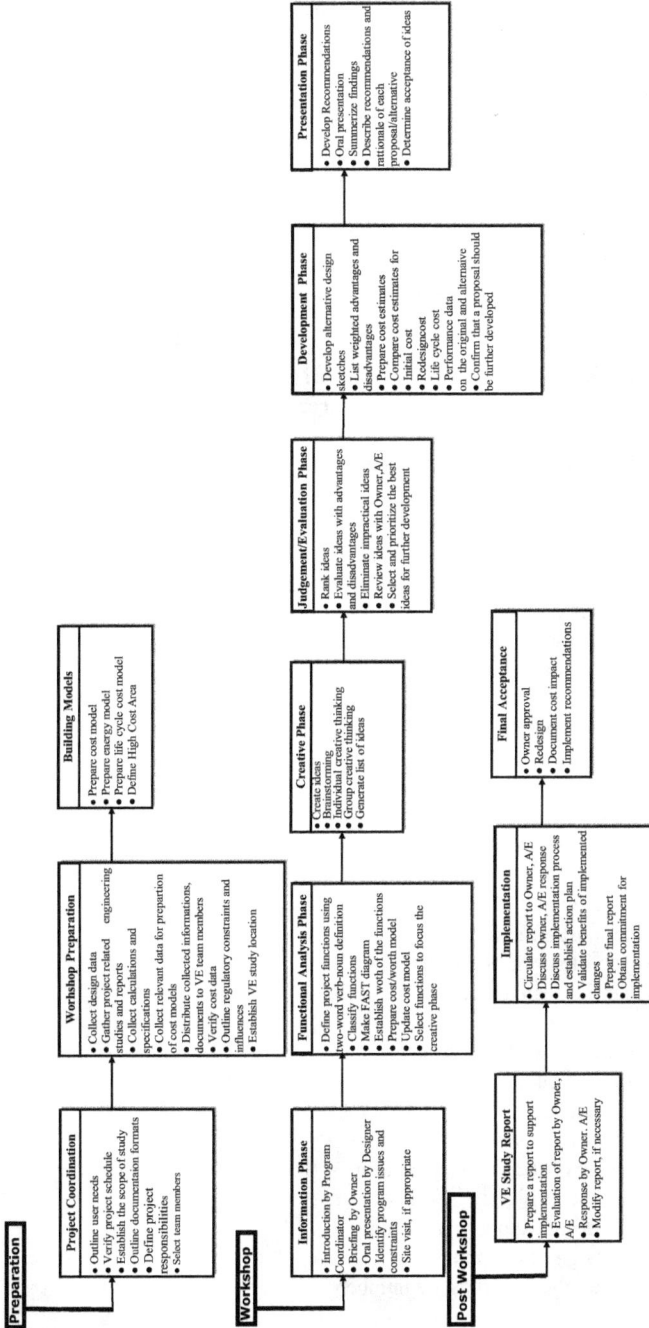

FIGURE 4.14 VE study process activities. (Abdul Razzak Rumane (2017), *Quality Management in Construction Projects*, Second Edition. Reprinted with permission from Taylor & Francis Group Company.)

TABLE 4.25
Major Points for Review of Preliminary Design

Serial Number	Description	Yes	No	Notes
A: Preliminary Design (General)				
A.1	Does the design support the owner's project goals and objectives?			
A.2	Does the design meet all the elements specified in TOR?			
A.3	Whether comments on concept design taken care while preparing schematic design?			
A.4	Whether Regulatory approvals obtained?			
A.5	Does the design meet all the performance requirements?			
A.6	Whether constructability has been taken care?			
A.7	Whether technical and functional capability considered?			
A.8	Whether health and safety requirements in the design are considered?			
A.9	Whether design confirm with fire and egress requirements?			
A.10	Whether design risks been identified, analyzed, and responses planned for mitigation?			
A.11	Whether environmental constraints considered?			
A.12	Does energy conservation is considered?			
A.13	Does sustainability requirements considered in design?			
A.14	Whether cost-effectiveness over the entire project life cycle is considered?			
A.15	Whether the design meets LEED requirements?			
A.16	Whether accessibility is considered?			
A.17	Does the design meet ease of maintenance?			
A.18	Does the design have provision for inclusion of facility management requirements?			
A.19	Does the design support proceeding to next design development stage?			
A.20	Whether all the drawings are numbered?			

TABLE 4.25 (Continued)
Major Points for Review of Preliminary Design

Serial Number	Description	Yes	No	Notes
B: Architectural				
B.1	Drawings coordinated with other discipline?			
B.2	Whether grid system is established?			
B.3	Whether zoning is taken care?			
B.4	Whether overall site plans showing all the major areas?			
B.5	Whether floor plans showing overall dimensions?			
B.6	Whether roof plans are shown?			
B.7	Whether preliminary elevations are shown?			
B.8	Whether all the rooms are numbered			
B.9	Whether entrances, stairways, lobbies, corridors identified?			
B.10	Whether services rooms, equipment rooms, plant rooms identified?			
B.11	Whether finishes schedule/requirements prepared?			
C: Structural				
C.1	Whether preliminary foundation plan shown?			
C.2	Whether structural systems identified?			
C.3	Whether preliminary building structure prepared?			
C.4	Whether preliminary framing plans for all floors and roof			
C.5	Whether relevant codes regarding seismic zone, wind speed considered for structural load calculations?			
C.6	Whether slab loading for equipment considered?			
D: Elevator				
D.1	Whether traffic analysis performed?			
D.2	Whether elevator locations shown?			
D.3	Whether equipment room location shown?			
E: Mechanical				
E.1	Whether preliminary plans for toilets/rest rooms, pantry shown?			
E.2	Whether main water supply, sanitary, and storm water system shown?			

(continued)

TABLE 4.25 (Continued)
Major Points for Review of Preliminary Design

Serial Number	Description	Yes	No	Notes
E.3	Whether riser diagram/single-line diagram for plumbing system shown?			
E.4	Whether plumbing fixtures identified?			
E.5	Whether fire protection system comply with regulatory requirements?			
E.6	Whether single-line diagram for fire protection system (sprinkler) prepared?			
E.7	Whether equipment room (plant room) location shown?			
E.8	Whether special fire suppression system for electrical substation, generator room considered?			
F: HVAC				
F.1	Whether single-line diagram is prepared for all related systems?			
F.2	Whether shaft locations and approximate pipe sizes, duct size shown?			
F.3	Whether location of plant room considered?			
F.4	Whether approximate load for chiller, pump sizes is considered on available data?			
F.5	Whether gross HVAC zoning and typical individual space zoning is considered while preparing the design?			
G: Electrical				
G.1	Whether Preliminary lighting plans prepared?			
G.2	Whether preliminary power layout plans prepared?			
G.3	Whether substation location is shown?			
G.4	Whether electrical room located?			
G.5	Whether cable tray and other raceways route shown?			
G.6	Whether riser shaft for cables, bus ducts considered?			
G.7	Whether system schematic diagram prepared?			
G.8	Whether fire alarm system complies with regulatory requirements?			
G.9	Whether preliminary plans for all low voltage systems (communication, public address, audio visual, access control, security system) prepared?			

TABLE 4.25 (Continued)
Major Points for Review of Preliminary Design

Serial Number	Description	Yes	No	Notes
G.10	Whether emergency diesel generator is considered?			
H: Landscape				
H.1	Whether landscape plans prepared?			
H.2	Whether plants selected?			
H.3	Whether irrigation system layout prepared?			
I: External				
I.1	Whether site layout plans showing roads, walkways, parking areas prepared?			
I.2	Whether street lighting plans prepared?			
I.3	Whether project site boundaries are properly marked and demarcated?			
J: Financial				
J.1	Project cost is properly estimated?			
K: Schedule				
K.1	Project schedule is practically achievable?			
L: Value Engineering				
L.1	Whether value engineering study performed and recommendations taken care?			
M: Reports				
M.1	Whether the narrative description complete and include adequate information about the project?			
M.2	Whether outline specifications include all the works?			
M.3	Whether preliminary contract documents have taken care of all the TOR requirements?			
N: Drawings, Sketches				
N.1	Whether drawings, sketches for all trades prepared as per TOR?			
O: Models				
O.1	Whether the models meet the design objectives?			
P: Submittals				
P.1	Whether numbers of sets prepared are as per TOR?			

TABLE 4.26
Total Quality Management Concept for Preliminary Design Phase Activities

Serial Number	Phase Activity	Stakeholders/ Team Members	Customer Specifications/ Requirements	System for Managing/Tool
1	Identification of Stakeholders/ Project Team Members	1. Project owner 2. Designer (Design manager) 3. Project manager	1. Project need and requirements	Please refer Table 4.23
2	Identification of Preliminary Design Requirements	1. Designer/Design manager 2. Project manager 3. Quality manager	1. TOR requirements 2. Project data/ information 3. Owner's requirements 4. Sustainability requirements 5. Regulatory requirements	Similar to Table 4.15 and Table 4.16 Please refer Section 4.3.2.5
3	Establish Preliminary Design Scope	1. Designer/Design manager 2. Project manager 3. Quality manager	1. TOR requirements 2. Project data/ information 3. Owner's requirements 4. Sustainability requirements 5. Regulatory requirements	Please refer Table 4.24
4	Develop Preliminary Management Plan	Design manager Planning manager Project manager	TOR requirements	Please refer Section 4.3.2.4
5	Develop Preliminary Design	1. Designer/ Design manager/ Design team 2. Project manager 3. Quality manager	Please refer Section 2.5.3 (Application of Six Sigma DMADV Tool for Design)	Please refer Figure 4.13 and Table 4.24
6	Develop Contract Terms and Documents	1. Designer/Design manager 2. Project manager 3. Quality manager 4. Quantity surveyor/ Contract administrator	1. TOR requirements 2. Regulatory requirements 3. Applicable construction contract documents 4. Applicable codes and standards	Please refer Section 4.3.2.5.3
7	Regulatory/ Authority Approval	Designer/Design manager Regulatory authority	Applicable codes and standards	Follow agreed upon communication system

TABLE 4.26 (Continued)
Total Quality Management Concept for Preliminary Design Phase Activities

Serial Number	Phase Activity	Stakeholders/ Team Members	Customer Specifications/ Requirements	System for Managing/Tool
8	Estimate Preliminary Schedule	1. Designer/Design manager 2. Planning manager 3. Project manager 4. Quality manager	Owner's needs and requirements	Schedule is based on concept of Figure 4.10
9	Estimate Project Cost	1. Designer/Design manager 2. Planning manager 3. Project manager 4. Quality manager	Owner's needs and requirements	Project cost is based on Table 4.19
10	Quality Management	1. Quality manager 2. Designer (Design manager)	1. Quality manager 2. Designer (Design manager)	
11	Project Resources	1. Designer (Design Manager) 2. Planning manager	Resources for construction project	
12	Project Risk	Designer/Design team members	Typical risks to complete construction project	
13	HSE Issues/ Requirements	Designer/Safety manager	Environmental and Safety management compliance	
14	Value Engineering	Designer Value Engineering Consultant	Optimization of project design	Please refer Figure 4.14
15	Review Preliminary Design	Designer	1. Owner's requirements 2. Regulatory requirements 3. Funding requirements 4. Project quality	Please refer Table 4.25
16	Finalize Preliminary Design	1. Designer 2. Project owner	1. Owner's requirements 2. Regulatory requirements 3. Funding requirements 4. Project quality	

4.3.3 TQM in Detail Design Phase

Detailed design is the third phase of the construction project life cycle. It follows the preliminary design phase and takes into consideration the configuration and the allocated baseline derived during the preliminary phase. Design development is also known as detailed design/detailed engineering. During this phase, all suggested changes are reevaluated to ensure that the changes will not de-tract from meeting the project design goals/objectives. Detailed design involves the process of successively breaking down, analyzing, and designing the structure and its components so that it complies with the recognized codes and standards of safety and performance while rendering the design in the form of drawings and specifications that will tell the contractors exactly how to build the facility to meet the owner's need. During this phase, detail design of the work, contract documents, detail plan, budget, estimated cash flow, regulatory approval, and tender/bidding documents are prepared. Depending on the type of contract the owner would like to have for completing the facility, the designer (consultant) can start preparing the detailed design. The success of a project is highly correlated with the quality and depth of the engineering plans prepared during this phase.

Figure 4.15 illustrates a typical flowchart for TQM applications in detailed design phase.

4.3.3.1 Identify Stakeholders

Following stakeholders have direct involvement in the project during detailed design phase:

- Owner
- Consultant
- Designer
- Regulatory authorities
- Project/construction manager (if the owner decides to engage them during this phase, depending on the type of project delivery system)

Table 4.27 illustrates the responsibilities of various participants (design–bid–build type of contracts) during Detailed Design Phase.

4.3.3.1.1 Identify Design Team Members

Generally, the project team members selected to develop schematic design continues during detail design phase. Additional design personnel are included to meet the workload to develop detail design.

Accuracy in the project design is a key consideration of the life cycle of the project; therefore, it is required that the designer/consultant be not only an expert in the technical field but also should have a broad understanding of engineering principles, construction methods, and value engineering. The designer must know the availability of the latest products in the market and use proven technology, methods, and materials to meet the owner's objectives. He or she must refrain from using a monopolistic product, unless its use is important or critical for proper functioning of the system. He

Initiating TQM Process for Detail Design Phase

Lifecycle Phases of Construction Project. Major Activities relating to Detail Design Phase (Refer Table 4.2)

Identify Detail Design Phase Activities/Elements

Identify/Select Responsible Stakeholders

1-Identify Organization Involvement/Team members/ Stakeholders related to Detail Design Phase Activities
2-Define Roles and Responsibilities

Identify Detail Design Requirements

Identify/Define TQM Criteria as per Owner's Need/ Requirements/TOR Requirements/Preliminary Design Comments

1-Identify quality, specification meeting owner's need, requirements
2- TOR Requirements
3-Regulatory compliance
4-Identify critical processes and Measures

1-Apply Tools and Techniques
2- Codes and Standards
3-Critical processes and measures

Establish Scope, Context and Criteria of the Activity to meet Owner's Need/Requirements, TOR Requirements

1-Apply Tools and Techniques
2- Codes and Standards
3-TOR requirements
4-Owner;s requirements
5- Construction Quality

Prepare Detail Design/ Documents

Review/Analyze the Conformance of Activity Requirements

1-Process and measure
2- Codes and Standards
3-Performce of Activity to meet the ownr's requirements
4-Authorities' requirements
5-Stakeholders' requirements
6- Construction Project Quality

Finalize The Activity

Proceed to Next Activity

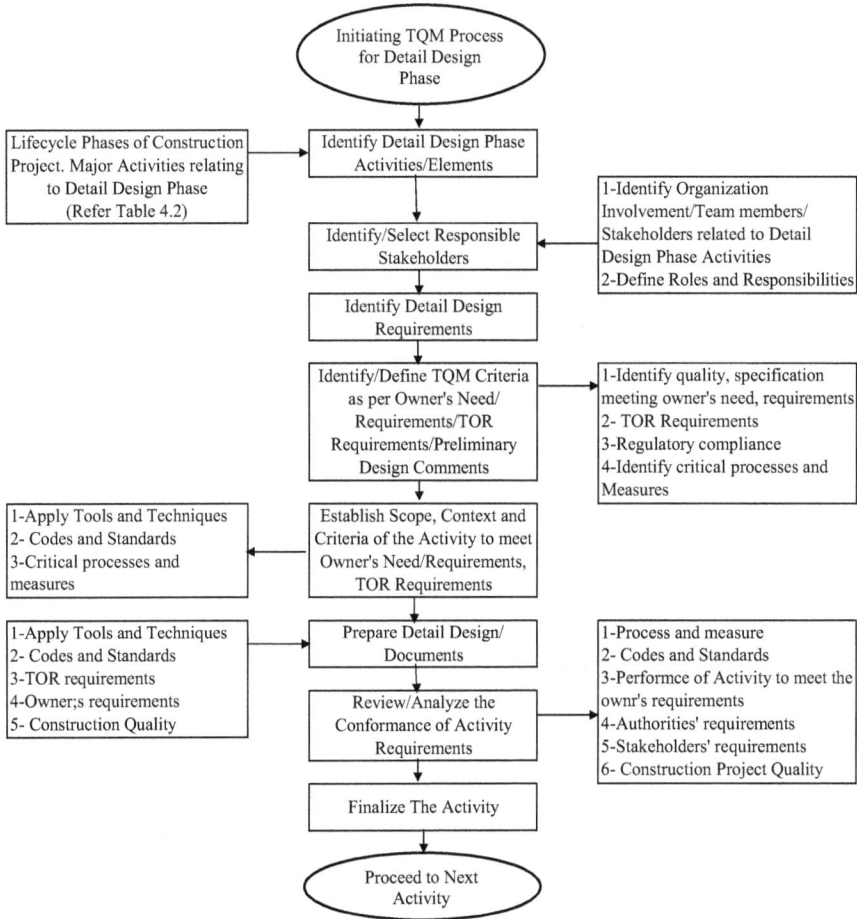

FIGURE 4.15 Typical flowchart for total quality management applications in quality of a detail design phase activity.

or she must ensure that at least two or three sources are available in the market producing the same type of product that complies with all its required features and intent of use. This will help the owner get competitive bidding during the tender stage.

Figure 2.14 discussed under Section 2.5.3 of Chapter 2 illustrates the Design Management Team and their major responsibilities.

Each of the managers has many other team members. These members are selected based on the organizational structure and suitable skills required to perform the job. These include

1. Team Leader (Principal Engineer) for each trade
2. Team Members (Engineers and CAD Technicians for each discipline)
3. Quantity Surveyor (Cost Engineer)

TABLE 4.27
Responsibilities of Various Participants (Design–Bid–Build Type of Contracts) during Detail Design Phase

| | **Responsibilities** | | |
Phase	**Owner**	**Designer**	**Regulatory Authorities**
Detail Design	• Approval of design • Approval of time schedule • Approval of budget	• Development of detail design • Submission for authorities approval • Detail plans • Schedule • Budget • BOQ • Specifications • Contract documents • Verification of design	• Review and approval of project submittals

4. Owner's Representative
5. End User

4.3.3.2 Identify Detail Design Requirements

Detail Design is an enhancement of the work carried out during preliminary design phase. During this phase a comprehensive design of works with detailed WBS of design, drawings, specifications, and contract documents are prepared. The detail design phase is realm of design professionals, including architects, interior designers, landscape architects, and several other disciplines such as civil, mechanical, electrical, and other engineering professionals as needed.

During this phase detailed plans, sections, and elevations are drawn to scales, principal dimensions are noted, and design calculations are checked to conform the accuracy of design and its compliance to the codes and standards.

4.3.3.2.1 Data/Information

In order to identify detail design requirements, the designer has to collect the following information:

1. Preliminary design comments
2. TOR requirements
3. Owner's requirements
4. Regulatory requirements
5. Sustainability requirements

4.3.3.3   Establish Detail Design Scope

Design development scope is developed taking into consideration requirements established under Section 4.3.3.2.1 which are as follows:

- Approved preliminary design documents
- TOR requirements
- Owner's preferred requirements
- Regulatory requirements
- Sustainability requirements
- Approved preliminary project schedule
- Approved preliminary cost estimate
- Codes and standards

4.3.3.3.1   Identify Detail Design Deliverables

Based on the scope for design development, following are the major deliverables to be developed during detail design phase:

1. Design drawings
    1.1   Site plans
    1.2   Architectural
    1.3   Interior
    1.4   Structural
    1.5   Conveying system
    1.6   Mechanical (public health and fire suppression)
        1.6.1   Clean water supply system
    1.7   HVAC
        1.7.1   Clean air supply system
    1.8   Electrical
        1.8.1   Alternate energy system
        1.8.2   Comfortable lighting
    1.9   Low voltage system
    1.10 Waste management system
    1.11 Pollution control system
    1.12 Landscape
    1.13 External
2. Specifications
3. Project schedule
4. Definitive cost estimate
5. Regulatory approvals

Table 4.28 lists Detail Design Deliverables to be developed during Detail Design phase.

4.3.3.4   Develop Project Management Plan

Preliminary project management plan developed during schematic phase is updated based on additional information collected during detail design phase. The contents of

TABLE 4.28
Detail Design Deliverables

Serial Number	Deliverables
1	General
1.1	a. Project Specifications
	b. Project Schedule
	c. Project Estimate
	d. Bill of Quantities
	e. Detailed Calculations (All trades)
	f. Schematic/Riser Diagram (All applicable systems)
	g. Site Investigations
	h. Design Report
	i. Model
2	Detail Design Drawings
2.1	Architectural
2.1.1	Overall site plans
2.1.2	Floor plans
2.1.3	Roof plans
2.1.4	Elevations
2.2	Structural
2.3	Elevator
2.4	Plumbing and Fire suppression
2.5	HVAC
2.6	Electrical
2.7	Waste management system
2.8	Pollution Control system
2.9	Landscape
3.0	External

Project Management Plan are same as that of preliminary design but additional information is added. Typical TOC of Project Management Plan is as given below:

1. Project Description
2. Project Objectives
3. Project Organization
   3.1 Organization chart
   3.2 Responsibility matrix
   3.3 Project directory
4. Project (Scope) Deliverables
5. Project Schedule
6. Project Budget
7. Project Quality plan
8. Project Resources
9. Risk Management

10. Communication Matrix
11. Contract Management
12. HSE Management
13. Project Finance Management
14. Claim Settlement

4.3.3.5   Develop Detail Design

Detail design activities are similar, although more in-depth than the design activities in the preliminary design stage. The size, shape, levels, performance characteristics, technical details, and requirements of all the individual components are established and integrated into the design. Design engineers of different trades have to take into consideration all these at a minimum while preparing the scope of works. The range of design work is determined by the nature of the construction project.

Figure 4.16 illustrates the stages to develop detail design.

The designer has to collect data/information and perform site investigations and verify the information is same as collected earlier.

4.3.3.5.1   Data Collection

The designer has to collect all the missing data to ensure that detailed design is developed without any errors and omissions.

4.3.3.5.2   Site Investigations

The designer has to revisit the site to ensure that there are not many changes to the earlier performed investigations.

4.3.3.5.3   Prepare Detail Design

Following detail drawings for building projects are generally prepared during this phase:

1. Architectural
    *   Site plans
    *   Floor plans
    *   Roof plans
    *   Partitions
    *   Sections
    *   Elevations
    *   Reflected ceiling plan
    *   Finishes schedule
    *   Door schedule
    *   Typical windows details
    *   Toilets
    *   Furnishings

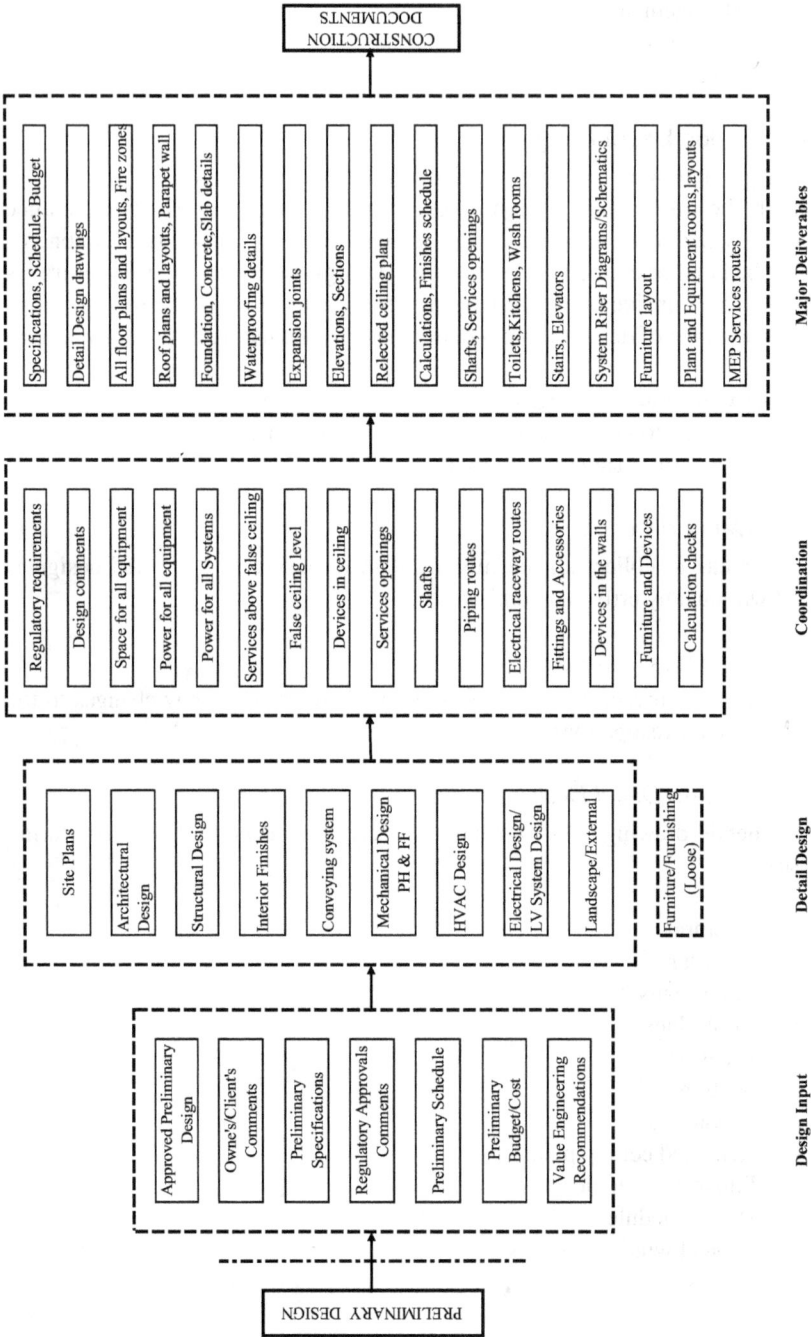

FIGURE 4.16 Detail design development stages. (Abdul Razzak Rumane. (2013). *Quality Tools for Managing Construction Projects.* Reprinted with permission from Taylor & Francis Group.)

2. Structural
   • Building structure
   • Foundation system
   • Footings
   • Stairs
   • Roof
   • Sections
   • Structural floor plans
3. Conveying system
   • Elevator location
   • Machine room
4. MEP
   • General arrangement of each system
   • Single-line riser diagrams
   • Electrical schematic diagram
   • Layout of equipment rooms
   • Layout of electrical substation
   • Electrical rooms
   • Vertical shafts
5. Low voltage systems
6. Waste management system
7. Pollution control system
8. External works
9. Landscape plans

4.3.3.5.3.1 Calculations

The designer has to submit following calculations along with the detail design:

• Calculation of all structural elements (Structural Works)
• Analysis and selection of HVAC system
• Analysis and selection of mechanical systems
• Calculation of lighting (Isolux calculations)
• Short-circuit calculations for cables
• Lightening protection system
• Earthing (grounding) system
• Sound system (audiovisual system)

4.3.3.5.4 Prepare Drawings

Detail design activities are similar, although more in-depth than the design activities in the preliminary design stage. The size, shape, levels, performance characteristics, technical details, and requirements of all the individual components are established and integrated into the design. Design engineers of different trades have to take into consideration all these at a minimum while preparing the scope of work. The range of design work is determined by the nature of the construction project.

4.3.3.5.4.1   Detail Design of Works

The following are the aspects of works to be considered by design professionals while preparing the detail design. These can be considered as a base for development of design to meet customer requirements and will help achieve the qualitative project.

Architectural Design

- Intent/use of building/facility
- Property limits
- Aesthetic look of the building
- Environmental conditions
- Elevations
- Plans
- Axis, grids, levels
- Room size to suit the occupancy and purpose
- Zoning as per usage/authorities requirements
- Identification of zones, areas, rooms
- Modules to match with structural layout/plan
- No. of floors
- Ventilation
- Thermal insulation details
- Stairs, elevators (horizontal and vertical transportation)
- Fire exits
- Ceiling height and details
- Reflected ceiling plan
- Internal finishes
- Internal cladding
- Partition details
- Masonry details
- Joinery details
- Schedule of doors and windows
- Utility services
- Toilet details
- Required electromechanical services
- External finishes
- External cladding
- Glazing details
- Finishes schedule
- Door schedule
- Windows schedule
- Hardware schedule
- Special equipment
- Fabrication of items, such as space frame, steel construction, retaining wall, having special importance for appearance/finishes
- Special material/product to be considered, if any
- Any new material/product to be introduced

- Conveying system core details
- Maintenance access for equipment/services requirements
- Ramp details
- Hard and soft landscape
- Parking areas
- Provision for future expansion (if required)

Concrete Structure

- Property limits/surrounding areas
- Type of foundation
- Energy-efficient foundation
- Design of foundation based on field and laboratory test of soil investigation which gives the following information:
    a) Subsurface profiles, subsurface conditions, and subsurface drainage
    b) Allowable bearing pressure and immediate and long-term settlement of footing
    c) Co-efficient of sliding on foundation soil
    d) Degree of difficulty for excavation
    e) Required depth of stripping and wasting
    f) Methods of protecting below grade concrete members against impact of soil and groundwater (water and moisture problems, termite control, and radon where appropriate)
    g) Geotechnical design parameters such as angle of shear resistant, cohesion, soil density, modulus of deformation, modulus of sub-grade reaction, and predominant soil type
    h) Design loads such as dead load, live load, wind load, and seismic load
        - Footings
        - Grade and type of concrete
        - Size of bars for reinforcement and the characteristic strength of bars
        - Clear cover for reinforcement for
    a) Raft foundation
    b) Underground structure
    c) Exposed to weather structure such as columns, beams, slabs, walls, and joists
    d) Not exposed to weather columns, beams, slabs, walls, and joists
        - Reinforcement bar schedule, stirrup spacing
        - Expansion joints
        - Concrete tanks (water)
        - Insulation
        - Services requirements (shafts, pits)
        - Shafts and pits for conveying system
        - Location of columns in coordination with architectural requirements
        - No. of floors
        - Height of each floor
        - Beam size and height of beam

- Openings for services
- Substructure

a) Columns
b) Retaining walls
c) Walls
d) Stairs
e) Beams
f) Slab

- Super structure

a) Columns
b) Stairs
c) Walls
d) Beams
e) Slabs

- Consideration of water proofing requirements for roof slab against water leakage
- Deflection which may cause fatigue of structural elements, crack or failure of fixtures, fittings or partitions or discomfort of occupants
- Movement and forces due to temperature
- Equipment vibration criteria
- Load sensors to measure deflection
- Reinforcement bar schedule, stirrup spacing
- Building services to fit in the building
- Environmental compatibility
- Parapet wall
- Excavation
- Dewatering
- Shoring
- Backfilling

Elevator Works

- Type of elevator
- Loading capacity
- Speed
- No. of stops
- Travel height
- Cabin, cabin accessories, cabin finishes, and car operating system
- Door, door finishes, and door system
- Safety features
- Drive, size, and type of motor
- Floor indicators, call button
- Control system
- Cab overhead dimensions
- Pit depth

- Hoist way
- Machine room
- Operating system

Fire Suppression System

The fire suppression system provides protection against fire to life and property. The system is designed taking into consideration the local fire code and NFPA standards. The system includes the following:

- Sprinkler system for fire suppression in all the areas of the building
- Hydrants (landing valve) for professional fighting
- Hose reel for public use throughout the building
- Gaseous fire protection system for communication rooms
- Fire protection system for diesel generator room
- Size of fire pumps and controls
- Water storage facility
- Interface with other related systems

Plumbing Works

- Maximum working pressure to have adequate pressure and flow of water supply
- Maximum design velocity
- Maximum probable demand
- Demand weight of fixture in fixture units for public uses
- Friction loss calculation
- Maximum hot water temperature at fixture outlet
- Water heater outlet hot temperature
- Providing isolating valves to ensure that the system is easily maintainable
- Hot water system
- Central water storage capacities
- Size of pumps and controls
- Location of storage tank
- Schematic diagram for water distribution system

Drainage System

While designing the drainage system, the schedule of foul drainage demand units and frequency factors for the following items should be considered for sizing the piping system, number of manholes, and capacity of sump pump, capacity of sump pit:

- Washbasins
- Showers
- Urinals
- Restrooms
- Kitchen sinks

- Other equipment such as dishwashers and washing machines
- Waste management system

HVAC Works

- Environmental conditions
  - Outdoor design conditions
  - Indoor design conditions
  - Indoor air quality
- Air conditioning calculations
  - Cooling load calculations
  - Heating load calculations
  - Space temperature and humidity at required set point
  - Occupancy load
  - Lighting load
- Room pressurizing and leakage calculations
- Energy consumption calculations
- Air conditioning calculations for IT equipment room(s) based on heat emission of equipment
- Air distribution system calculations
- Smokes extract ventilation calculations
- Exhaust ventilation calculations
- Ductwork sizing calculations
- Selection of the ductwork components such as balancing dampers, constant volume boxes, variable air volume boxes, attenuators, grilles and diffusers, fire dampers, pressure relief dampers, and so on
- Pipe work sizing calculations
- Selection of the inline pipe work components, for example, valves, strainers, air vents, commissioning sets, flexible connections, and sensors
- Selection of boilers, pressurization units, air conditioning calculations
- Pipework and duct work insulation selection
- Details of grilles and diffusers, control valves, and so on
- Selection of the duct work systems plant and equipment, for example, air handling units, fan coil units, filters, coils, fans, humidifiers, and duct heaters
- Selection of chillers, cooling towers
- Selection of pumps
- Selection of fans
- Equipment system calculations
- Space requirements for chillers, cooling towers, pumps, and other equipment (plant room)
- Mechanical room location and access
- Preparation of the plan and section layouts and plant room drawings
- Electrical load calculations
- Comparison of electrical consumption with electrical conservation code
- Preparation of equipment schedules

- HVAC-related electrical works
- Control details
- Starter panels, MCC panels, schematic diagram of MCC
- Selection of program equipment
- Preparation of point schedule for building management system
- Schematic diagram for BMS (Building Management System)

Electrical System

- Lighting calculations for different areas based on illumination level recommended by CIE/CEN/CIBSE and Isolux diagrams
- Selection of light fittings, type of lamps
- Daylighting system
- Selection of control gear for light fixture
- Environmental consideration for selection of light fixture and control gear for comfortable lighting
- EXIT/EMERGENCY lighting system
- Circuiting references, normal as well as emergency
- Sizing of conduits
- Power for wiring devices
- Power supply for equipment (HVAC, PH&FF, conveying system, and others)
- Sizing of cable tray
- Sizing of cable trunking
- Selection (type and size) of wires and cable
- Voltage drop calculations for wires and cables
- Selection of upstream and downstream breakers
- Derating factor
- Sensitive of breakers (degree of protection)
- Selection of isolators
- IP ratings (degree of ingress protection) of panels, boards, isolators
- Schedule of distribution boards, switch boards, and main low tension boards
- Cable entry details
- Location of distribution boards, switch boards, and low tension panels
- Short-circuit calculations
- Sizing of diesel generator set for emergency power supply
- Sizing of ATS (Automatic Transfer Switch)
- Generator room layout
- Sizing of capacitor bank
- Provision for solar system integration
- Schematic diagrams
- Sizing of transformers
- Substation layout
- Calculations for grounding (earthing) system
- Grounding system layout
- Calculations for lightning protection system

- Lightning protection system layout
- Renewable energy/alternate energy system

Fire Alarm System

A fire alarm system is designed taking into consideration the local fire code and NFPA standards. The system includes the following:

- Conduiting and raceways
- Type of system: analog/digital/addressable
- Type of detectors based on the area and spacing between the detectors and the walls
- Break glass/pull station
- Type of horns/bells
- Voice evacuation system, if required
- Type of wires and cables
- Mimic panel, if required
- Repeater panel, if required
- Main control panel
- Interface with other systems such as HVAC, elevator
- Riser diagram

Solar System

The designer should consider the following points while developing solar power system:

- Avoid shading from trees, buildings, and so on (especially during peak sunlight hours)
- Check the proposed plan for the proposed site to ensure that future neighboring construction will not cast shade on the array
- Determine where solar array can be placed (roof, carport, facade, curtain wall, boundaries, double skin, or elsewhere)
- Keep the south-facing section obstruction-free if possible. If the roof is sloped, the south-facing section will optimize the system performance
- Minimize rooftop equipment to maximize available open area for solar collector placement
- Ensure that the type of roof will have adequate space to install solar system at later stage to optimize cost of installing solar system at later stage.
- Ensure the roof is capable of carrying the load of the solar equipment.
- Analyze wind loads on rooftop solar equipment in order to ensure that the roof structure is sufficient.
- Add additional safety equipment for solar equipment access and installation

Information and Communication Technology

- Structured cabling considering type and size of cable: copper, fiber optic
- Type and size of the cables
- Racks

- Wiring accessories/devices
- Access/distribution switches
- Internet switches
- Core switch
- Access gateway
- Router
- Network management system
- Servers
- Telephone handsets

Public Address System

- Conduiting and raceways
- Type of system: analog/digital/IP-based
- Types of wires and cables
- Types of speakers
- Distribution of speakers
- Required noise level in different areas
- Calculations for sound pressure level
- Zoning of system, if required
- Size and type of pre-mixer
- Size and type of amplifier
- Microphones
- Paging system
- Message recorder/player
- Interface with other systems

Audiovisual System

- Conduiting and raceways
- Type of system: analog/digital/IP-based
- Types of wires and cables
- Racks
- Type, size, and brightness of projectors
- Type and size of speakers and sound pressure level
- Type and size of screens
- Microphones
- Cameras (visualizers)
- CD/DVD players-recorders
- Control processors
- Video switch matrix
- Mounting details of equipment

Security System/CCTV

- Type of system: digital/IP-based
- Conduiting and raceways

- Wires and cabling network
- Level of security required
- Type and size of cameras
- Types of monitors/screens
- Video/event recording
- Video servers
- Database server
- System software
- Schematic diagram
- System console

Security System/Access Control

- Conduiting and raceways
- Wires and cabling network
- Proximity RFID reader
- Fingerprint and proximity combine reader
- Magnetic lock
- Release button
- Door contact
- RFID card
- Reader control panel
- Server
- Multiplexer
- Monitors
- Workstation
- Metal detector

Landscape Works

As a landscape architect, the following points are to be considered while designing the landscape system:

- Property boundaries
- Size and shape of the plot
- Shape and type of dwelling
- Integration with surrounding areas
- Orientation to the sun and wind
- Climatic/environmental conditions
- Ecological constraints (soil, vegetation, etc.)
- Location of pedestrian paths and walkways
- Pavement
- Garage and driveway
- Vehicular circulation
- Location of sidewalk
- Play areas and other social/community requirements
- Outdoor seating

- Location of services, positions of both under- and aboveground utilities and their levels
- Location of existing plants, rocks, or other features
- Site clearance requirements
- Foundation for paving, including front drive
- Top soiling, or top soil replacement
- Soil for planting
- Planting of trees, shrubs, and ground covers
- Grass area
- Sowing grass or turfing
- Lighting poles/bollard
- Special features, if required
- Signage, if required
- Surveillance, if required
- Installation of irrigation system
- Marking out the borders
- Storage for landscape maintenance material

External Works (Infrastructure and Road)

External works are part of the contract requirements of a project that involves the construction of a service road and other infrastructure facilities to be connected to the building and also includes care of existing services passing through the project boundary line. The designer has to consider the following while designing external works:

- Grading material
- Asphalt paving for road or street
- Pavement
- Pavement marking
- Precast concrete curbs
- Curbstones
- External lighting
- Cable routes
- Piping routes for water, drainage, storm water system
- Sump pump(s) for drainage, storm water
- Trenches or tunnels
- Bollards
- Manholes and hand holes
- Traffic marking
- Traffic signals
- Boundary wall/retaining wall, if required

Bridges

Designers should use relevant authorities' design manual and standards and consider the following points while designing bridges.

- Soil stability
- Alignment with road width, property lines
- Speed
- Intersections/interchanges
- No. of lanes, width
- Right-of-way lines
- Exits, approaches, and access
- Elevation datum
- Superelevation
- Clearance with respect to railroad, roadway, navigation (if applicable)
- High and low levels of water (if applicable)
- Utilities passing through the bridge length
- Slopes
- Number and length of span
- Live loads, bearing capacity
- Water load, wind load, earthquake effect (seismic effect)
- Bridge rails, protecting screening, guard rails, barriers
- Shoulder width
- Footings, columns, and piles
- Abutment
- Beams
- Substructure
- Super structure, deck slab
- Girders
- Slab thickness
- Reinforcement
- Supporting components, deck hanger, tied arch
- Expansion and fixed joints
- Retaining walls, crash wall
- Drainage
- Lighting
- Aesthetic
- Sidewalk, pedestrian, and bike facilities
- Signage, signals
- Durability
- Sustainability

Highways

Designer should use relevant authorities' design manual and standards and consider the following points while designing highways:

- Type of highway
- Soil stability
- Speed
- No. of lanes, width

- Shoulder width
- Gradation
- Type of pavement and thickness
- Right-of-way lines
- Exits, approaches, access, and ramp
- Superelevation
- Slopes, curvature, turning
- Median, barriers, curb
- Sidewalks, driveways
- Pedestrian accommodation
- Bridge roadway width
- Drainage
- Gutter
- Special conditions, such as snow and rain
- Pump(s) for drainage, rainwater
- Lighting
- Signage, signals
- Durability
- Sustainability

Furnishings/Furniture (Loose)

In building construction projects, loose furnishings/furniture are tendered as a special package and are normally not part of main contract. In order to express all the features of the furnishing/furniture products in the specification, the descriptive feature of the product is not enough. In order to give enough information and understanding of the product, the product specifications are accompanied with the pictorial view/cut out sheet/photo of the product and the furniture layout.

Table 4.29 illustrates a tool for mistake proofing to eliminate design errors.

4.3.3.6   Develop Contract Documents

Preparation of detailed documents and specifications per master format is one of the activities performed during this phase of the construction project. The contract documents must specify the scope of works, location, quality, and duration for completion of the facility. As regards the technical specifications of the construction project, master format specifications are included in the contract documents. The master format is a master list of section titles and numbers for organizing information about construction requirements, products, and activities into a standard sequence. It is a uniform system for organizing information in project manuals, for organizing cost data, for filling product information and other technical data, for identifying drawing objects, and for presenting construction market data. MasterFormat™ (1995 edition) consisted of 16 divisions; however, MasterFormat (2004 edition) consists of 48 divisions (49 is reserved). MasterFormat contract documents produced jointly by the Construction Specifications Institute (CSI) and Construction Specifications Canada (CSC) are widely accepted as standard practice for the preparation of contract documents.

TABLE 4.29
Mistake Proofing for Eliminating Design Errors

Serial Number	Items	Points to Be Considered to Avoid Mistakes
1	Information	1. Terms of Reference (TOR) 2. Client's preferred requirements matrix 3. Data collection 4. Regulatory requirements 5. Codes and standards 6. Historical data 7. Organizational requirements
2	Mismanagement	1. Compare production with actual requirements 2. Inter disciplinary coordination 3. Application of different codes and standards 4. Drawing size of different trades/specialist consultants
3	Omission	1. Review and check design with TOR 2. Review and check design with client requirements 3. Review and check design with regulatory requirements 4. Review and check design with codes and standards 5. Check for all required documents
4	Selection	1. Qualified team members 2. Available material 3. Installation methods

Source: Abdul Razzak Rumane. (2013). *Quality Tools for Managing Construction Projects*. Reprinted with permission from Taylor & Francis Group.

Table 4.30 lists division numbers and titles of MasterFormat® 2016 published by the CSI and CSC.

Particular specifications consist of many sections related to a specific topic. Detailed requirements are written in these sections to enable the contractor understand the product or system to be installed in the construction project. The designer has to interact with the project team members and owner while preparing the contract documents.

Typical sections are as follows:
Section No.
Title

Part 1—General
1.01—General reference/related sections
1.02—Description of work
1.03—Related work specified elsewhere in other sections
1.04—Submittals

TABLE 4.30
MasterFormat® System Organization (© 2022)

Division Numbers and Titles

PROCUREMENT AND CONTRACTING REQUIREMENTS GROUP
Division 00 Procurement and Contracting Requirements

SPECIFICATIONS GROUP

General Requirements Subgroup
Division 01 General Requirements

Facility Construction Subgroup
Division 02 Existing Conditions
Division 03 Concrete
Division 04 Masonry
Division 05 Metals
Division 06 Wood, Plastics, and Composites
Division 07 Thermal and Moisture Protection
Division 08 Openings
Division 09 Finishes
Division 10 Specialties
Division 11 Equipment
Division 12 Furnishings
Division 13 Special Construction
Division 14 Conveying Equipment
Division 15 Reserved
Division 16 Reserved
Division 17 Reserved
Division 18 Reserved
Division 19 Reserved

Facility Services Subgroup
Division 20 Reserved
Division 21 Fire Suppression
Division 22 Plumbing
Division 23 Heating, Ventilation, and Air
     Conditioning (HVAC)
Division 24 Reserved
Division 25 Integrated Automation
Division 26 Electrical
Division 27 Communications
Division 28 Electronic Safety and Security
Division 29 Reserved

Site and Infrastructure Subgroup
Division 30 Reserved
Division 31 Earthwork
Division 32 Exterior Improvements
Division 33 Utilities
Division 34 Transportation
Division 35 Waterway and Marine
     Construction
Division 36 Reserved
Division 37 Reserved
Division 38 Reserved
Division 39 Reserved

Process Equipment Subgroup
Division 40 Process Interconnections
Division 41 Material Processing and
     Handling Equipment
Division 42 Process Heating Cooling,
     and Drying Equipment
Division 43 Process Gas and Liquid
     Handling, Purification, and
     Storage Equipment
Division 44 Pollution and Waste Control
     Equipment
Division 45 Industry-Specific
     Manufacturing Equipment
Division 46 Water and Wastewater
     Equipment
Division 47 Reserved
Division 48 Electric Power Generation
Division 49 Reserved

Source: The Construction Specifications Institute. Reprinted with permission from CSI.

1.05—Delivery, handling, and storage
1.06—Spare parts
1.07—Warranties

In addition to the foregoing, a reference is made for items such as preparation of mock up, quality control plan, and any other specific requirements related to the product or system specified herein.

Part 2—Product
2.01—Materials
2.02—List of recommended manufacturers

Part 3—Execution
3.01—Installation
3.02—Site quality control

Shop Drawing and Materials Submittals

The detailed procedure for submitting shop drawings, materials, and samples is specified under the section titled "SUBMITTAL" of contract specifications. The contractor has to submit the specifications to the owner/consultant for review and approval. The following are the details of preparation of shop drawings and materials.

A—Shop Drawings
The contractor is required to prepare shop drawings taking into account the following partial list of considerations:

1. Reference to contract drawings. This helps the A&E (consultant) to compare and review the shop drawing with the contract drawings
2. Detail plans and information based on the contract drawings
3. Notes of changes or alterations from the contract documents
4. Detailed information about fabrication or installation of works
5. Verification of all dimensions at the job site
6. Identification of product
7. Installation information about the materials to be used
8. Type of finishes, color, and textures
9. Installation details relating to the axis or grid of the project
10. Roughing in and setting diagram
11. Coordination certification from all other related trades (subcontractors)

The shop drawings are to be drawn accurately to scale and shall have project-specific information in it. They should not be reproductions of contract drawings.

Immediately after approval of individual trade shop drawings, the contractor has to submit builder's workshop drawings, composite/coordinated shop drawings taking into consideration the following at a minimum.

A1—Builder's Workshop Drawings

Builder's workshop drawings indicate the openings required in the civil or architectural work for services and other trades. These drawings indicate the size of openings, sleeves, and level references with the help of detailed elevation and plans.

A2—Composite/Coordination Shop Drawings

The composite drawings indicate the relationship of components shown on the related shop drawings and indicate the required installation sequence. Composite drawings should show the interrelationship of all services with one another and with the surrounding civil and architectural work. Composite drawings should also show the detailed coordinated cross sections, elevations, reflected plans, and so on resolving all conflicts in levels, alignment, access, space, and so forth. These drawings are to be prepared taking into consideration the actual physical dimensions required for installation within the available space.

B—Materials

Similarly, the contractor has to submit the following, at a minimum, to the owners/consultants to get their review and approval of materials, products, equipment, and systems. The contractor cannot use these items unless they are approved for use in the project.

B1—Product Data

The contractor has to submit the following details:

1. Manufacturer's technical specifications related to the proposed product
2. Installation methods recommended by the manufacturer
3. Relevant sheets of manufacturer's catalogs
4. Confirmation of compliance with recognized international quality standards
5. Mill reports (if applicable)
6. Performance characteristics and curves (if applicable)
7. Manufacturer's standard schematic drawings and diagrams to supplement standard information related to project requirements and configuration of the same to indicate product application for the specified works (if applicable)
8. Compatibility certificate (if applicable)
9. Single-source liability (this is normally required for systems approval when different manufacturers' items are used)

B2—Compliance Statement

The contractor has to submit a specification comparison statement along with the material transmittal.

The consultant reviews the transmittals and actions as follows:

a. Approved
b. Approved as noted
c. Approved as noted, resubmit
d. Not approved

In certain projects, the owner is involved in the approval of materials.

In case of any deviation from specifications, the contractor has to submit a schedule of such deviations listing all the points not conforming to the specification.

B3—Samples
The contractor has to submit (if required) the samples from the approved material to be used for the work. The samples are mainly required to

1. Verify color, texture, and pattern
2. Verify that the product is physically identical to the proposed and approved material
3. Comparison with products and materials used in the works
4. At times it may be specified to install the samples in such a manner as to facilitate review of qualities indicated in the specifications.

4.3.3.7 Regulatory Approvals

Government agency regulatory requirements have a considerable impact on precontract planning. Some agencies require that the design drawings be submitted for their preliminary review and approval to ensure that the designs are compatible with local codes and regulations. These include submission of drawings to electrical authorities showing the anticipated electrical load required for the facility, approval of fire alarm and firefighting system drawings, and approval of drawings for water supply and drainage system. Technical details of the conveying system are also required to be submitted for approval from the concerned authorities.

4.3.3.8 Develop Project Schedule

The project schedule is developed using bottom-up planning details using key events. It is also known as Class 1 schedule or Schedule Level 3.

In order to improve the understanding and the communication among stakeholders involved with preparing, evaluating, and using project schedule, AACE International has published the guideline to classify schedules into five classes and five levels. Figure 4.17 illustrates Schedule Classifications versus Schedule Levels that schedule can be developed and/or presented.

Table 4.31 illustrates the Generic Schedule Classification Matrix.

4.3.3.9 Estimate Project Cost

The cost estimate during this phase is based on elemental parametric methodology. It is also known as Detailed Costing (Detailed Estimate). Please refer to Table 4.19 discussed under Section 4.3.1.14 for cost estimation levels.

4.3.3.10 Manage Design Quality

In order to reduce errors and omissions, it is necessary to review and check the design for quality assurance by the quality control personnel from the project team through itemized review checklists to ensure that design drawings fully meet the owner's objectives/goals. It is also required to review the design with the owner to ensure a

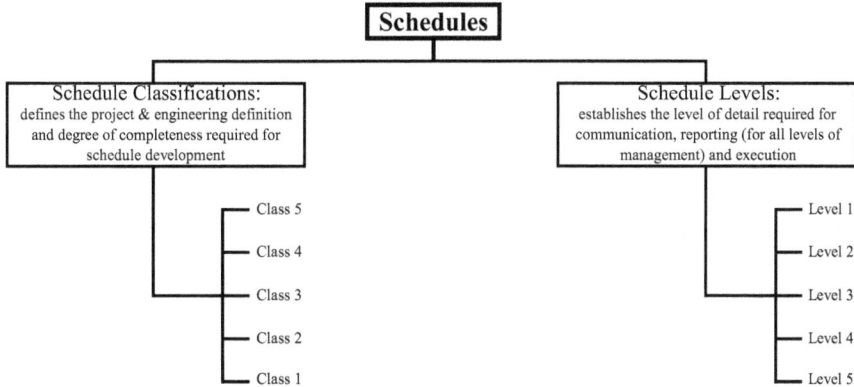

FIGURE 4.17 Schedule: Classifications versus levels. (Adapted from AACE International Recommended Practice No. 27R-03, Copyright @ 2010 AACE International. Reprinted with permission from AACE International, 726 East Park Ave., #180, Fairmont, WV 26554. Email: info@aacei.org; Phone: +1.304.296.8444; website: web.aacei.org. All AACE International content is copyright protected ©2010, All rights reserved.)

mutual understanding of the build process. The designer has to ensure that the installation/execution specification details are comprehensively and correctly described and also the installation quality requirements for systems are specified in detail.

Figure 4.18 illustrates the quality management procedure for detail design phase. The designer has to plan quality (planning of design work), perform quality assurance, and control quality for preparing detail design. This will mainly consist of the following:

4.3.3.10.1   Plan Quality
- Review comments on preliminary design
- Determine number of drawings to be produced
- Establish scope of work for preparation of detail design
- Identify requirements listed under TOR
- Identify quality standards and codes to be complied
- Establish design criteria
- Identify regulatory requirements
- Identify environmental requirements
- Establish quality organization with responsibility matrix
- Develop design (drawings and documents) review procedure
- Establish submittal plan
- Establish design review procedure

4.3.3.10.2   Quality Assurance
- Collect data
- Investigate site conditions
- Prepare design drawings

TABLE 4.31
Generic Schedule Classification Matrix

SCHEDULE CLASS	Primary Characteristic Degree of Project Definition (Expressed as % of Complete Definition)	Secondary Characteristic End Usage	Scheduling Methods Used
Class 5	0% to 2%	Concept Screening	Top-down planning using high level milestones and key project events
Class 4	1% to 15%	Feasibility Study	Top-down planning using high level milestones and key project events. Semi-detailed
Class 3	10% to 40%	Budget, Authorization, or Control	"Package" top down planning using key events. Semi-detailed
Class 2	30% to 75%	Control or Bid/Tender	Bottom-up planning. Detailed
Class 1	65% to 100%	Bid/Tender	Bottom-up planning. Detailed

Source: The table is from AACE International Recommended Practice No. 27R-03 Copyright @ 2010 by AACE International. Reprinted with permission from AACE International, 726 East Park Ave., #180, Fairmont, WV 26554. Email: info@aacei.org; Phone: +1.304.296.8444; website: web. aacei.org. All AACE International content is copyright protected ©2010, All rights reserved.

- Prepare detailed specifications
- Prepare contract documents
- Prepare bill of quantities
- Ensure functional and technical compatibility
- Ensure the design is constructible
- Ensure operational objectives are met
- Ensure drawings are fully coordinate with all disciplines
- Ensure the design is cost-effective
- Ensure selected/recommended material meet owner objectives
- Ensure that design fully meets the owner's objectives/goals

4.3.3.10.3   Control Quality
- Check quality of design drawings
- Check accuracy and correctness of design

Start Detail Design

Review Project Charter, Terms of Reference (TOR)

Identify Project Goals and Objectives

1-Review Comments on Preliminary Design
2-Collect Data/Information

Establish Detail Design Scope of Work

Stakeholders Requirements

Identify Project Team/ Organization Chart

Organizational Quality Management System

Design Basis

Design Input from Specialist (if applicable)

Coordination Among All Disciplines

Information/Data Input

Technical, Functional, Operational Study

Constructability

Specifications/ Documents

Material/Systems Selection

Cost-effective Design

Plan Quality

Perform Quality Assurance

Quality Control

Check Design Drawings

Check Design for Regulatory Requirements

Review Specifications/Documents

Check Project Schedule

Check Cost Estimate

Interdisciplinary coordination

BOQ/Detail Design Drawings

Design Criteria

Codes and Standards to be Followed

Regulatory Requirements

Quality Metrics and Measures

Project Risks

HSE Issues and Requirements

Design Submittal Plan

Design Review Procedure

Corrective and Preventive Action

Remedial Action

Quality Audit

Submit Design

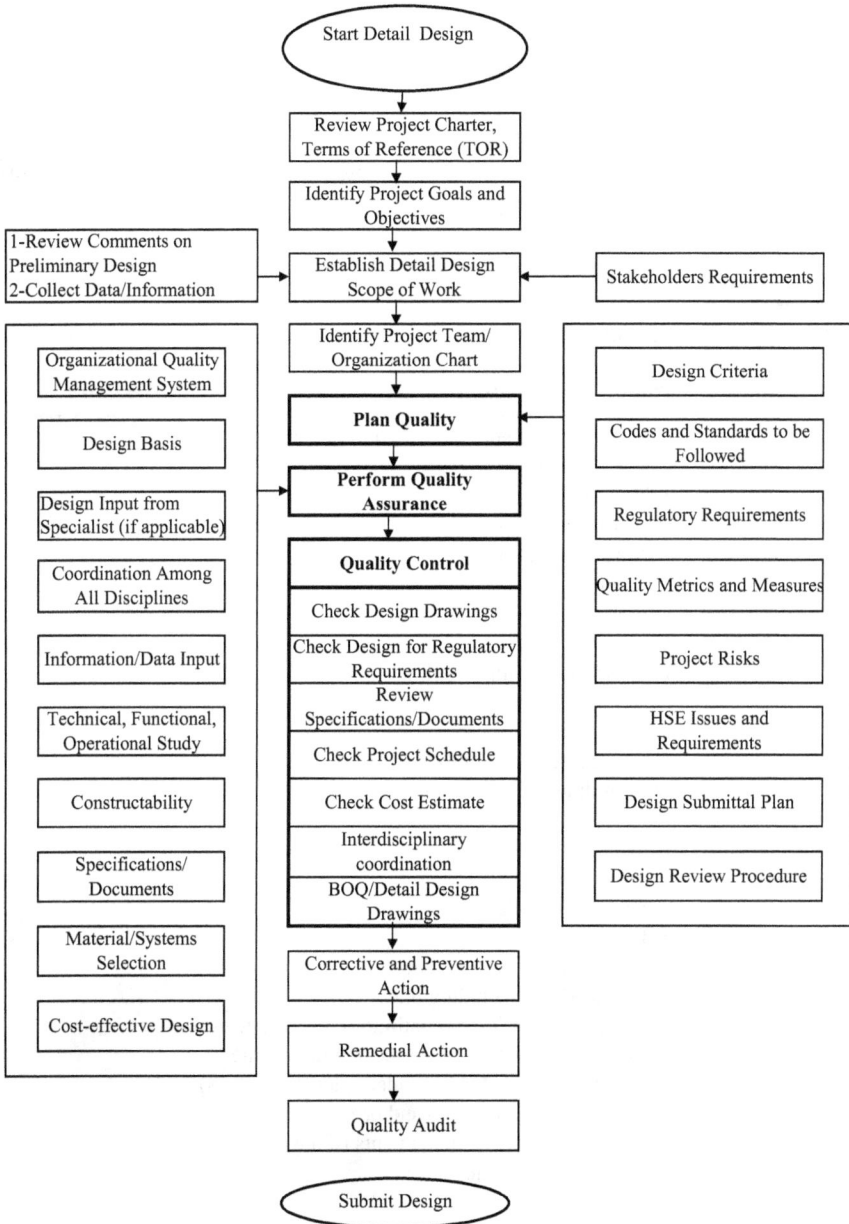

FIGURE 4.18   Quality management procedure for detail design phase.

- Verify bill of quantities
- Check specifications
- Check contract documents
- Check for regulatory compliance

- Check project schedule
- Check project cost
- Check interdisciplinary requirements
- Check required number of drawings are prepared

Table 4.32 lists items to be verified and checked internally by the designer before submission to the owner/project manager.

TABLE 4.32
Quality Check for Design Drawings

Serial Number	Points to Be Checked	Yes/No
1	Check for use of approved version of AutoCAD	
2	Check drawing for	
	• Title frame	
	• Attribute	
	• North orientation	
	• Key plan	
	• Issues and revision number	
3	Client name and logo	
4	Designer (Consultant name)	
5	Drawing title	
6	Drawing number	
7	Contract reference number	
8	Date of drawing	
9	Drawing scale	
10	Annotation	
	• Text size	
	• Dimension style	
	• Fonts	
	• Section and elevation marks	
11	Layer standards including line weights	
12	Line weights, line type (continuous, dash, dot, etc.)	
13	Drawing continuation reference and match line	
14	Plot styles (CTB color-dependent plot style tables)	
15	Electronic CAD file name and project location	
16	XREF (X Reference) attachments (if any)	
17	Image reference (if any)	
18	Section references	
19	Symbols	
20	Legends	
21	Abbreviations	
22	General notes	
23	Drawing size as per contract requirements	
24	List of drawings	

Source: Abdul Razzak Rumane. (2013). *Quality Tools for Managing Construction Projects*. Reprinted with permission from Taylor & Francis Group.

4.3.3.11   Estimate Resources

Designer has to estimate the resources required to complete the project. During this phase, detailed information is available to estimate manpower resources during construction phase.

The human resource planning process includes organizational planning taking into consideration the requirements of the project. While estimating the resources, the designer has to ensure the availability of resources to perform/execute the assigned particular activity/activities without affecting or delaying the estimated schedule.

4.3.3.12   Manage Risks

Following are typical risks which normally occur during detail design phase:

- Preliminary design deliverables and review comments are not taken into consideration while preparing the detail design
- Regulatory authorities' requirements are not taken into consideration
- Detail design scope of work is not properly established and is incomplete
- The related project data and information collected are incomplete
- The related project data and information collected are likely to be incorrect and wrongly estimated
- Site investigations for existing conditions not verified
- Fire and safety considerations recommended by the authorities not incorporated in the design
- Environmental consideration
- Incomplete design drawings and related information
- Inappropriate construction method
- Conflict with different trades.
- Interdisciplinary coordination not done
- Wrong selection of materials and systems
- Undersize HVAC equipment selection
- Incorrect water supply requirements
- Estimated total electrical load is much lower than expected actual consumption
- Traffic study for conveying system not verified taking into consideration final load
- Prediction of possible changes in design during construction phase
- Inadequate and ambiguous specifications
- Project schedule not updated as per detailed data and project assumptions
- Errors in detail cost estimation
- Number of drawings not as per TOR requirements

Designer has to take into account above-mentioned risk factors while developing the detail design.

Further, the designer has to consider the following risks while planning the duration for completion of design development phase:

- Impractical detail design preparation schedule
- Duration to obtain authorities' approval

4.3.3.13   HSE Issues and Requirements

The designer has to develop HSE plan and identify issues to manage the activities during detail design stage. Designer has to do Environmental Impact Assessment study and take into consideration while developing detail design.

While developing the design, the designer has to consider the following:

- Hazardous properties of the materials and products used in the project
- Hazardous emissions
- Pollution and its impact
- Impact on health
- Safety in design (safe design)
- Safety in operation of MEP systems
- Waste management system
- Environmental compatibility
- Regulatory and other environmental protection agencies' requirements

The designer has to prepare Environmental Impact Assessment Report for submission to the owner along with other documents while submitting the concept design package. The typical contents of Environmental Impact Assessment Report are as follows:

1. Purpose of report
2. Description of project (goals and objectives)
3. References (relevant codes, standards, owner references)
4. Scope of work (during all the life cycle phases of the project)

The designer is also required to submit HSE Management Plan along with other documents during this phase.

4.3.3.14   Review Detail Design

It is unlikely that the design of a construction project will be right in every detail the first time. Effective management and design professionals who are experienced and knowledgeable in the assigned task will greatly reduce the chances of error and oversight. However, so many aspects must be considered, especially for designs involving multiple disciplines and enfaces, and changes will be inevitable. The design should be reviewed taking into consideration requirements of all the disciplines before the release of design drawings for a construction contract. Engineering design has significant importance to the construction projects and must meet the customer's requirement at the start of project implementation. Engineering design has significant importance for construction projects. Engineering weakness can adversely impact the quality of design to such an extent that marginal changes can easily increase costs beyond the budget, which may affect schedule. Some areas are deemed critical to the proper design of a product; therefore, explicit design, material specification, and grades of the material specified in documentation have great importance. Most of the products used in construction projects are produced by other construction-related industries/manufacturers; therefore, the designer, while specifying the products, must specify related codes, standards, and technical compliance of these products.

The success of a project is highly correlated with the quality and depth of the engineering design prepared during this phase. Coordination and conflict resolution are important factors during the development of design to avoid omissions and errors. The designer has to review the detail design for accuracy of drawings, interdisciplinary coordination, and documents before these are submitted to the owner/project manager for subsequent preparation of construction documents.

4.3.3.14.1   Review Drawings
Figure 4.19 illustrates design review steps for detail design.

4.3.3.14.2   Perform Interdisciplinary Coordination
Table 4.33 illustrates major points to perform interdisciplinary coordination.

4.3.3.14.3   Review Documents
The designer has to review contract documents and ensure all the requirements listed under TOR are taken care. The contract documents are prepared for the approved type of project delivery system and contract pricing method.

Table 4.34 lists items to be checked for detail design review by the designer before submission to the owner/project manager.

4.3.3.15   Finalize Detail Design
Final detail design is prepared incorporating the comments, if any, found during analysis, review, and interdisciplinary coordination of the drawings and documents for submission to the client/owner.

Table 4.35 summarizes the TQM concept for detailed design phase activities.

4.3.4   TQM in Construction Document Phase

Construction development phase is the fourth phase of construction project life cycle. During this phase, the drawings and specifications prepared during detail design phase are further developed into the working drawings. All the drawings, specifications, documents, and other related elements necessary for construction of the project are assembled and subsequently released for bidding and tendering.

Figure 4.20 illustrates a typical flowchart for TQM applications in construction documents phase.

4.3.4.1   Identify Stakeholders
Following stakeholders have direct involvement in the construction document phase:

- Owner
- Consultant

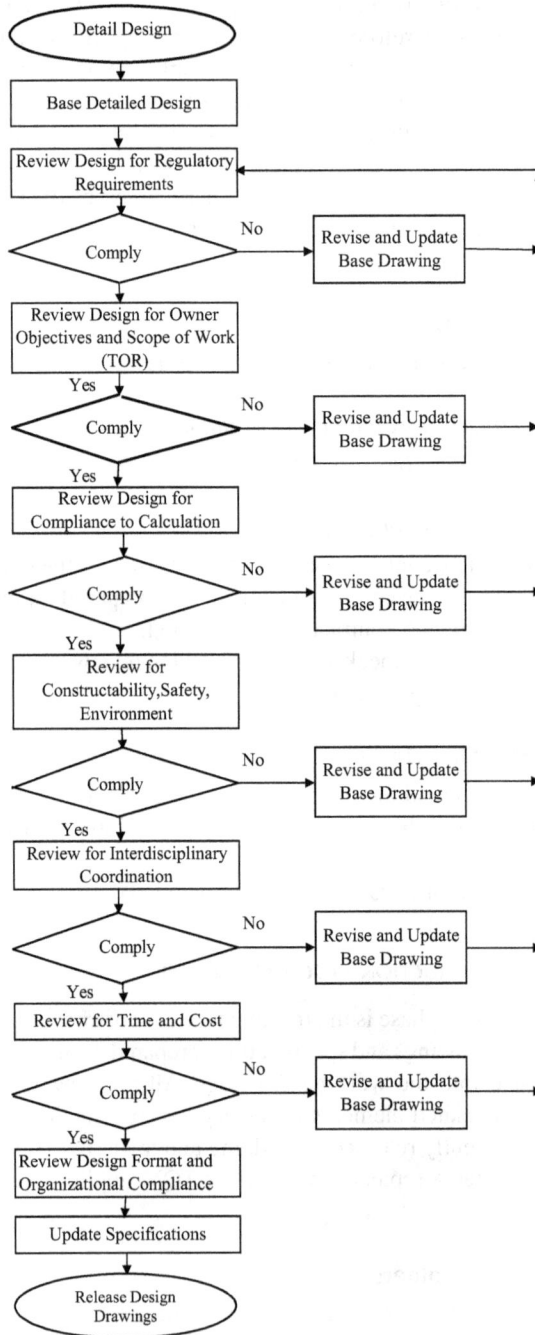

FIGURE 4.19 Design review steps. (Abdul Razzak Rumane. (2013). *Quality Tools for Managing Construction Projects*. Reprinted with permission from Taylor & Francis Group.)

TABLE 4.33
Interdisciplinary Coordination

				Discipline			
Serial Number	Discipline	Architectural	Structural	Mechanical	HVAC	Electrical	External Works
1	**ARCHITECTURAL**		1. Structural framing plans	1. Pump room location and size of room	1. Plant room location and size	1. Location and size of substation and door sizing	1. Property limits
			2. Axis, grids, levels	2. Void above false ceiling for piping	2. Void above false ceiling for HVAC equipment, duct and piping	2. Trenches for cables in substation, electrical room and generator room	2. Location of outdoor equipment
			3. Location of columns, beams	3. Sprinkler in false ceiling	3. Access for maintenance of equipment	3. Location and size of electrical room and closets	3. Location of plants
			4. modules to match with structural plan	4. Location of sanitary fixtures and accessories	4. K-Value of thermal insulation, type of external glazing and U-value	4. Location of electrical devices	4. Location of seating/relax area

(continued)

TABLE 4.33 (Continued)
Interdisciplinary Coordination

Serial Number	Discipline	Architectural	Structural	Mechanical	HVAC	Electrical	External Works
						Discipline	
			5. Location of stairs, fire exits	5. Location of fire hydrant, cabinet and landing valves	5. Location of louvers, grills, diffusers	5. Location of light fittings in the false ceiling	5. Location of maintenance room/area
			6. Expansion joints	6. Location of water tank	6. Location of thermostat and other devices	6. Void above false ceiling for cable tray and trunking	6. Location of manholes
			7. Building dimensions	7. Shaft for water supply, sanitary and drainage pipes	7. Stair case pressurization system with respect to HVAC	7. Cable tray, cable trunking route	7. Location of generator exhaust pipe
				8. Location of fuel filling point for fuel carrying tanker	8. Location of HVAC equipment on roof	8. Location and size of low voltage rooms	
				9. Location of manholes	9. HVAC shaft requirement	9. Location and size of generator room	

2 STRUCTURAL			10. Ventilation of substation and generator room	
1. Structural framing plans	1. Opening for pipe crossing in the walls and slab	1. Shaft for pipings and duct	1. Base for transformers	1. Manholes
2. Axis, grids, levels	2. Shaft for pipe risers (water supply, sanitary, drainage)	2. Openings/sleeves for duct and pipings	2. Base for generator	2. Foundation for light poles
3. Location of columns, beams	3. Opening for roof drain	3. Operating weight of all HVAC equipment	3. Trenches for electrical cables	3. Manhole/Foundation for electrical panels
4. Modules to match with structural plan	4. Openings/sleeves for piping	4. Floor height to accommodate equipment	4. Openings/sleeves for cable tray, electrical bus duct	4. Manhole/Foundation for feeder pillars
5. Location of stairs, fire exits	5. Opening for main circulation drain	5. Expansion joints requirements	5. Shaft for cable trays	5. Underground services tunnel
6. Expansion joints	6. Water tank inlet location	6. Pump room equipment loads with HVAC Equipment	6. Foundation for light poles	

(continued)

TABLE 4.33 (Continued)
Interdisciplinary Coordination

Serial Number	Discipline	Discipline					
		Architectural	Structural	Mechanical	HVAC	Electrical	External Works
3	MECHANICAL	7. Building dimensions		7. Sanitary manholes		7. Manhole/ Foundation for electrical panels	
		1. Pump room location and size of room	1. Opening for pipe crossing in the walls and slab		1. Make up water requirements for HVAC	1. Power supply for pumps and other equipment	1. Irrigation system with external Works
		2. Void above false ceiling for piping	2. Shaft for pipe risers (water supply, sanitary, drainage)		2. Connection of chilled water for plumbing works	2. Location of isolators for power supply	2. Area drain, road gully with external/ asphalt work
		3. Sprinkler in false ceiling	3. Opening for roof drain		3. Interface with building management system	3. Interface with fire alarm system	3. Storm water manholes with external works
		4. Location of sanitary fixtures and accessories	4. Openings/sleeves for piping		4. HVAC/AHU drain with drainage system		4. External services to be hooked up with municipality route

		5. Location of fire hydrant, cabinet and landing valves 6. Location of water tank 7. Shaft for water supply, sanitary and drainage pipes 8. Location of fuel filling point for fuel carrying tanker 9. Location of manholes	5. Opening for main circulation drain 6. Water tank inlet location 7. Sanitary manholes				
4	HVAC	1. Plant room location and size 2. Void above false ceiling for HVAC equipment, duct and piping	1. Shaft for ducts and piping 2. Opening for ducts and pipings in the wall and roof	1. Make up water requirements for HVAC 2. Connection of chilled water for plumbing works	1. Power supply for chillers, pumps, AHUs, and other equipment 2. Location of isolators for power supply	1. Access for underground services 2. Location of exhaust for underground ventilation system

(continued)

TABLE 4.33 (Continued)
Interdisciplinary Coordination

Serial Number	Discipline	Discipline					
		Architectural	Structural	Mechanical	HVAC	Electrical	External Works
		3. Access for maintenance of equipment		3. Interface with building management system		3. Heat dissipation from lighting and other electrical panels	
		4. K-Value of thermal insulation, type of external glazing and U-value		4. HVAC/AHU drain with drainage system		4. Three phase/ Single phase power requirements	
		5. Location of louvers, grills, diffusers				5. Power supply load during summer/winter	
		6. Location of thermostat and other devices				6. Electrical power supply for equipment connected to generator	
		7. Stair case pressurization system with respect to HVAC				7. Interface with fire alarm system	

	8. Location of HVAC equipment on roof			8. Interface with building management system	
	9. HVAC shaft requirement				
5 ELECTRICAL	1. Location and size of substation and door sizing	1. Base for transformers	1. Power supply for pumps and other equipment	1. Power supply for chillers, pumps, AHUs, and other equipment	1. Location of lighting poles
	2. Trenches for cables in substation, electrical room, and generator room	2. Base for generator	2. Location of isolators for power supply	2. Location of isolators for power supply	2. Location of earth pits
	3. Location and size of electrical room and closets	3. Trenches for electrical cables	3. Interface with fire alarm system	3. Heat dissipation from lighting and other electrical panels	3. Location of electrical manholes, handholes
	4. Location of electrical devices	4. Openings/sleeves for cable tray, electrical bus duct		4. Three phase/ Single phase power requirements	4. Underground cable routes

(continued)

TABLE 4.33 (Continued)
Interdisciplinary Coordination

Serial Number	Discipline	Discipline					
		Architectural	Structural	Mechanical	HVAC	Electrical	External Works
		5. Location of light fittings and other devices in the false ceiling	5. Shaft for cable trays		5. Power supply load during summer/winter		5. Location of bollards
		6. Void above false ceiling for cable tray and trunking	6. Foundation for light poles		6. Electrical power supply for equipment connected to generator		6. Location of electrical panels, feeder pillars
		7. Cable tray, cable trunking route	7. Manhole/Foundation for electrical panels		7. Interface with fire alarm system		
		8. Location and size of low voltage rooms			8. Interface with building management system		
		9. Location and size of generator room					
		10. Ventilation of substation and generator room					

6	LAND SCAPE/ EXTERNAL	1. Property limits	1. Manholes	1. Manholes	1. Irrigation system with external works	1. Access for underground services	1. Location of lighting poles
			2. Location of outdoor equipment	2. Foundation for light poles	2. Area drain, road gully with external/asphalt work	2. Location of exhaust for underground ventilation system	2. Location of earth pits
			3. Location of plants	3. Manhole/ Foundation for electrical panels	3. Storm water manholes with external works		3. Location of electrical manholes, handholes
			4. Location of seating/relax area	4. Manhole/ Foundation for feeder pillars	4. External services to be hooked up with municipality route		4. Underground cable routes
			5. Location of maintenance room/area	5. Underground services tunnel			5. Location of bollards
			6. Location of manholes				6. Location of electrical panels, feeder pillars
			7. Location of generator, exhaust pipe				7. Location of generator, exhaust pipe

Source: Abdul Razzak Rumane. (2013). *Quality Tools for Managing Construction Projects*. Reprinted with permission from Taylor & Francis Group.

TABLE 4.34
Check List for Detail Design Review

Serial Number	Items to Be Checked
1	Whether design meets owner requirements and complete scope of work (TOR)?
2	Whether designs were prepared using authenticated and approved software?
3	Whether design calculation sheets are included in the set of documents?
4	Whether design is fully coordinated for conflict between different trades?
5	Whether design has taken into consideration relevant collected data requirements?
6	Whether reviewer's comments on preliminary design responded?
7	Whether regulatory approval obtained and comments, if any, incorporated and all review comments responded?
8	Whether design has environmental compatibility?
9	Whether energy efficiency measures are considered?
10	Whether sustainability requirements are considered while selecting the equipment/systems?
11	Whether design constructability is considered?
12	Whether design matches with property limits?
13	Whether legends match the layout?
14	Whether design drawings are properly numbered?
15	Whether design drawings have owner logo, designer logo as per standard format?
16	Whether the design format of different trades have uniformity?
17	Whether project name and contract reference are shown on the drawing?

Source: Abdul Razzak Rumane. (2013). *Quality Tools for Managing Construction Projects.* Reprinted with permission from Taylor & Francis Group.

- Designer
- Project/Construction Manager (if the owner decided to engage them during this phase, depending on the type of project delivery system)

Table 4.36 illustrates the responsibilities of various participants during construction document phase.

4.3.4.1.1   Identify Construction Documents Team Members

Following project team members have direct involvement in the construction document phase:

- Owner

TABLE 4.35
Total Quality Management Concept for Detail Design Phase Activities

Serial Number	Phase Activity	Stakeholders/Team Members	Customer Specifications/Requirements	System for Managing/Tool
1	Identification of Stakeholders/Project Team Members	1. Project owner 2. Designer (design manager) 3. Project manager	1. Project need and requirements	Please refer Table 4.27
2	Identification of Detail Design Requirements	1. Designer/design manager 2. Project manager 3. Quality manager	1. TOR requirements 2. Project data/information 3. Owner's requirements 4. Sustainability requirements 5. Regulatory requirements	Update requirements similar to Table 4.15 and Table 4.16 Please refer Section 4.3.3.2
3	Establish Detail Design Scope	1. Designer/design manager 2. Project manager 3. Quality manager	1. TOR requirements 2. Project data/information 3. Owner's requirements 4. Sustainability requirements 5. Regulatory requirements	Please refer Table 4.28
4	Develop Project Management Plan	1. Design manager 2. Planning manager 3. Project manager	TOR requirements	Please refer Section 4.3.3.4
5	Develop Detail Design	1. Designer/design manager/ design team 2. Project manager 3. Quality manager	Please refer Section 2.5.3 (application of Six Sigma DMADV tool for design)	Please refer Figure 4.16, Section 4.3.3.5.3, 4.3.3.5.4 and Table 4.32
6	Develop Contract Terms and Documents	1. Designer/design manager 2. Project manager 3. Quality manager 4. Quantity surveyor/contract administrator	1. TOR requirements 2. Regulatory requirements 3. Applicable construction contract documents 4. Applicable codes and standards	Please refer Section 4.3.3.6 and Table 4.30

(continued)

TABLE 4.35 (Continued)
Total Quality Management Concept for Detail Design Phase Activities

Serial Number	Phase Activity	Stakeholders/Team Members	Customer Specifications/Requirements	System for Managing/Tool
7	Regulatory/Authority Approval	1. Designer/design manager 2. Regulatory authority	Applicable codes and standards	Follow agreed upon communication system
8	Estimate Project Schedule	1. Designer/design manager 2. Planning manager 3. Project manager 4. Quality manager	Owner's needs and requirements	Schedule is based on concept of Figure 4.17 and Table 4.31
9	Estimate Project Cost	1. Designer/design manager 2. Planning manager 3. Project manager 4. Quality manager	Owner's needs and requirements	Project cost is based on Table 4.19
10	Quality Management	1. Quality manager 2. Designer (design manager)	1. Quality manager 2. Designer (Design manager)	Please refer Table 4.32
11	Project Resources	1. Designer (design manager) 2. Planning manager	Resources for construction project	
12	Project Risk	Designer/Design team members	Typical risks to complete construction project	
13	HSE Issues/Requirements	Designer/safety manager	Environmental and safety management compliance	
14	Review Detail Design	Designer	1. Owner's requirements 2. Regulatory requirements 3. Funding requirements 4. Project quality	Please refer Figure 4.18, Table 4.33. and Table 4.34
15	Finalize Detail Design	1. Designer 2. Project owner	1. Owner's requirements 2. Regulatory requirements 3. Funding requirements 4. Project quality	

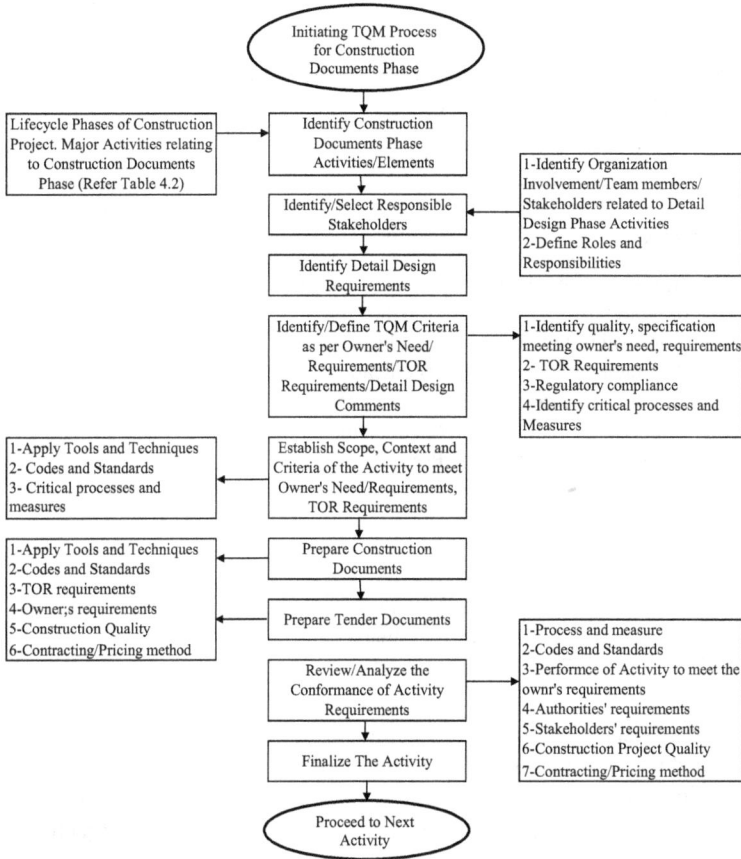

FIGURE 4.20 Typical flowchart for total quality management applications in quality of construction documents phase activity.

- Consultant
- Designer
- Quantity surveyor (contract administrator)
- Planning and control manager
- Project/construction manager (if the owner decided to engage them during this phase, depending on the type of project delivery system)

During this phase, Quantity Surveyor/Contract Administrator has great responsibilities. His/her team under the leadership of project manager (design) is responsible for coordinating and assembling all the required documents.

4.3.4.2 Identify Construction Documents Requirements

Construction document phase provides a complete set of working drawings of all the disciplines, site plans, technical specifications, Bill of Quantities (BOQ), schedule,

TABLE 4.36
Responsibilities of Various Participants (Design–Bid–Build Type of Contracts) during Construction Documents Phase

Phase	Responsibilities		
	Owner	Designer	Regulatory Authorities
Construction Document	• Approval of working drawings • Approval of tender documents • Approval of time schedule • Approval of budget	• Development of working drawings • Development of specifications • Development of contract documents • Project schedule • Project budget • BOQ • Development of tender Documents • Review of construction Documents	• Review and approval of project submittals

(except the standards specifications and documents for insertions normally added during bidding and tendering phase), and related graphic and written information to bid the project. It is necessary that utmost care is taken to develop and assemble all the documents and ensure the accuracy and correctness to meet the owner's objectives.

4.3.4.2.1 Data/Information

In order to identify requirements to assemble contract documents, the designer has to gather the comments on the submitted detail design by the owner/project manager, collect TOR requirements, regulatory requirement, identify owner requirements, and all other related information to ensure nothing is missed.

4.3.4.2.2 Gather Detail Design Comments

The designer has to review the comments on detail design and coordinate with all the disciplines/trades and incorporate the same while preparing the construction documents.

4.3.4.2.3 Identify TOR Requirement

Identify all the requirements listed under TOR. This includes

- Final drawings to be prepared to the required scales, format with necessary logo, client name, location map, north orientation, project name, designer name, drawing title, drawing number, contract reference number, date of drawing, revision number, drawing scale, duly signed by the designer.

- BOQ and schedule of rates
- Contract documents
- Project schedule
- Technical specifications
- Cost estimates
- Summary report

4.3.4.2.4   Collect Owner's Requirements

The designer has to discuss with the owner to ascertain there are no changes to the earlier established owner's requirement and if there are any additional requirements or changes then the scope is to be updated and incorporated in the final design, documents before sending for tendering.

4.3.4.2.5   Identify Regulatory Requirements

The designer has to ensure that there are no changes to existing regulatory requirements. If there are updates to regulatory requirements, then the designer has to incorporate the same as any changes during construction have adverse effect on the project.

4.3.4.2.6   Identify Project Delivery System/Contracting Requirements

The designer has to prepare contracts documents taking into consideration the project delivery system as per the agreed upon organization's procurement policy and the contracting/pricing method. (Please refer to Section 2.5 of Chapter 2 for more details).

4.3.4.2.7   Identify Environmental Requirements

Environmental agencies always update their requirements to protect the environment. The designer has to verify that there are no changes to the requirements/assumptions considered during detail design development phase.

4.3.4.2.8   Identify Sustainability Requirements

The designer has to also identify sustainability requirements for development of construction documents. Following are the basic elements that have to be taken care of by the designer to develop construction documents:

- Meeting/satisfying owner's requirements
- Project value within approved budget
- Efficient resources
- Efficient and durable material
- Working drawings with all the details
- Schematics of MEP systems
- Energy-efficient MEP systems, equipment
- Green building concept
- Aesthetic
- Landscape compatible with surrounding area
- Regulatory requirements

- Applicable codes and standards
- Contract documents clearly written in simple language that is unambiguous and easy to understand

4.3.4.3   Establish Construction Documents Scope

The scope for development of construction documents is prepared taking into consideration requirements established under Sections 4.3.4.2 which are as follows:

- Approved detail design phase documents
- TOR requirements
- Owner's preferred requirements
- Project delivery system, contracting/pricing method
- Regulatory requirements
- Environmental requirements
- Approved project schedule
- Approved detail cost estimate
- Codes and standards

The purpose of construction documents is to provide sufficient information and detail to ensure that the bidders will be able to submit the definitive cost for the project. There is no ambiguity in the drawings and specifications, and the work to be performed by the contractor is properly identified and correctly addressed taking necessary measures to mitigate errors and omissions in the design. The scope of work during construction document phase mainly comprises of

- Preparation of working (final) drawings
- Technical specifications
- BOQ
- Project schedule
- Definitive cost estimate
- Authorities approvals
- Existing site conditions/site plans
- Site surveys
- Design calculations

4.3.4.3.1   Identify Construction Documents Deliverables

Table 4.37 lists construction document deliverables to be developed during this phase.

4.3.4.4   Develop Construction Documents

The construction documents are developed taking into consideration all the items discussed under Sections 4.3.4.1 and 4.3.4.2 and any other related information. Following items are mainly developed during construction documents phase:

1. Working drawings
2. Technical specifications
3. Documents

TABLE 4.37
Construction Documents Deliverables

Serial Number	Deliverables
1	Document I
1.1	Tendering procedure
	i. Invitation to tender
	ii. Instructions to bidders
	iii. Forms for tender and appendix
	iv. List of equipment and machinery
	v. List of contractor's staff
	vi. Contractor's certificate of work statement
	vii. List of subcontractor(s) or specialist(s)
	viii. Initial bond
	ix. Final bond
	x. Forms of agreement
2	Document II
2.1	Conditions of contract
	II-1 General conditions
	II-2 Particular conditions
	II-3 Public tender laws
3	Document III
	III-1 General specifications
	III-2 Particular specifications
	III-3 Drawings with schematics
	III-4 Schedule of rates and bill of quantities
	III-5 Analysis of prices
	III-6 Addenda
	III-7 Tender requirements (if any) and any other instructions issued by the owner

4.3.4.4.1 *Working Drawings*

All the drawings prepared during detail design phase are reviewed to ensure all the related information and adjustments are carried out. Following is the list of major disciplines in building construction projects for which working drawings are developed:

1. Architectural Design
2. Concrete Structure
3. Elevator
4. Fire Suppression
5. Plumbing
6. Drainage
7. HVAC Works
8. Electrical System (Light and power)
9. Fire Alarm System
10. Information and Communication System

11. Public Address System
12. Audio Visual System
13. Security System/CCTV
14. Security System/Access Control
15. Satellite/Main Antenna System
16. Integrated Automation System
17. Landscape
18. External Works (Infrastructure and Road)
19. Furnishings/Furniture (Loose)

Each of these trade drawings shall have

- Detail drawings produced at different scales and format
- Plans
- Sections
- Elevations
- Schedule
- Drawing index

Following information will be included in all the drawings

- Client name
- Client logo
- Location map
- North orientation
- Project name
- Drawing title
- Drawing number
- Date of drawing
- Revision number
- Drawing scale
- Contract reference number
- Signature block
- Signed by the designer for check and approval

4.3.4.4.2 Specifications

Designer has to prepare comprehensive technical specifications as per the Division and Section taking into consideration related drawings. It is essential to have close coordination between working drawings and specifications. MasterFormat® specification documents are used to prepare specifications for building project. These Divisions and Sections are divided into numbers of Volumes for ease of reference. BOQ for the project activities is prepared corresponding to these Divisions and Sections.

4.3.4.4.3 BOQ

BOQ developed during detail design phase is reviewed to ensure it matches with the working drawings.

Table 4.38 is an example BOQ form.

TABLE 4.38
Bill of Quantities (BOQ)

Owner Name
Project Name

Project Number:
SAMPLE BOQ

ITEM	DESCRIPTION	QTY	UNIT	UNIT RATE	TOTAL AMOUNT	REMARKS
	DIVISION 3 – CONCRETE					
	The Contractor is referred to the Specifications and Drawings for all details related to this section of the Works and he is to include for complying with all the requirements contained therein, whether or not they are specifically mentioned within the item descriptions					
	03300: CAST IN-PLACE CONCRETE					
	SUBSTRUCTURE					
	Plain concrete 17.5MPa using Sulphate Resisting Cement (Type V) including formwork and additives					
A	100mm Blinding		m3			
	Plain concrete 17.5MPa using Ordinary Portland Cement (Type I) including formwork and additives					
B	Concrete filling		m3			
C	To kerb foundation		m3			

(*continued*)

TABLE 4.38 (Continued)
Bill of Quantities (BOQ)

Owner Name
Project Name

Project Number:
SAMPLE BOQ

ITEM	DESCRIPTION	QTY	UNIT	UNIT RATE	TOTAL AMOUNT	REMARKS
	Reinforced concrete 28MPa using sulphate resisting cement (Type V) including formwork, reinforcement, water stops, expansion and construction joints, joint filler and additives					
D	Raft foundation		m3			
E	300mm Walls		m3			
F	400mm Ditto		m3			
	CARRIED TO COLLECTION					
	03480: PRECAST CONCRETE SPECIALTIES					
	Depressed curbs including reinforcement, finish, anchors, fixings, grouting, fair face finish and painting					
A	250 x 180mm Barrier kerb		m			
B	Ditto, curve on plan		m			
C	350 x 150mm Flush kerb		m			
	Wheel stoppers, fair face finish and painting					
D	2000 x 140 x 100mm Overall size, Parking Area		no			

TABLE 4.38 (Continued)
Bill of Quantities (BOQ)

Owner Name
Project Name

Project Number:
SAMPLE BOQ

ITEM	DESCRIPTION	QTY	UNIT	UNIT RATE	TOTAL AMOUNT	REMARKS
	03520: LIGHTWEIGHT CONCRETE					
	Lightweight concrete laid in bays of 25m2 minimum, maximum length of bay 7m and expansion joints filled with impregnated compressible foam and resilient packing					
E	50mm Thick, laid to falls, to roof		m2			
F	170mm Thick, laid to falls, to pathways		m2			
	COLLECTION					
	Total of page No. 1/2					
	Total of page No. 2/2					
	CARRIED TO SUMMARY					

4.3.4.4.4   Contract Documents

Contract documents are prepared taking into consideration contract format suitable for specific type of project from any of the following organizations that are producing different types of contracting systems:

1. EJCDC—The Engineers Joint Contract Documents Committee, USA
2. FIDIC—Federation Internationale Des Ingenieurs (International Federation of Consulting Engineers)

3. MasterFormat®—Construction Specifications Institute (CSI) and Construction Specifications Canada (CSC)
4. NEC—New Engineering Contract (NEC) or NEC Engineering and Construction Contract, UK (Institution of Civil Engineers)

4.3.4.5 Develop Tender Documents

Following documents are prepared during this phase:

1. Complete set of working (construction) drawings duly coordinated with other disciplines and technical specifications
2. Detailed BOQ
3. Technical specifications for all the activities shown on the drawings
4. Schedule
5. Cost estimate
6. Legal and contractual information
7. Contractor bidding requirements
8. Contract conditions
9. General specifications
10. Reports

The above-listed documents are used as guidelines by the designer to prepare Tender Documents. These are as follows:

Document-I

1. Tendering procedure consisting of
   i.   Invitation to tender
   ii. Instruction to bidders
   iii. Forms for tender and appendix
   iv. List of equipment and machinery
   v.   List of contractor's staff
   vi. Contractor certificate of work statement
   vii. List of subcontractor (s) or specialist(s)
   viii.Initial bond
   ix. Final bond
   x.   Form of agreement
   xi. List of tender documents

Document-II

1. II-1 General Conditions
2. II-2 Particular Conditions
3. II-3 Public Tender Laws

Document-III

1. III-1 General Specifications
2. III-2 Particular Specifications (Division 1-49)
3. III-3 Drawings
4. III-4 Schedule of Rates and BOQ
5. III-5 Analysis of Prices

4.3.4.6   Develop Project Schedule

The project schedule is developed using bottom-up planning details using key activities/events. It is also known as Class 1 schedule or Schedule Level 4 (please refer to Figure 4.34 and Table 4.34).

4.3.4.7   Estimate Project Cost

The cost estimate during this phase is based on detailed costing methodology. During this phase all the project activities are known, and detailed BOQ is available for costing purpose. It is also known as Detailed Costing (Definitive Estimate). (Please refer to Table 4.19 discussed under Section 4.3.1.14 for cost estimation levels).

4.3.4.8   Manage Construction Documents Quality

In order to reduce errors and omissions it is necessary to review and check the design for quality assurance by the quality control personnel from the project team through itemized review checklists to ensure that working drawings are suitable for construction. The designer has to ensure that the installation/execution specification details are comprehensively and correctly described and coordinated with working drawings and also the installation quality requirements for systems are specified in detail.

The designer has to plan quality, perform quality assurance, and control quality for preparing contract documents. This will mainly consist of the following:

1. Plan Quality:
    *   Review comments on detail design package
    *   Determine number of drawings to be produced
    *   Establish scope of work for preparation of construction documents
    *   Identify requirements listed under TOR
    *   Identify quality standards and codes to be complied
    *   Identify regulatory requirements
    *   Identify environmental requirements
    *   Establish quality organization with responsibility matrix
    *   Develop review procedure for the produced working drawings
    *   Develop review procedure for the specifications and contract documents
    *   Establish submittal plan for construction documents

2. Quality Assurance:
    - Prepare working drawings
    - Prepare detailed specifications
    - Prepare contract documents
    - Prepare BOQ and schedule of rates
    - Ensure functional and technical compatibility
    - Ensure the design is constructible
    - Ensure operational objectives are met
    - Ensure drawings are fully coordinate with all disciplines
    - Ensure the design is cost-effective
    - Ensure selected/recommended materials meet owner objectives
    - Ensure that design fully meets the owner's objectives/goals
    - Ensure that construction documents match with approved project delivery system
    - Ensure type of contracting/pricing as per adopted methodology

3. Control Quality:
    - Check quality of design drawings
    - Check accuracy and correctness of design
    - Verify BOQ for correctness as per working drawings
    - Check complete specifications are prepared and coordinated to match working drawings and BOQ
    - Check contract documents as per Project Delivery System
    - Check for regulatory compliance
    - Check project schedule
    - Check project cost
    - Check calculations
    - Review studies and reports
    - Check accuracy of design
    - Check interdisciplinary requirements
    - Check required number of drawings are prepared

Before the drawings are released for bidding and tendering, it is necessary to check the drawings for formatting, annotation, and interpretation. The designer has to check the item listed under Table 4.32 (refer Section 4.3.3.10) for quality check (correctness) of design drawings.

4.3.4.9 Estimate Resources

At this stage, the designer can estimate resources having accuracy as more details are available to estimate exact resources.

4.3.4.10 Manage Risks

Following are typical risks which normally occur during construction document phase:

- Design development deliverables and review comments are not taken into consideration while preparing the construction documents
- Scope of work to produce construction documents not properly established and is incomplete
- Documents not matching as per project delivery system
- Documents not as per the type of contract/pricing methodology
- Regulatory authorities' requirements are not taken into consideration
- Latest environmental consideration not considered
- Conflict with different trades
- Conflict between working drawings and specifications
- Prediction of possible changes in design during construction phase
- Inadequate and ambiguous specifications
- Project schedule not updated as per detailed data and project assumptions
- Errors in definitive cost estimation
- Number of drawings not as per TOR requirements
- It is likely that owner-supplied items, if any, are not included in the documents

Designer has to take into account above-mentioned risk factors while construction documents.

Further, the designer has to consider the following risks while planning the duration for completion of construction document phase.

- Impractical construction document preparation.

4.3.4.11 HSE Issues and Requirements
The designer has to consider HSE issues and requirements while developing the construction documents.

4.3.4.12 Review Construction Documents (Designer)
The designer has to review construction documents. Table 4.39 illustrates items to be reviewed for constructability of design (working) drawings.

4.3.4.13 Review Construction Documents (Owner)
The owner has to review the construction documents prior to releasing them for bidding and tendering.

4.3.4.14 Finalize Construction Documents
The final construction documents package is prepared taking into consideration review comments and identified risks by the designer and comments from the owner/ project manager.

4.3.4.15 Release for Bidding and Tendering
Normally following items are submitted to the owner for their review and approval in order to proceed for bidding and tendering phase of the project:

TABLE 4.39

Constructability Review for Design (Construction-Working) Drawings

Serial Number	Items to be Reviewed	YES/NO
1	Are there construction elements that are impossible or impractical to build?	
2	Does the design follow industry standards and practices?	
3	Is structural design per site conditions, soil conditions and bearing capacity?	
4	Are the site conditions verified and suitable with respect to access, availability of utility services?	
5	Will all the specified material be available during construction phase?	
6	Is the specified material available from single source or multiple sources and brands?	
6	Is the design suitable for construction using the specified, material, equipment?	
8	Is the design suitable for construction using recommended method statement?	
9	Are the available labor resources capable of building the facility as per contract drawings and contracted methods and practices?	
10	Is the design fully coordinated with technical specifications and CSI-Format divisions?	
11	Is the specifications cover the material considered in the design?	
12	Does the design fully meet regulatory requirements?	
13	Is the drawings coordinated with all the trades and cross references are indicated wherever applicable?	
14	Is the design coordinated with adjacent land and its accessibility?	
15	Are requirements of general public and persons of special needs considered?	
16	Are construction schedule and milestone practical to achieve?	
17	Can application of QA/QC requirements be complied with?	
18	Has environmental impact and its mitigation considered?	
19	Is there space for temporary office facilities and parking space for workforce vehicles?	
20	Has availability of storage space for construction material considered?	
21	Is the design is sustainable?	

Source: Abdul Razzak Rumane. (2013). *Quality Tools for Managing Construction Projects*. Reprinted with permission from Taylor & Francis Group.

1. Working drawings
2. Project specifications
3. BOQ
    a. Priced
    b. Unpriced
4. Project schedule
5. Definitive cost estimate
6. Project summary report
7. Tender documents comprising of
    I. Document-I Tendering Procedure
    II. Document-II Condition of Contract
    III. Document-III consisting of
        III-1 General Specifications
        III-2 Particular Specifications
        III-3 Drawings
        III-4 BOQ and Schedules of Rates
        III-5 Analysis of Prices
8. Soft copy of construction documents (DVD)

Table 4.40 summarizes the TQM concept for construction documents phase activities.

4.3.5 TQM in Bidding and Tendering Phase

Construction development phase is the fifth phase of construction project life cycle of major projects. During this phase, tender documents are released for bid, and contract is awarded to successful bidder.

Most of the cost of the construction project is expended during the construction phase. In most cases, the contractor is responsible for procurement of all the materials, providing construction equipment and tools, and supplying the manpower to complete the project in compliance with the contract documents.

In many countries, it is a legal requirement that government-funded projects employ the competitive bidding method. This requirement gives an opportunity to all qualified contractors to participate in the tender, and normally the contract is awarded to the lowest bidder. Private-funded projects have more flexibility in evaluating the tender proposal. Private owners may adopt the competitive bidding system, or the owner may select a specific contractor and negotiate the contract terms. Negotiated contract systems have flexibility in pricing arrangement as well as the selection of the contractor based on his expertise or the owner's past experience with the contractor successfully completing one of his or her projects. The contracting system is based on the owner's procurement policy. (Please refer to Section 2.2.5 for contract pricing systems applicable to construction projects.)

It is the owner's desire that his or her facility be of good quality and the price reasonable. In order to achieve this, the owner has to share risks and/or provide incentives and safeguards to enhance the quality of construction. The risks involved in various types of contracts based on forms of payment are as follows:

TABLE 4.40
Total Quality Management Concept for Construction Documents Phase Activities

Serial Number	Phase Activity	Stakeholders/ Team Members	Customer Specifications/ Requirements	System for Managing/Tool
1	Identification of Stakeholders/ Project Team Members	1. Project owner 2. Designer (design manager) 3. Project manager	1. Project need and requirements	Please refer Table 4.36
2	Identification of Construction Documents Requirements	1. Designer/design manager 2. Project manager 3. Quality manager	1. TOR requirements 2. Project data/ information 3. Owner's requirements 4. Sustainability requirements 5. Regulatory requirements	Update requirements similar to Table 4.15 and Table 4.16 Please refer Section 4.3.4.2
3	Establish Construction Documents Scope	1. Designer/design manager 2. Project manager 3. Quality manager 4. Quantity surveyor/ contract administrator	1. TOR requirements 2. Project data/ information 3. Owner's requirements 4. Sustainability requirements 5. Regulatory requirements	Please refer 4.3.4.3 and Table 4.37
4	Develop Construction Documents	1. Designer/ design manager/ design team 2. Project manager 3. Quality manager	Please refer Section 2.5.3 (application of Six Sigma DMADV tool for design)	Please refer 4.3.4.4
5	Develop Tender Documents	1. Designer/design manager 2. Project manager 3. Quality manager 4. Quantity surveyor/ contract administrator	1. TOR requirements 2. Regulatory requirements 3. Applicable construction contract documents 4. Applicable codes and standards	Please refer Section 4.3.4.5
6	Estimate Project Schedule	1. Designer/design manager 2. Planning manager 3. Project manager 4. Quality manager	Owner's needs and requirements	Schedule is based on concept of Figure 4.17 and Table 4.31 and class 1 level 4

TABLE 4.40 (Continued)
Total Quality Management Concept for Construction Documents Phase Activities

Serial Number	Phase Activity	Stakeholders/ Team Members	Customer Specifications/ Requirements	System for Managing/Tool
7	Estimate Project Cost	1. Designer/design manager 2. Planning manager 3. Project manager 4. Quality manager	Owner's needs and requirements	Project cost is based on Table 4.19
8	Manage Construction Documents Quality	1. Quality manager 2. Designer (Design manager)	1. Quality manager 2. Designer (design manager)	Please refer Table 4.32 and Table 4.39
9	Project Resources	1. Designer (design manager) 2. Planning manager	Resources for construction project	
10	Manage Project Risk	Designer/design team members	Typical risks to complete construction project	
11	HSE Issues/ Requirements	Designer/safety manager	Environmental and safety management compliance	
12	Review Construction Documents (Designer)	Designer/designer team members	1. Owner's requirements 2. Regulatory requirements 3. Funding requirements 4. Project quality	Please refer Table 4.39
13	Review Construction Documents (Owner)	1. Owner 2. Project manager/ construction manager	1. Owner's requirements 2. Regulatory requirements 3. Funding requirements 4. Project quality	
14	Finalize Construction Documents	1. Designer 2. Project owner	1. Owner's requirements 2. Regulatory requirements 3. Funding requirements 4. Project quality	
15	Release for Bidding and Tendering	1. Owner 2. Designer	As per corporate policy	

1. Cost plus—high risk
2. Reimbursement—intermediate
3. Fixed price—low risk

In order to maintain a climate of mutual cooperation during construction, the owner has to develop an understanding with the contractor. The contract needs to be adapted through mutual agreement with the contractor. The contract strategy needs to provide incentives and safeguards to deal with the risks.

In the case of a competitive bidding system, it is necessary that the detailed design and specifications for the project be prepared by the designer for bidding purposes. Under the competitive bidding system, normally there are four stages in tendering of a construction project:

1. Selection of tenderer (pre-qualification)
2. Invitation to bid
3. Tender preparation and submission
4. Meetings
5. Appraisal of tenders, negotiation, and decision
6. Award of contract

For most construction projects, selection of a tenderer is based on the lowest tender price. Tenders received are opened and evaluated by the owner/owner's representative. Normally, tender results are declared in the official gazette or by some sort of notifications. The successful tenderer is informed of the acceptance of the proposal and is invited to sign the contract. The tenderer has to submit the performance bond before the formal contract agreement is signed. If a successful tenderer fails to submit the performance bond within the specified period or withdraws his tender, then the contractor loses the initial bond and may be subjected to other regulatory applicable conditions.

The signing of a contract agreement between the owner/owner's representative and the contractor binds both parties to fulfill their contractual obligations.

Figure 4.21 illustrates a typical flowchart for TQM applications in bidding and tendering phase.

4.3.5.1 Identify Stakeholders

Following stakeholders have direct involvement in the bidding and tendering phase:

- Owner
- Tender Committee
- Designer (Consultant)
- Project/Construction Manager (if the owner decided to engage them during this phase, depending on the type of project delivery system)
- Bidders
- Table 4.41 illustrates the responsibilities of various participants during construction document phase.

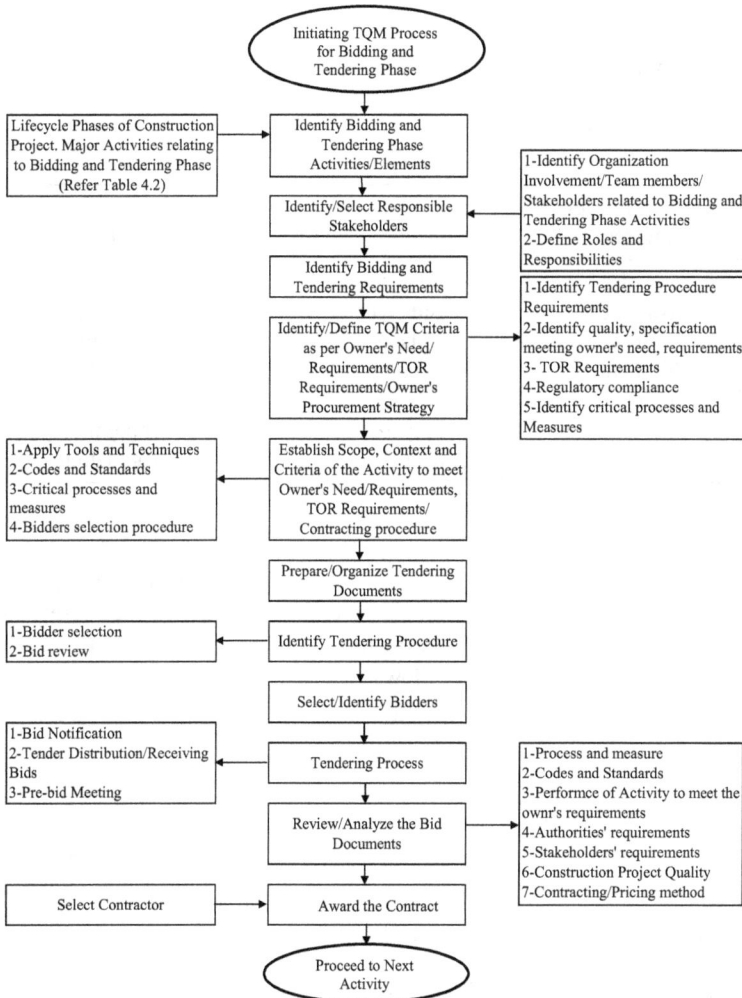

FIGURE 4.21 Typical flowchart for total quality management applications in quality of bidding and tendering phase activity.

4.3.5.2   Organize Tendering Documents

The owner hand over the approved construction documents/tender documents to the tender committee for further action. The bid documents are prepared as per the procurement method and contract strategy adopted during the early stage of the project. Tendering procedure documents submitted by the designer are updated and necessary owner-related information is inserted in the tender documents. The bid advertisement material is prepared and upon approval from the owner, the bid notification is announced through different media as per the organization's/agency's policy.

TABLE 4.41

Responsibilities of Various Participants (Design–Bid–Build Type of Contracts) during Bidding and Tendering Phase

	Responsibilities		
Phase	Owner/Tender Committee	Consultant (Designer)	Bidder/Contractor
Bidding and Tendering	• Advertise bid • Distribute bid • Collect bid (proposal) • Negotiation • Approve contractor • Award contract	• Review/evaluate bid • Bid conference/meeting • Bid clarification • Recommend successful bidder • Prepare contract documents	• Collection of bid documents • Preparation of proposal • Submission of proposal • Negotiation

4.3.5.3 Identify Tendering Procedure

The owner has to identify the tendering procedure. There are different types of tendering methods based on the organization's procurement strategy. Please refer to Sections 2.2.2 and 2.2.3.1 under Chapter 2, the most common procurement methods for selection of project contractor are as follows:

1. Low bid
2. Best value
    a. Total cost
    b. Fees
3. QBS

Figure 4.22 illustrates bidding and tendering process.

4.3.5.3.1 Define Bidder Selection Procedure

The selection of project team (contractor) is mainly done as follows:

1. Screening of qualified contractors (pre-qualification of contractor)
2. Selecting contractor using contracting methods such as;
    a. Competitive bidding
    b. Competitive negotiations
    c. Direct negotiation
3. Awarding contract

While engaging/selecting a contractor, the following selection criteria are to be considered as a minimum to pre-qualify the contractor:

- Available skill level
- Relevant/past performance on similar type of work
- Reputation about their works

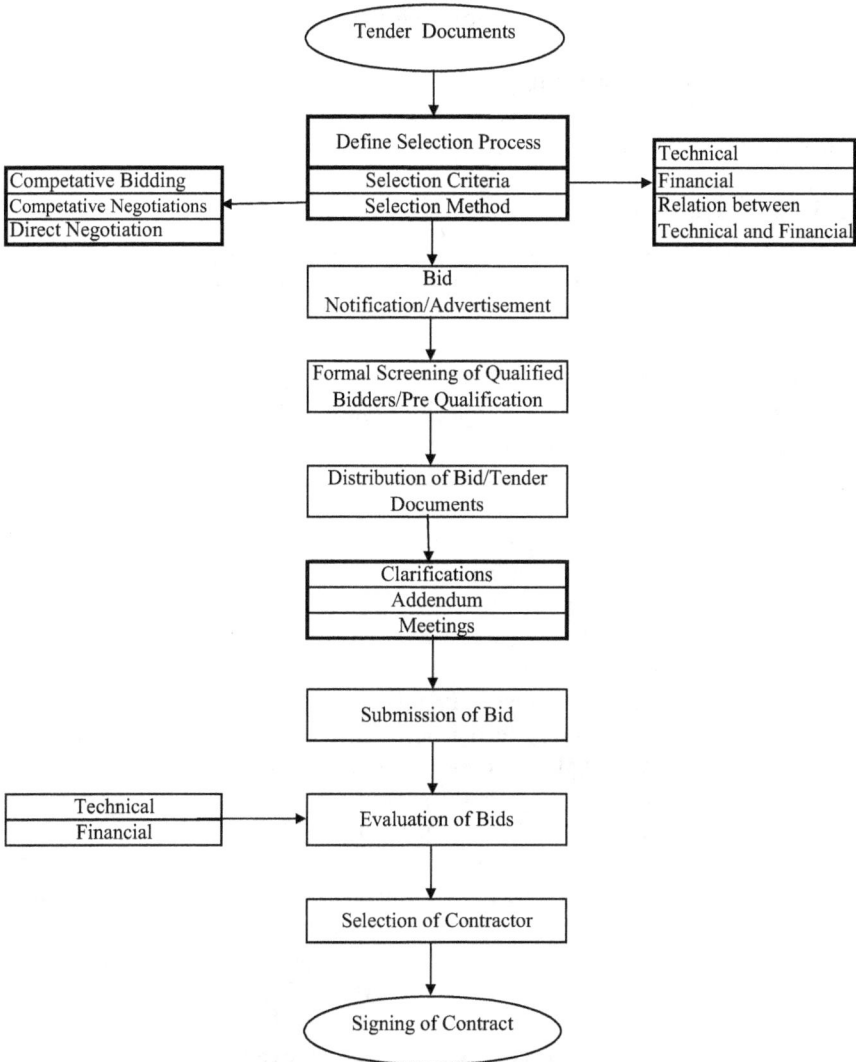

FIGURE 4.22 Bidding and tendering process.

- Number of projects (works) successfully completed
- Technical competence
- Knowledge about the type of projects (works) for which likely to be engaged
- Available resources
- Commitment to creating best value
- Commitment to containing sustainability
- Rapport/behavior
- Communication

The above information is gathered through a Request for Information, Request for Pre-Qualification, or Pre-Qualification Questionnaires from the prospective agency, consultant, and contractor. Figure 4.23 illustrates shortlisting/selection of bidders/contractors, and Table 4.42 lists pre-qualification questionnaires to select contractor for design–bid–build type of contract.

4.3.5.3.2 *Identify Bidders*

Upon receiving the information relating to the qualification data, the documents are evaluated as per the selection criteria determined earlier by the project owner. Figure 4.24 illustrates sample contractor selection criteria.

4.3.5.4 Establish Bid Review Process

Most government and public sector projects follow Low Bid selection method. There are three international bidding procedures which may be selected by the project owner to suit the nature of project procurement. These are

1. Single stage–One Envelope: In this procedure, the bidders submit bids in one envelope containing both technical and financial proposals. The bids are evaluated by the selection committee which send their recommendation to the owner. Following the review and concurrence by the owner, the contract is awarded to the lowest bidder.
2. Single stage–Two Envelopes: In this procedure, the bidders submit two envelopes, one containing technical proposal and the other the financial proposal. Initially, technical proposals are evaluated without referring to the price. Bidders whose proposal does not conform to the requirements may be rejected/not accepted. Following the technical proposal evaluation, the financial proposals of technically responsive bidders are reviewed. Upon review and concurrence by the owner, the contract is awarded to the lowest bidder.
3. Two Stages: In this procedure, during the first stage, the bidders submit their technical offers on the basis of operating and performance requirements, but without price. The technical offers are evaluated by the selection committee. Any deviations to the specified performance requirements are discussed with the bidders who are allowed to revise or adjust the technical offer and resubmit the same.

During the second stage, the bidders, whose technical offers are accepted, are invited to submit final technical proposal and financial proposal. Both the proposals are reviewed by the selection committee and following review and concurrence by the owner, the contract is awarded to the competitive bidder as per the procurement policy.

The designer (consultant) establishes bid evaluation procedure in consultation with the owner. Table 4.43 lists the items to be reviewed prior to the evaluation of bid documents.

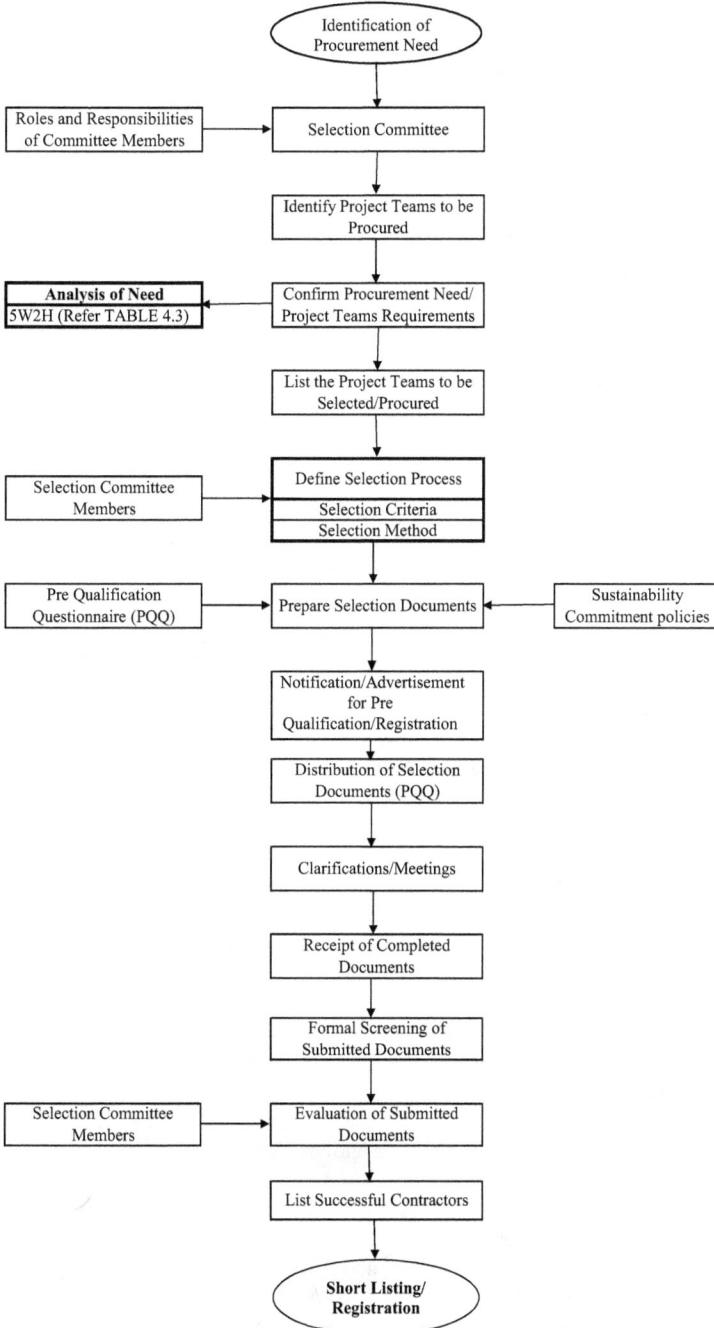

FIGURE 4.23 Short listing/registration of contractors.

TABLE 4.42
Pre-Qualification Questionnaires (PQQ) for Selecting Contractor
(Design–Bid–Build)

Serial Number	Question	Answer
1	Name of the Organization and Address	
2	Organization's Registration and License Number	
3	ISO Certification	
4	Registration/Classification Status of the Organization	
5	Joint Venture with Any International Contractor	
6	Total Turnover Last 5 years	
7	Audited Financial Report for Last 3 years	
8	Insurance and Bonding Capacity	
9	Total Experience (years) in Construction of Following Type of Projects	
	9.1 Residential	
	9.2 Commercial (Mix use)	
	9.3 Institutional (Governmental)	
	9.4 Industrial	
	9.5 Infrastructure	
10	Size of Project (Maximum Amount Single Project)	
	10.1 Residential	
	10.2 Commercial (Mix use)	
	10.3 Institutional (Governmental)	
	10.4 Industrial	
	10.5 Infrastructure	
11	List Successfully Completed Projects	
	11.1 Residential	
	11.2 Commercial (Mix use)	
	11.3 Institutional (Governmental)	
	11.4 Industrial	
	11.5 Infrastructure	
12	List Similar Type (Type to Be mentioned) of Projects completed	
	12.1 Project Name and Contracted Amount	
	12.2 Project Name and Contracted Amount	
	12.3 Project Name and Contracted Amount	
	12.4 Project Name and Contracted Amount	
	12.5 Project Name and Contracted Amount	
13	List of Subcontractors	
14	Resources	
	14.1 Management	

TABLE 4.42 (Continued)
Pre-Qualification Questionnaires (PQQ) for Selecting Contractor
(Design–Bid–Build)

Serial Number	Question	Answer
	14.2 Engineering	
	14.3 Technical	
	14.4 Foreman/Supervisor	
	14.5 Skilled Manpower	
	14.6 Unskilled Manpower	
	14.7 Plant and Equipment	
15	Current Projects	
16	Quality Management Policy	
17	Health, Safety and Environment Policy	
	17.1 Number of Accidents during Last 3 Years	
	17.2 Number of Fires at Site	
18	Staff Development Policy	
19	List of Delayed Projects	
21	List of Failed Contract	
22	List of Professional Awards	
23	Litigation (Dispute, Claims) on Earlier Projects	

4.3.5.5   Manage Tendering Process

The tendering process involves the following activities:

- Bid notification
- Distribution of tender documents
- Prebid meeting(s)
- Issuing addendum, if any
- Bid submission

4.3.5.5.1   Bid Notification/Advertisement

The tender is announced in different types of media such as newspaper, magazines, and electronic media, as per the organization's/agency's policy.

4.3.5.5.2   Distribute Tender Documents

Normally tender documents are distributed to eligible bidders against payment of fee announced in the bid notification, which is nonrefundable.

4.3.5.5.3   Conduct Prebid Meeting

The owner conducts prebid meeting to provide an opportunity for the contractors bidding on the project to review and discuss the construction documents and to discuss

Evaluation Criteria	Weightage	Key Points for Consideration	Review Result
1. General Information			
1 Company information		Company's current position-a MUST information	Yes OR No
2. Financial	**25%**		
1 Total Turn Over (last 5 years)	25%	Sum of the Turn over for the last five uears	
2 Values of Current work-in-hand	25%	Project value / Value of current work-in-hand	
3 Audite Financial Reports	10%	To confirm the ratio given in point three	
4 Financial Standing	30%		
33% Assets		Current Assets / Current Liabilities	
34% Liabilities		Total Liabilities - Total Equity / Total assets	
33% Profit/Loss		Net Profit before Tax / Total Equity	
5 Bonding and Insurance Limit	10%		
60% a. Performance & Bonding Capacity		Provided or Not provided	
40% b. Insurance		Provided or Not provided	
3. Organization Details	**20%**		
3a. Business			
1 Company's Core Area of Business	30%	Degree of satisfactory answer	
2 Experience of years in business	30%	No. of Years	
3 ISO Certification	15%	Yes or No	
4 Registration/Classification Status	15%	Grade or Classification	
5	10%	Key staff indicated (Name/Title), Balanced Resources, Departmental(Specialization) Diversity, Lines Of Communication	
Organizational Chart			
6 Dispute/Claims		Degree of satisfactory answer	
3b. Experiene	**30%**		
50% a. Projects' value		No. of projects with comparable value	
50% b. Projects' type (similar type and complexity)		No. of projects with similar complexity	
4. Resources	**20%**		
1 Personnel	60%		
30% Management		No. of Managerial staff	
30% Engineers		No. of engineers and project staff	
30% Technicaians		No. of cad technicians and foreman	
10% Staff Development		% turnover spent on training	
2 Technology	10%	% of turnover spent on acquiring latest construction technology	
3 Plant & Equipment	30%	List of plant and equipment	
5. GENERAL	**5%**		
1 Bank Refrences	30%	Provided or Not provided	
2 Project Refences	30%	Provided or Not provided	
2 Health, Safety and Enviromment narration	40%	Degree of Satisfactory answer	

FIGURE 4.24 Contractor selection criteria.

TABLE 4.43
Check List for Bid Evaluation

Serial Number	Description	Yes	No	Notes
A: Documents				
A.1	Bid submitted before closing time on the date specified in the bid documents			
A.2	Bidders identification is verified			
A.3	Bid is properly signed by the authorized person			
A.4	Bid bond is included			
A.5	Required certificates are included			
A.6	Bidders confirmation to the validity period of bid			
A.7	Confirmation to abide by the specified project schedule			
A.8	Bid documents have no reservation or conditions (limitation or liability)			
A.9	Preliminary method statement			
A.10	List of equipment and machinery			
A.11	List of proposed core staff as listed in the tender documents			
A.12	Complete responsiveness to the commercial terms and conditions			
A.13	All the required information is provided (completeness of information)			
A.14	All the supporting documents required to determine technical responsiveness is submitted			
B: Financial				
B.1	All the items are priced			
B.2	Bid amount clearly mentioned			
B.3	Prices of provision items			

- General scope of the project
- Any particular requirements of bidders that may have been difficult to specify
- Explain details of complex matters
- Engagement of subcontractor, specialist subcontractors
- Particular risks
- Any other matters that will contribute to the efficient delivery of project

The meeting is attended by the designer (consultant), bidders (contractors), project/construction manager, and tender committee member. Queries from the contractors pertaining to construction documents are noted, and the designer (consultant) provides written responses to these queries by clarifying all the points. The bidders have to consider the clarification points and incorporate the requirements while calculating the bid price. The responses recorded in the meeting become the part of contract documents (part of addendum) which are signed by the owner and successful bidder.

4.3.5.5.4 Bid Clarification

Figure 4.25 illustrates the Bid Clarification Form which becomes part of contract documents.

4.3.5.5.5 Site Visit

Site visit is arranged for the bidders to familiarize with the site conditions.

4.3.5.5.6 Addendum Any

Addendum, if any, is issued to bidders to be considered for submission of the bid.

4.3.5.6 Submit/Receive Bids

Bids are received in accordance with the Instructions to Bidders section of the tender documents. The bid should be accompanied by an initial bond in favor of the owner/ tender committee to be valid for a period mentioned in tendering procedures. All the bids received are documented and notified. The tender, which is submitted as a sealed document, is opened as mentioned in tendering procedures. Figure 4.26 illustrates typical submittal procedure by the bidder/contractor.

Serial Number	Name of Contractor	Item No. and Clause Reference	Queries	Owner/Condultant's Clarification	Remarks
			Project Name **Project Number** **Bid Clarification Form**		
			SAMPLE FORM		
			SAMPLE FORM		

FIGURE 4.25 Bid clarification form. (Abdul Razzak Rumane. (2016). *Handbook of Construction Management.* Reprinted with permission from Taylor & Francis Group.)

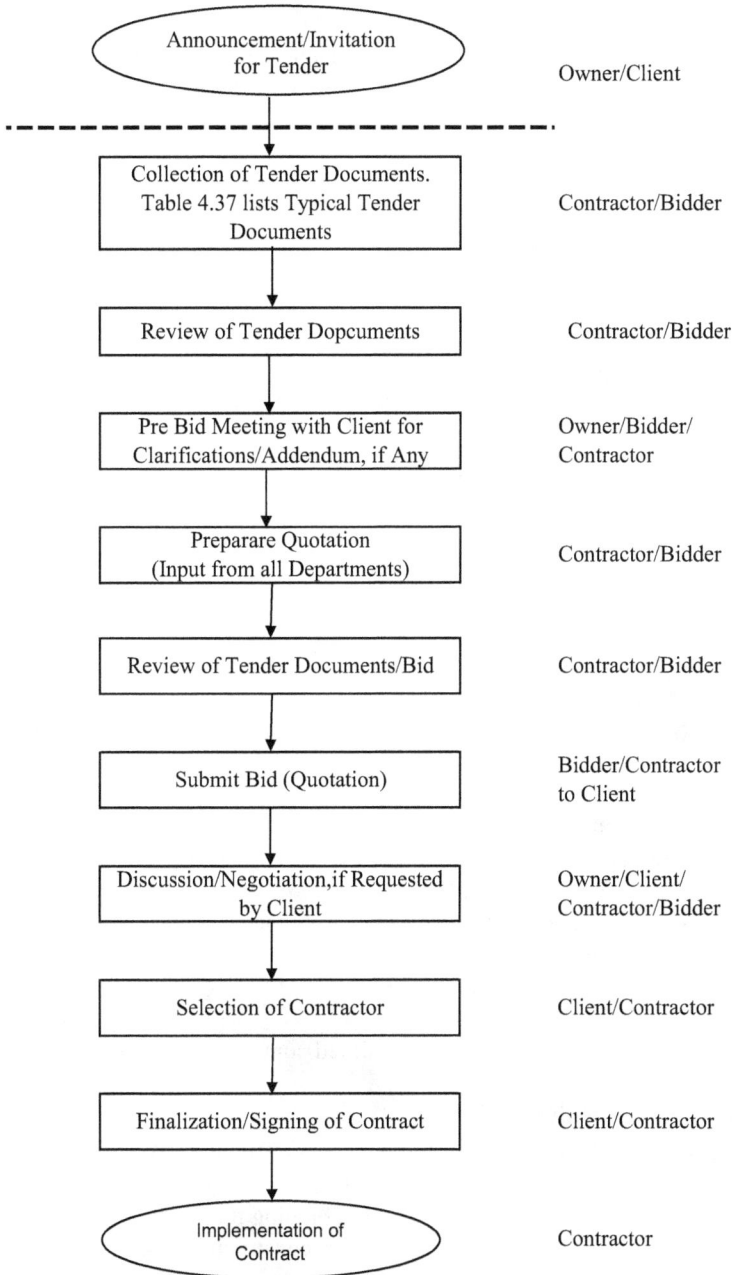

FIGURE 4.26 Typical tender submittal procedure by contractor.

4.3.5.7 Manage Bidding and Tendering Quality

In order to minimize errors during bidding and tendering phase, minimize omissions, the designer has to plan quality (planning of bidding work), perform quality assurance, and control quality for preparing bidding and tendering activities. This will mainly consist of the following:

1. Plan Quality:
   - Organize tender documents
   - Identify regulatory requirements
   - Identify bidders
   - Bid notification
   - Tender documents distribution system
   - Establish responsibility matrix
   - Bid clarification method
   - Bid collection method
   - Bid evaluation procedure
   - Contract award procedure
2. Quality Assurance:
   - Collect regulatory information
   - Owner approval for bidding
   - Distribution of bid documents
   - Pretender meetings
   - Site visits
   - Bid clarifications
   - Bid collection
3. Control Quality:
   - Bid documents to prequalified/shortlisted/registered bidders
   - Contents of bid documents as required
   - Bid document distribution as per announce date
   - Check for specified number of documents
   - Check bid bond amount
   - Bid evaluation
   - Award contract to selected/qualified contractor

4.3.5.8 Manage Tendering Risk

Following are typical risks likely to occur during this phase:

- Not all the qualified bidders take part in bidding for the project
- Bidders noticing errors and omissions in construction documents resulting in delay in the submission of bids
- BOQ not matching with design drawings
- Amendment to construction documents
- Addendum
- Delay in submission of bids than the notified one

TABLE 4.44
Major Risk Factors Affecting Contractor

Serial Number	Risk Factor
1	**Bidding/Tendering**
	1.1 Low bid
	1.2 Poor definition of scope of work
	1.3 Overall understanding of project
	1.4 Review of contract specs with bill off quantities
	1.5 Errors in resource estimation
	1.6 Errors in resource productivity
	1.7 Errors in resource availability
	1.8 Errors in material price
	1.9 Improper schedule
	1.10 Quality standards
	1.11 Exchange rate
	1.12 Review of contract document requirements with regulatory requirements
	1.13 Unenforceable conditions or contract clauses

- Bid value exceeding the estimated definitive cost
- Successful bidder fails to submit performance bond

The owner/designer has to consider these risks and plan the phase duration accordingly.

Table 4.44 lists the risks the contractor has to manage during bidding and tendering phase.

4.3.5.9   Review Bid Documents

The designer (consultant) reviews the bids for compliance to tender requirements.

4.3.5.9.1   Evaluate Bids

The submitted bids are reviewed and evaluated by the contractor selection committee members for full compliance with the tender requirements.

The main purpose of bid evaluation is to determine that the bid responses are complete in all respects in accordance with the evaluation and selection methodology specified in the tender documents.

4.3.5.10   Select Contractor

After the bids have been reviewed and evaluated by the consultant, they are sent to the owner for approval and further action. In certain cases, the bids are reviewed by the tender approving agency. The comments raised by the agency are to be taken care of before the finalization of bid. In most cases, the bidder with lowest bid amount wins the contract. In case the bid amount is more than the approved project budget, the owner has to update the budget amount or negotiate with the contractor to meet owner's expectations. Upon approval from the relevant agency, the owner awards the contract to one bidder.

4.3.5.11   Award Contract

Figure 4.27 illustrates the contract award process.

Table 4.45 summarizes the TQM concept for bidding and tendering phase activities.

FIGURE 4.27   Contract award process. (Abdul Razzak Rumane. (2016). *Handbook of Construction Management.* Reprinted with permission from Taylor & Francis Group.)

TABLE 4.45
Total Quality Management Concept for Bidding and Tendering Phase Activities

Serial Number	Phase Activity	Stakeholders/ Team Members	Customer Specifications/ Requirements	System for Managing/Tool
1	Identification of Stakeholders/ Project Team Members	1. Project owner 2. Designer (design manager) 3. Project manager 4. Tendering committee members 5. Bidders/ contractors	1. Project need and requirements 2. Procurement strategy	Please refer Table 4.41
2	Organize Tender Documents	1. Designer/design manager 2. Project manager 3. Quality manager 4. Tendering committee members	1. TOR requirements 2. Owner's requirements 3. Regulatory requirements 4. Pricing/contracting method	Please refer Section 4.3.5.2
3	Identify Tendering Process	1. Designer/design manager 2. Project manager 3. Quality manager 4. Quantity surveyor/contract administrator 5. Tendering committee members	1. TOR requirements 2. Owner's requirements 3. Regulatory requirements	Please refer Sections 2.2.2, 2.2.3.1 and Figure 4.22
4	Identify Bidders	1. Designer/ design manager/ design team 2. Project manager 3. Quality manager 4. Tendering committee members	Owner's procurement strategy	Please refer Section 4.3.5.3.1, 4.3.5.3.2 and Figure 4.23 and Table 4.43
5	Establish Bid Review Process	1. Designer/design manager 2. Project manager 3. Quality manager 4. Quantity surveyor/contract administrator 5. Tendering committee members	1. TOR requirements 2. Owner's procurement strategy	Please refer Section 4.3.5.4 and Table 4.43

(continued)

TABLE 4.45 (Continued)
Total Quality Management Concept for Bidding and Tendering Phase Activities

Serial Number	Phase Activity	Stakeholders/ Team Members	Customer Specifications/ Requirements	System for Managing/Tool
6	Manage Tendering Process	1. Designer/design manager 2. Project manager 3. Quality manager 4. Tendering committee members	As per agreed upon tendering process	Please refer Section 4.3.5.5 and Figure 4.24
7	Submit/Receive Bids	1. Designer/design manager 2. Tendering committee members 3. Bidders/ contractors	As mentioned in the tendering documents	
8	Manage Bidding and Tendering Quality	1. Quality manager 2. Designer (design manager)	1. Owner's requirements 2. Compliance to tendering procedure	Please refer Section 4.3.5.7
9	Manage Tendering Risk	1. Designer/design team members 2. Quality manager 3. Bidders/ contractors	Typical risks during tendering phase	Please refer Section 4.3.5.8 and Table 4.44
10	Review Bid Documents	1. Designer/ designer team members 2. Quantity surveyor/contract administrator	1. Compliance to owner's procurement strategy 2. Regulatory requirements	Please refer Section 4.3.5.9
11	Select Contractor	1. Owner 2. Tendering committee members 3. Bidders/ contractors	1. Owners procurement strategy	Please refer Section 4.3.5.10
12	Award Contract	1. Owner 2. Tendering committee members 3. Successful bidder/contractor	1. Owner's procurement strategy	Please refer Figure 4.27

4.3.6 CONSTRUCTION PHASE

Construction is translating owner's goals and objectives, by the contractor, to build the facility as stipulated in the contract documents, plans, and specifications within budget and on schedule. Construction is the sixth phase of construction project life cycle and is an important phase in construction projects. A majority of total project budget and schedule is expended during construction. Similar to costs, the time required to construct the project is much higher than the time required for preceding phases. Construction usually requires a large number of work force and variety of activities. Construction activities involve erection, installation, or construction of any part of the project. Construction activities are actually carried out by the contractor's own work force or by subcontractors. Construction therefore requires more detailed attention of its planning, organizations, monitoring, and control of project schedule, budget, quality, safety, and environment concerns. The construction phase consists of various activities such as Mobilization, Submittals, Planning and Scheduling, Management of Resources/Procurement, Execution of Works, Control and Monitoring, Quality, and Inspection.

Figure 4.28 illustrates a typical flowchart for TQM applications in bidding and tendering phase.

Once the contract is awarded to successful bidder (contractor), then it is the responsibility of the contractor to respond to the needs of client (owner) by constructing the project as specified in the contract documents, drawings, and specifications within the specified time and budget.

A letter from the client/owner is issued to the contractor to begin the project work subject to the conditions of the contract. This letter is known as "Notice to Proceed" letter.

Figure 4.29 is a sample Notice to Proceed.

Once the "Notice to Proceed" letter is given to the contractor, a "Kickoff Meeting" is held between the owner/client and contractor. It is also called preconstruction meeting. The Kickoff Meeting provides opportunity to all project team members to interact and knowing each other.

Figure 4.30 is an example Kickoff Meeting Agenda.

4.3.6.1 Identify Stakeholders

Following project stakeholders/team members have involvement in the project during construction phase:

1. Owner
   - Owner's representative/Project Manager
2. Construction Supervisor
   - Construction Manager (if applicable)
   - Consultant (Designer)
   - Specialist consultant
3. Contractor
   Main contractor
   Subcontractor
   Supplier

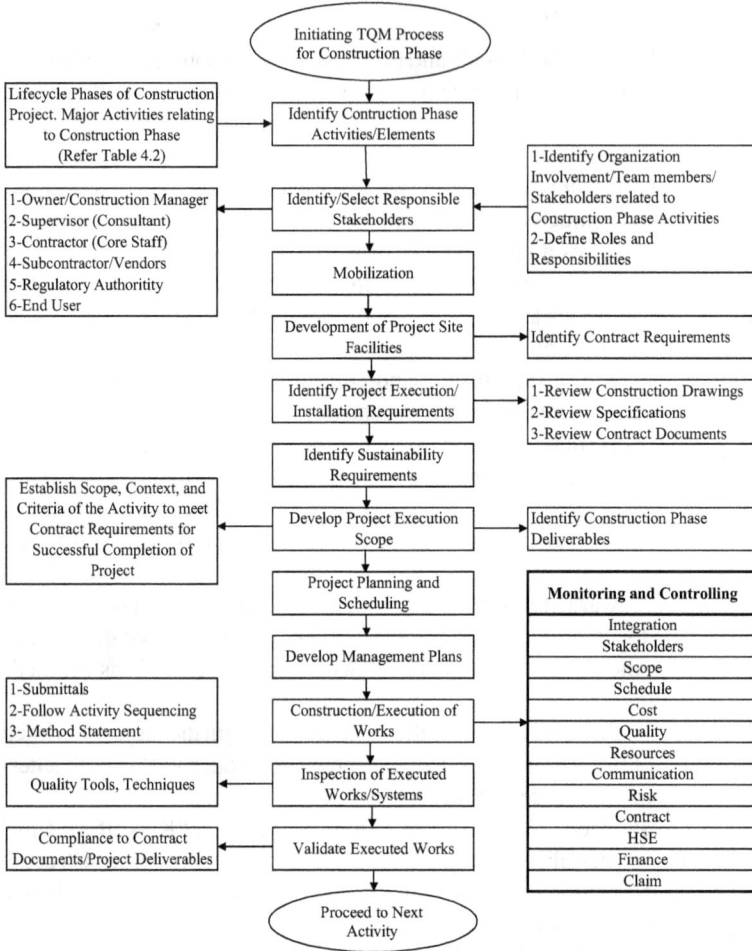

FIGURE 4.28 Typical flowchart for total quality management applications in quality of construction phase activity.

4. Regulatory Authorities
5. End User

4.3.6.1.1   Identify Owner's Representative/Project Manager

The owner from his/her office deputes or hire from outside, Owner's Representative (OR) to administer the overall project. The OR should have relevant experience, knowledge about the construction processes of similar nature of project. He/she should be able to manage the project with the help of supervision team members. In FIDIC terminology, this person is known as "Engineer" who is appointed by the "Employer" (Owner).

LETTER HEAD

NOTICE TO PROCEED

To,

Contractor Name: SAMPLE LETTER

Address:

Subject: Contract Number------------

Attention: -------------------

Sir/Madam

You are hereby authorized to proceed with Project No. ------------ in accordance with

construction contract dated ----------- This contract calls all the contracted works to be

completed within --------- calendar days. The date of enterprise shall be -------------.

Sincerely,

Enclosures:

CC:

FIGURE 4.29   Notice to proceed.

4.3.6.1.2   Identify Supervision Team

In a traditional type of contract, the client selects the same firm which has designed the project. The firm known as "Consultant" is responsible to supervise the construction process and achieving project quality goals. The firm appoints a representative, who is acceptable and approved by the owner/client, to be on-site and is often called Resident Engineer (R.E.). The R.E. along with supervision team members is responsible to supervise, monitor and control, and implement the procedure specified in the contract documents and ensure completion of project within specified time, budget, and per defined scope of work.

In order to ensure smooth flow of supervision activities, R.E. has to follow the organization's supervision manual and contractual requirements. Depending on the type and size of the project, the supervision team usually consists of following personnel:

PROJECT NAME

Contract Number:			
Type of Meeting:		Date of Meeting:	
Place of Meeting:		Time of Meeting:	
Owner:			
PMC:			
Contractor:			
Others (As Applicable)			

AGENDA   SAMPLE AGENDA

1.0 Points to be Discussed

1.1 Intoduction

1.2 Project goals and objectives

1.3 Scope of work

1.4 Permit, Bonds, Insurance

1.5 Site handover procedure

1.6 Mobilization

1.7 Contractor's organization chart (Design, Construction)

1.8 Construction Schedule

1.9 Communication and Correspondance

1.10 Transmittals and Submittal Procedure

1.11 Meetings (Progrees, Coordination)

1.12 Construction Management Plan

1.13 Quality Management Plan

1.14 Risk Management Plan

1.15 HSE Management Plan

1.16 Payment

1.7 Nominated sub contractors

2.0 Any other business

Signed by: ....   Position: ....

Deate: ....

FIGURE 4.30   Kickoff meeting agenda.

1. Resident Engineer
2. Contract Administrator/Quantity Surveyor
3. Planning/Scheduling Engineer
4. Engineers from different trades such as Architectural, Structural, Mechanical, HVAC, Electrical, Low Voltage System, Landscape, Infrastructure
5. Quality Manager
6. HSE Officer

7. Inspectors from different trades
8. Interior Designer
9. Document Controller
10. Office Secretary

The construction phase consists of various activities such as Mobilization, Execution of Works, Planning and Scheduling, Monitoring and Controlling, Management of Resources/Procurement, Quality, and Inspection. Table 4.46 illustrates major activities to be performed by the supervisor during construction phase.

The owner may engage construction manager (if required) to supervise complex and major projects. As agency CM, the construction manager is responsible to perform services related to following areas of activities:

- Performance bond by contractor
- Worker's insurance

TABLE 4.46
Responsibilities of Supervision Consultant

Sr. No.	Description
1	Achieving the quality goal as specified
2	Review contract drawings and resolve technical discrepancies/errors in the contract documents
3	Review construction methodology
4	Approval of contractor's construction schedule
5	Regular inspection and checking of executed works
6	Review and approval of construction materials
7	Review and approval of shop drawings
8	Inspection of construction material
9	Monitoring and controlling construction expenditure
10	Monitoring and controlling construction time
11	Maintaining project record
12	Conduct progress and technical coordination meetings
13	Coordination of owner's requirements and comments related to site activities
14	Project-related communication with contractor
15	Coordination with regulatory authorities
16	Processing of site work instruction for Owner's action
17	Evaluation and processing of variation order/change order
18	Recommendation of contractor's payment to owner
19	Evaluating and making decisions related to unforeseen conditions
20	Monitor safety at site
21	Supervise testing, commissioning, and handover of the project
22	Issue substantial completion certificate

Source: Abdul Razzak Rumane. (2017). *Quality Management in Construction Projects, Second Phase.* Reprinted with permission from Taylor & Francis Group.

- Selecting and recommending contractor's core staff
- Selecting and recommending subcontractor
- Establishing and implementing procedures for processing and approval of shop drawings
- Material approvals
- Managing contractor's Request for Information (RFI)
- Change order management
- Construction schedule approval
- Construction supervision
- Quality management
- Coordination of on-site, off-site inspection
- Inspection of works
- Construction contract administration
- Conduction periodic progress meetings
- Preparation of minutes of meeting and distribution as per agreed upon matrix
- Document control
- Technical correspondence between contractor
- Managing submittals
- Monitor daily progress
- Monitoring contractor's performance and ensure that the work is performed as specified, as per approved shop drawings, and as per applicable codes
- Scope control
- Construction scheduling and monitoring
- Cost tracking and management
- Review, evaluation, and documentation of claims
- Maintain project progress record
- Evaluation of payment request and recommending progress payments
- Monitor project risk
- Monitoring contractor's HSE plans (Health, Safety, and Environment)
- Coordination of works by multiple contractors
- Coordination for delivery and storage of owner-supplied materials and systems
- Testing of systems
- Punch list

4.3.6.1.3   Identify Contractor's Core Staff

Contract documents normally specify a list of minimum number of core staff to be available at site during the construction period. The absence of any of these staff may result in penalty to be imposed on contractor by the owner.

Following is a typical list of contractor's minimum core staff needed during construction period for execution of work of a major building construction project.

1. Project Manager
2. Site Senior Engineer for Civil Works
3. Site Senior Engineer for Architectural Works
4. Site Senior Engineer for Electrical Works
5. Site Senior Engineer for Mechanical Works
6. Site Senior Engineer for HVAC Works

  7. Site Senior Engineer for Infrastructure Works
  8. Planning Engineer
  9. Senior Quantity Surveyor/Contract Administrator
 10. Civil Works Foreman
 11. Architectural Works Foreman
 12. Electrical Works Foreman
 13. Mechanical Works Foreman
 14. HVAC Works Foreman
 15. Laboratory Technician
 16. Quality Control Engineer
 17. Safety Officer

4.3.6.1.4 Identify Regulatory Authorities

In certain countries there is a regulation to submit electrical, mechanical, HVAC drawings for review and approval by the authorities. Contractor has to identify which drawings/documents are to be submitted to authorities during construction phase. A necessary letter to the regulatory authorities/agencies is issued by the owner upon request for such letter from the contractor.

4.3.6.1.5 Identify Subcontractors

In most construction projects, the contractor engages special subcontractors to execute certain portion of contracted project works. Areas of subcontracting are generally listed in the Particular Conditions of the contract document. Generally, the contractor has to submit subcontractors/specialist contractors to execute following type of works:

  1. Precast Concrete Works
  2. Metal Works
  3. Space Frame, Roofing Works
  4. Wood Works
  5. Aluminum Works
  6. Internal Finishes such as painting, false ceiling, tiling, cladding
  7. Furnishings
  8. Waterproofing and Insulation Works
  9. Mechanical Works
 10. HVAC Works
 11. Electrical Works
 12. Low Voltage Systems/Smart Building System
 13. Landscape
 14. External Works
 15. Any other specialized works

The contractor has to submit their names for approval to the owner prior to their engagement to perform any work at the site. Table 4.47 is an example subcontractor selection questionnaire.

TABLE 4.47
Subcontractor Pre-qualification Questionnaire

Instructions

Please type or write all your replies legibly. Attach additional sheets, if required.

PART I

I.1 Company Information

I.1.1   Name of Organization:
I.1.2   Commercial Registration no.:
I.1.3   Year of Establishment:
I.1.4   Type of Company:
I.1.5   Company Address:
I.1.6   Affiliate company name(s) and Address:

I.2 Subcontract Works (Please tick mark all interested)

Sr.No.	Work Description	Detail Design	Preparation of Shop Drawing	Construction
1	Architectural Work	☐		
2	Structural Work			
3	Precast Work			
4	Internal Finishes			
5	External Finishes			
6	Plumbing			
7	Drainage			
8	Fire Fighting			
9	HVAC			
10	Elevator			
11	Escalator			
12	Electrical Work(Power)			
13	Electrical Work (Low Voltage)- Specify			
14	Instrumentation			
15	Building Management System			
16	Irrigation			
17	Landscape Work			
18	Pavements			
19	Streets/Roads			
20	Water Proofing			

TABLE 4.47 (Continued)
Subcontractor Pre-qualification Questionnaire

PART II

II.1 Financial Information

II.1.1   Provide copy of audited Balance Sheet:
II.1.2   Provide Bonding Capacity:
II.1.3   Provide Insurance Capacity:
II.1.4   Provide Bank Reference:

PART III

III.1 Organization Details

III.1.1.   Core Business Area:
III.1.2.   Organization Chart:
III.1.3.   ISO Certification:
III.1.4.   Year of Experience:

III.2 Project Details

III.2.1 Project History for last 10 years

Sr.No.	Name of Project	Type of Work	Value	Peak Workforce	Start Date	End Date
1						
2						
3						
4						

III.2.2 Current Projects

Sr.No.	Name of Project	Type of Work	Value	Peak Workforce	Start Date	End Date
1						
2						
3						

(*continued*)

TABLE 4.47 (Continued)
Subcontractor Pre-qualification Questionnaire

PART IV

IV.1 Management Staff

IV.1.1 Provide list of Project Managers, Project Engineers, Engineers

IV.2 Workforce

Sr.No.	Work Description	TechniciansForeman	Skilled	UNSkilled
1	Architectural Work			
2	Structural Work			
3	Precast Work			
4	Internal Finishes			
5	External Finishes			
6	Plumbing			
7	Drainage			
8	Fire Fighting			
9	HVAC			
10	Elevator			
11	Escalator			
12	Electrical Work (Power)			
13	Electrical Work (Low Voltage)- Specify			
14	Instrumentation			
15	Building Management System			
16	Irrigation			
17	Landscape Work			
18	Pavements			
19	Streets/Roads			
20	Water Proofing			

PART V

V.1 Quality Management System
V.1.1 Provide copy of ISO Certificate:
V.I.2 Person incharge of QA/QC Activities:
V.I.3 Number of Quality Auditors:

TABLE 4.47 (Continued)
Subcontractor Pre-qualification Questionnaire

PART VI

VI.1 HSE System

VI.1   Does company has HSE Policy?
VI.1.2 Provide Site Accident records for last 2 years
Declaration
We hereby declare that the information provided herein is true to our knowledge.
Note:
All relevant documents attached.

Signature of Authorized Person

Source: Abdul Razzak Rumane (2013), *Quality Tools for Managing Construction Projects*. Reprinted
          with permission from Taylor & Francis Group Company.

4.3.6.1.5.1   Nominated Subcontractors

In certain projects, the owner/client selects a contractor, known as nominated sub-contractor, to carry out certain part of the work on behalf of the owner/client. Normally nominated subcontractor is imposed on the main contractor after the main contractor is appointed. Nominated subcontractor(s) may be a specialist supplier, specialist contractor nominated in accordance with the contract to be employed by the contractor for the supply of materials or services, or to execute the project work.

4.3.6.1.6   Establish Stakeholder's Requirements

A distribution matrix based on the interest and involvement of each stakeholder is prepared and the appropriate documents are sent for their action/information as per the agreed upon requirements. Table 4.48 shows an example matrix for site adminis-tration of a building construction project.

4.3.6.1.7   Develop Responsibility Matrix

Table 4.49 illustrates the contribution of various participants during construc-tion phase.

TABLE 4.48
Matrix for Site Administration and Communication

Sr. No.	Description of Activities	Contractor	Consultant	Owner
1	General			
	1.1 Notice to Proceed	-	-	P
	1.2 Bonds and Guarantees	P	R	A
	1.3 Consultant Staff Approval	-	P	A
	1.4 Contractor's Staff Approval	P	R/B	A
	1.5 Payment Guarantee	P	R	A
	1.6 Master Schedule	P	R	A
	1.7 Stoppage of Work	-	P	A
	1.8 Extension of Time	-	P	A
	1.9 Deviation from Contract Documents	P	R	A
	a. Material			
	b. Cost			
	c. Time			
2	Communication			
	2.1 General Correspondence	P	P	P
	2.2 Job Site Instruction	D	P	C
	2.3 Site Works Instruction	D	P/B	A
	2.4 Request for Information	P	A	C
	2.5 Request for Modification	P	B	A
3	Submittals			
	3.1 Subcontractor	P	B/R	A
	3.2 Materials	P	A	C
	3.3 Shop Drawings	P	A	C
	3.4 Staff Approval	P	B	A
	3.5 Pre-meeting Submittals	P	D	C
4	Plans and Programs			
	4.1 Construction Schedule	P	R	C
	4.2 Submittal Logs	P	R	C
	4.3 Procurement Logs	P	R	C
	4.4 Schedule Update	P	R	C
5	Monitor and Control			
	5.1 Progress	D	P	C
	5.2 Time	D	P	C
	5.3 Payments	P	R/B	A
	5.4 Variations	P	R/B	A
	5.5 Claims	P	R/B	A
6	Quality			
	6.1 Quality Control Plan	P	R	C
	6.2 Check Lists	P	D	C
	6.3 Method Statements	P	A	C
	6.4 Mock up	P	A	B

TABLE 4.48 (Continued)
Matrix for Site Administration and Communication

Sr. No.	Description of Activities	Contractor	Consultant	Owner
	6.5 Samples	P	A	B
	6.6 Remedial Notes	D	P	C
	6.7 Non Conformance Report	D	P	C
	6.8 Inspections	P	D	C
	6.9 Testing	P	A	B
7	Site Safety			
	7.1 Safety Program	P	A	C
	7.2 Accident Report	P	R	C
8	Meetings			
	8.1 Progress	E	P	E
	8.2 Coordination	E	P	C
	8.3 Technical	E	P	C
	8.4 Quality	P	C	C
	8.5 Safety	P	C	C
	8.6 Close Out	-	P	
9	Reports			
	9.1 Daily Report	P	R	C
	9.2 Monthly Report	P	R	C
	9.3 Progress Report	-	P	A
	9.4 Progress Photographs	-	P	A
10	Close Out			
	10.1 Snag List	P	P	C
	10.2 Authorities Approvals	P	C	C
	10.3 As-Built Drawings	P	D/A	C
	10.4 Spare Parts	P	A	C
	10.5 Manuals and Documents	P	R/B	A
	10.6 Warranties	P	R/B	A
	10.7 Training	P	C	A
	10.8 Hand Over	P	B	A
	10.9 Substantial Completion Certificate	P	B/P	A

P-Prepare/Initiate
B-Advise/Assist
R-Review/Comment
A-Approve
D-Action
E-Attend
C-Information

Source: Abdul Razzak Rumane. (2017). *Quality Management in Construction Projects*, Second Edition. Reprinted with permission from Taylor & Francis Group.

TABLE 4.49
Responsibilities of Various Participants (Design–Bid–Build Type of Contracts) during Construction Phase

Phase	Responsibilities		
	Owner	Supervisor	Contractor
Construction	• Approve subcontractor(s) • Approve contractor's core staff • Legal/regulatory clearance • SWI • V.O. • Payments	• Supervision • Approve plans • Monitor work progress • Approve material • Approve shop drawings • Approval of submittals • Monitor schedule • Approve executed/installed works • Control budget • Recommend payment • Coordination between owner and contractor	• Execution of work • Contract management • Selection of subcontractor(s) • Planning • Resources • Procurement • Quality • Safety

4.3.6.2 Mobilization

The activities to be performed during the mobilization period are defined in the contract documents. During this period, the contractor is required to perform many of the activities before the beginning of actual construction work at the site. Necessary permits are obtained from the relevant authorities to start the construction work at the site. After being granted access to the construction site by the owner, the contractor starts mobilization work, which consists of preparation of site offices/field offices for the owner, supervision team (consultant), and for the contractor himself. This includes all the necessary on-site facilities and services necessary to carry out specific tasks. Mobilization activities usually occur at the beginning of a project but can occur anytime during a project when specific on-site facilities are required. During this time the project site is handed over to the contractor. The contractor performs site survey and testing of soil to facilitate start of construction work.

In anticipation of the award of contract, the contractor begins the following activities much in advance, but these are part of contract documents, and the contractor's action is required immediately after signing of the contract in order to start construction:

- Mobilization of construction equipment and tools
- Workforce to execute the project

4.3.6.2.1   Bonds, Permits, and Insurance

In order to proceed with project execution activities and as per contract documents, the contractor has to

1.   Submit advance payment guarantee
2.   Permit from local authority (municipality)
3.   Insurance policies covering following areas:
     a.   Contractor's all risks and third-party insurance policy
     b.   Contractor's plant and equipment insurance policy
     c.   Workmen's compensation insurance policy
     d.   Site storage insurance policy

Normally the submitted originals are retained by the owner and the copies are kept with the R.E. or project/construction manager (consultant).

4.3.6.3   Development of Project Site Facilities

The requirements to set up temporary facilities are specified in the contract. The contractor has to submit layout plans, dimensions, and other pertinent details for temporary facilities to be constructed. Upon approval of plans, the contractor proceeds with construction of temporary facilities and necessary utilities. These include

1.   Site offices for owner, supervisor (consultant), regulatory authority (if applicable)
2.   Storage facilities
3.   Toilets and washrooms
4.   Sanitary and drainage system
5.   Drinking water facility
6.   Safety and healthcare facilities
7.   Site electrification
8.   Temporary firefighting system
9.   Site fence
10. Site access road
11. Necessary utilities for construction
12. Communication system
13. Signage
14. Fuel storage area
15. Guard room
16. Testing laboratory
17. Waste dumping area
18. Area for storage of hazardous material
19. Project signboard

The contractor has to identify and designate the authority approved dumping area for waste and hazardous material.

4.3.6.4   Identify Project Execution/Installation Requirements

Once the contract is awarded to successful bidder (contractor), then it is the responsibility of the contractor to respond to the needs of client (owner) by building the facility as specified in the contract documents, drawings and specifications within the schedule and budget.

4.3.6.4.1   Gather Contract Document Requirements

In order to develop contract requirements, the contractor has to review all the construction documents which are part of the contract that the contractor has signed with the owner of the project. This includes

1. Construction drawings
2. Specification
    • General specifications
    • Particular specification
    • BOQ
3. Contract documents
    • General conditions
    • Particular conditions
4. Other related documents

4.3.6.4.1.1   Review Construction Drawings

The contractor has to review the construction/working drawings to understand the project requirements. Each trade engineer has to review the drawing and understand the construction procedure to be followed to avoid omission and rework. The trade engineers have to

• Prepare a list of issue to be resolved and information needed from the designer (consultant)
• Prepare a list of conflicting items in the drawings
• Compare the estimated bid activities and actual material for proper execution of project
• Identify high risk items
• Check constructability as per the specified method statement
• Identify the execution process which can be simplified
• Identify discrepancies between drawings and specifications
• Identify any missing drawing required to execute the project
• Identify value engineering change proposal
• Identify complete scope of work
• Prepare construction material takeoff (BOQ)

4.3.6.4.1.2   Review Specifications
Contractor has to review specifications and check for following major points:

- Matching of construction drawings and specification requirements
- Any missing specifications
- Issues that need to be resolved
- If there are discrepancies between specifications and drawings
- Codes and Standards to be followed
- Submittal requirements
- Quality requirements
- Safety requirements
- Environmental requirements
- Installation procedure for owner-supplied items

4.3.6.4.1.3   Review Contract Documents
Contractor has to carefully study all the clauses of contract and identify

- High risk clauses
- Items having price difference between bid price and the actual price for the items to be installed for specified performance of work/system
- Items need to be resolved by raising Request for Information (RFI)
- Clauses having conflict with regulatory requirements
- Schedule constraints
- Ambiguous clauses
- Clauses that favor owner
- Coordination among all the parties involved in the project
- Any hidden clause that will entitle owner for compensation claim from the contractor
- Priorities among various construction documents
- Discrepancies and conflicting clauses
- Change order process
- Claim for extra work
- Payment procedure
- Damage and penalty for delay clauses
- Force Majeure clause

4.3.6.5   Identify Sustainability Requirements
The contractor has to also identify sustainability requirements for construction/execution of project. Following are the basic elements that have to be taken care by the contractor to develop construction project:

- Meeting/satisfying owner's requirements
- Completion of execution/installation works within agreed schedule
- Completion of execution/installation of project works within budget
- Quality management

- Use of energy-efficient equipment, systems
- Water conservation systems
- Use of local resources (manpower)
- Efficient and durable material
- Ecofriendly material
- Material handling and utilization
- Optimized material storage
- Supply chain management
- Just-in-time delivery
- Follow systematic work process
- Use efficient method of execution/installation.
- Follow green building concept
- Aesthetic
- Install energy-efficient MEP systems
- Waste treatment system
- Stormwater management
- Hazardous material management
- Clean air/ventilation system
- Daylighting
- Sensor-controlled lighting in certain areas
- Install waste treatment plant
- Risk management
- Landscape compatible with surrounding area
- Plantation in project area
- Regulatory requirements
- Applicable Codes and standards

4.3.6.6   Develop Project Execution Scope

Following is the scope of works to be executed during construction phase:

A. Execution of Work

- Site work such as cleaning and excavation of project site
- Construction of foundations including footings and grade beams
- Construction of columns and beams
- Forming, reinforcing, and placing the floor slab
- Laying up masonry walls and partitions
- Installation of roofing system
- Finishes
- Furnishings
- Conveying system
- Installation of fire suppression system
- Installation of water supply, plumbing, and public health system
- Installation of heating, ventilating, and air conditioning system

- Integrated automation system
- Installation of electrical lighting and power system
- Emergency power supply system
- Fire alarm system
- Communication system
- Electronic security and access control system
- Landscape works
- External works

B. Monitoring and Control during Construction

- Stakeholder's requirements
- Project schedule
- Project cost
- Project quality
- Safety during construction

C. Inspection of executed works

4.3.6.6.1   *Identify Construction Phase Deliverables*

The construction phase deliverables are established based on project execution scope discussed under Sections 4.3.6.4 and 4.3.6.5. These are

A. Construction-installation/execution and inspection of following works:

1. Structural work
2. Architectural
3. Internal finishes
4. External finishes
5. Conveying system
6. MEP work
7. Information and communication
8. Low voltage systems
9. Landscape work
10. External work
11. Furnishings

B. Record drawings/updated working drawings (shop drawings)

C. Updated specifications

4.3.6.7   Project Planning and Scheduling

Project planning is a logical process to determine what work must be done to achieve project objectives and ensure that the work of project is carried out:

- In an organized and structured manner
- Reducing uncertainties to minimum

- Reducing risk to minimum
- Establishing quality standards
- Achieving results within scheduled time and budget

Table 4.50 lists the reasons for planning, and Table 4.51 lists advantages of planning and scheduling.

Planning is a mechanism that conveys or communicates to project participants what activity to be done, how, and in what order to meet the project objectives by scheduling the same. Project planning is required to bring the project to completion on schedule, within budget, and in accordance with the owner's needs as specified in the contract. The planning process considers all the individual tasks, activities, or jobs that make up the project and must be performed. It takes into account all the resources available, such as human resources, finances, materials, plants, and equipment. It also considers the works to be executed by the subcontractors.

Planning and scheduling are often used synonymously for preparing construction program because both are performed interactively. Planning is the process of identifying the activities necessary to complete the project, while scheduling is the process of determining the sequential order of the planned activities and the time required to carry out and complete the activities. Scheduling is the mechanical process of formalizing the planned functions, assigning the starting and completion dates to each part or activity of the work in such a manner that the whole work proceeds in a logical sequence and in an orderly and systematic manner.

TABLE 4.50
Reasons for Planning

1.	To execute work in an organized and structured manner
2.	To eliminate or reduce uncertainty
3.	To reduce risk to the minimum
4.	To reduce rework
5.	To improve the efficiency of the process
6.	To establish quality standards
7.	To provide basis for monitoring and control of project work
8.	To know duration of each activity
9.	To know cost associated with each activity
10.	To establish benchmark for tracking the quantity, cost and timing of work required to complete the project
12.	To know responsibility and authority of people involved in the project
13.	To establish timely reporting system
14.	Integration of project activities for smooth flow of project work

Source: Abdul Razzak Rumane. (2013). *Quality Tools for Managing Construction Projects.* Reprinted with permission from Taylor & Francis Group.

TABLE 4.51
Advantages of Project Planning and Scheduling

Serial Number	Advantages
1	It facilitates management by objectives.
2	It facilitates execute the work in an organized and structured manner.
3	It eliminates or minimizes uncertainties.
4	It reduces risk to the minimum.
5	It helps proper coordination.
6	It helps integration of project activities for smooth flow of project work.
7	It help reduce rework.
8	It improves the efficiency of the process and increase productivity.
9	It improves communication.
10	It establishes timely reporting system.
11	It establishes the duration of each activity.
12	It provides basis for monitoring and controlling of project work.
13	It helps to establish benchmark for tracking the quantity, cost and timing of work required to complete the project.
14	It helps foresee problems at early stage.
15	It helps to know the responsibility and authority of people involved in the project.

4.3.6.7.1 Development of Construction Schedule

Upon signing of the contract, the contractor has to submit mainly the following to PMC/supervisor (consultant) for their review and approval:

1. Contractor's construction schedule
2. S-curve
3. Contractor's quality control plan (CQCP)

Among these plans, construction schedule is most important. This is the first and foremost program contractor has to submit for approval. The contractor cannot proceed with construction unless the preliminary construction schedule is approved. In certain cases, the progress payment is having relation with approval of contractor's construction schedule. The contractor is not paid unless the contractor's construction schedule (CCS) is approved.

Prior to the start of execution of project or immediately after the actual project starts, the contractor prepares the project construction plans based on the contracted time schedule of the project. Detailed planning is needed at the start of construction in order to decide how to use the resources such as labors, plants, materials, finance, and subcontractors economically and safely to achieve the specified objectives. The plan shows the periods for all sections of the works and activities indicating that every-thing can be completed by the date specified in the contract and ready for use or for installation of equipment by other contractors.

Effective project management requires planning, measuring, evaluating, forecasting, and controlling all aspects of a project quality and quality of work, cost, and schedules. The purpose of the project plan is to successfully control the project to ensure completion within the budget and schedule constraints. Project planning is the evolution of the time and efforts to complete the project.

Construction projects are unique and non-repetitive in nature. Construction projects consist of many activities directed to the accomplishment of a desire objective. Activities are those operations of the plan which take time to carry out and on which resources are expended. In order to achieve quality objectives of the project, each activity has to be completed within specified limit, using spe-cified product and approved method of installation. Construction projects consist of number of related activities which are dependent of other activities and cannot be started until others are completed and some that can run in parallel. The most important point while starting the planning is to establish all the activities which constitute the project.

Planning involves defining objectives of the project; listing of tasks or jobs that must be performed; determining gross requirements for material, equipment, and manpower; and preparing costs and durations for the various jobs or activities to bring about the satisfactory completion of the project. The techniques for planning vary depending on the projects size, complexity, duration, personnel, and owner's requirements. Techniques used during the construction phase of the project should make possible the evaluation of the project progress against the plan. There are many different analytical and graphical techniques that are commonly used for planning of the project. These are

1. The Bar Chart
2. Critical Path Method (CPM)
3. Progress Curves
4. Matrix Schedule

The most widely used forms of program are bar charts and network diagram. Bar chart is the oldest planning method used in project management. Bar chart is graph-ical representation of estimated duration of each activity and the planned sequence of activity and the planned sequence of activity performance. The horizontal axis represents the time schedule whereas list of project activities are shown along the vertical axis.

Network diagrams such as program evaluation and review technique (PERT) and CPM are used for scheduling complex projects. PERT/CPM diagrams consist of nodes and links and represent the entire project as a network of arrows (activities) and nodes (events). In order to draw network diagram, work activities have to be iden-tified, relationship among the activities to be specified, and precedence relationship between the activities in a particular sequence need to be established.

The most widely used scheduling technique is the CPM. The CPM analysis represents the set of sequence of predecessor/successor activities which will take the longest time to complete. The duration of the critical path is the sum of all the activities duration along the path. Thus, the critical path is the longest possible path of the project activities network. The duration of the critical path represents the minimum time required to complete the project.

There are many computer-based programs available for preparing the network and critical path of activities for construction projects. These programs can be used to analyze the use of resources, review of project progress, and forecast the effects of changes in the schedule of works or other resources. Most computer programs automate preparation and presentation of various planning tools such as bar chart, PERT, and CPM analysis. The programs are capable of storing huge data and help process and update the program quickly. It manipulates data for multiple usages from the planning and scheduling perspectives.

In order to manage and control the project at different levels in most effective manner, the project is broken down into a group of smaller subprojects/subsystems and then to small well-defined activities. This break down is necessary because of the size and complexity of construction project and is referred to as WBS. To begin the preparation of detailed construction program, the contractor prepares a WBS. Its purpose is to define various activities that must be executed to complete the project. WBS help construction project planner to

1. Plan and schedule the work
2. Estimate costs and budget
3. Control schedule, cost, and quality

Depending on the size of the project, the project is divided into multiple zones and relevant activities are considered for each zone to prepare the construction program. While preparing the program the relationships between project activities and their dependency and precedence is considered by the planner. These activities are connected to their predecessor and successor activity based on the way the task is planned to be executed. There are four possible relationships that exist between various activities. These are finish–to-start relationship, the start-to-start relationship, the finish-to-finish relationship, and start-to-finish relationship.

Once all the activities are established by the planner and estimated duration of each activity has been assigned, the planner prepares a detailed program fully coordinating all the construction activities.

The CPM calculates the minimum completion time for a project along with the possible start and finish times for the project activities. The critical path is the longest in the network, whereas the other paths may be equal or shorter than that path. Therefore, there is a possibility that some of the events and activities can be completed before they are actually needed and accordingly it is possible to develop number of activity schedules from the CPM analysis to delay the start of each activity as long as possible but still finish the project with minimum possible time without extending the completion date of the project. To develop such schedule, it is required to find out when each activity needs to start and when it needs to be finished. There may be some activities in the project with some leeway in which some activity can start and finish. This is called slack time or float in an activity. For each activity in a project, there are four points in time: early start, early finish, late start, late finish. The early start and early finish is the earliest time an activity can start and be finished. Similarly, the late start and late finish are the latest times the activities can start and finish. The difference between the late start time and early start time is the slack time or float.

With the advent of powerful computer-based programs like Primavera and Microsoft Project, it is possible that the details of the work breakdown are fed to these software programs. The software is capable of producing network diagram and schedules and limitless the number of different reports, which also help in the efficient monitoring of the project schedule by comparing actual with planned progress. The software can be used to analyze the project for use of resources, forecasting the effects of changes in the schedule and cost control.

First step in preparation of construction program is to establish the activities, the next step is to establish estimated time duration of each activity. The deadline for each activity is fixed, but it is often possible to reschedule by changing the sequence in which the tasks are performed, while retaining the original estimated.

The activities to be performed during execution of project are grouped in a number of categories. Each of these categories has number of activities. Following are the major categories of construction projects schedule:

A. General Activities
   1. Mobilization
B. Engineering
   1. Subcontractor Submittal and Approval
   2. Materials Submittal and Approval
   3. Shop Drawing Submittal and Approval
   4. Procurement
C. Site Activities
   1. Site Earth Works
   2. Dewatering and Shoring
   3. Excavation and Backfilling
   4. Raft Works
   5. Retaining Wall Works
   6. Concrete Foundation and Grade Beams
   7. Water Proofing
   8. Concrete Columns and Beams
   9. Casting of Slabs
   10. Wall Partitioning
   11. Interior Finishes

      12. Furnishings
      13. External Finishes
      14. Equipment
      15. Conveying Systems Works
      16. Plumbing and Public Health Works
      17. Fire Fighting works
      18. HVAC Works
      19. Electrical Works
      20. Fire Alarm System Works
      21. Communication System Works
      22. Low voltage Systems Works
      23. Landscape Works
      24. External Works
   D. Close Out
      1. Testing and Commissioning
      2. Completion and Handover

Figure 4.31 illustrates the schedule development process.

Section 2.5.4 in Chapter 2 is an example procedure to develop CCS using the systematic approach of Six Sigma methodology concept of DMADV analytic tool set.

4.3.6.7.2 Develop Project S-Curve

S-Curve is a graphical display of cumulative cost, resources, or other quantities plotted against time. S-Curve is used for forecasting cash flow which is based on the work (activities) the contractor is expected to complete and how much amount (payment) will be received and predicts how much amount will be spent over the established project schedule (time). The S-Curve is used to measure project performance and predict the expenses over project duration. The S-Curve helps owner/client to know the project funding requirements. It is also an indication of the progress of work to be completed in a project. Funding requirements, total and periodic, are derived from the S-Curve. This also represents the planned progress of a project. Figure 4.32 illustrates planned project S-Curve prepared by the contractor which is based on the construction schedule.

4.3.6.7.2.1 Cost Loaded Curve

Conceptually, cash flow is a simple comparison of when revenue will be received and when the financial obligations must be paid. This is obtained by loading each activity in the approved schedule with the budgeted cost in the BOQ. The process of inputting schedule of values is known as cost loading. The graphical representation of the above is obtained as a curve and is known as cost loaded curve. Figure 4.33 illustrates cost loaded S-Curve prepared by the contractor.

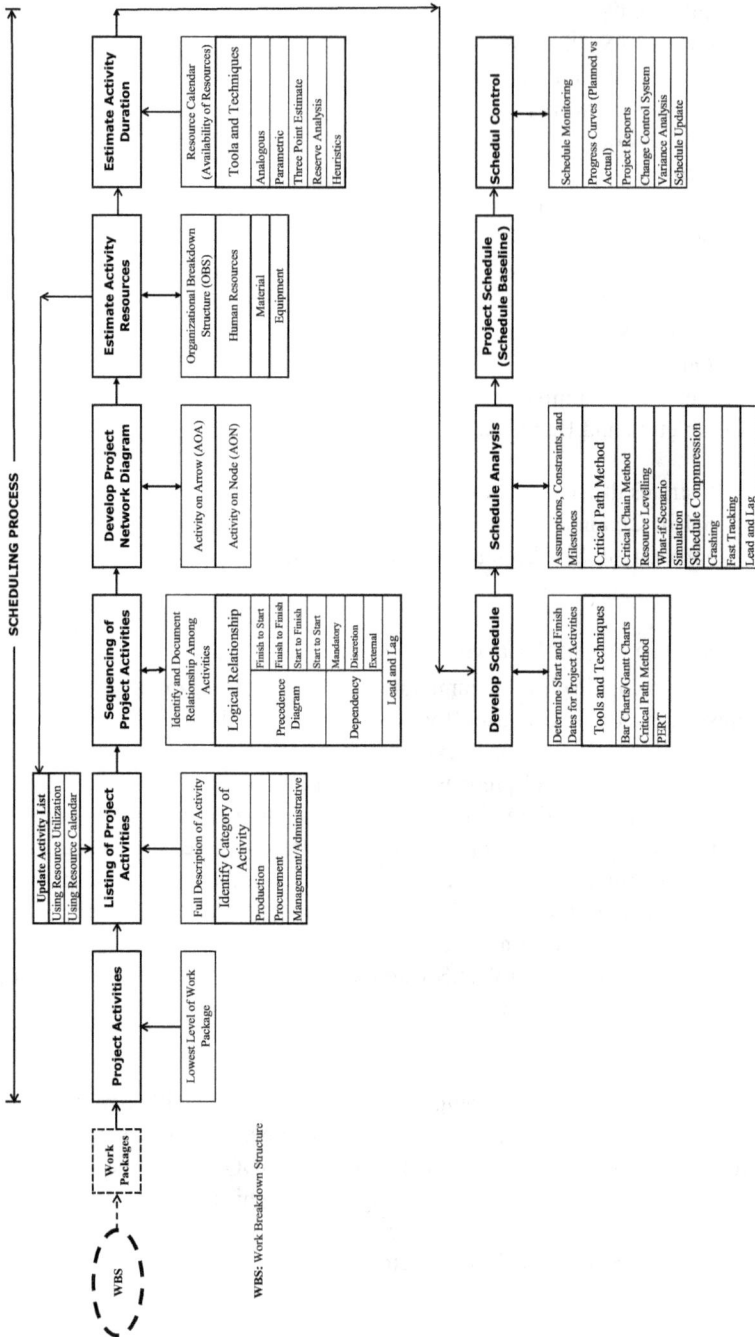

FIGURE 4.31 Schedule development process. (Abdul Razzak Rumane. (2016). *Handbook of Construction Management: Scope, Schedule, and Cost Control.* Reprinted with permission from Taylor & Francis Group.)

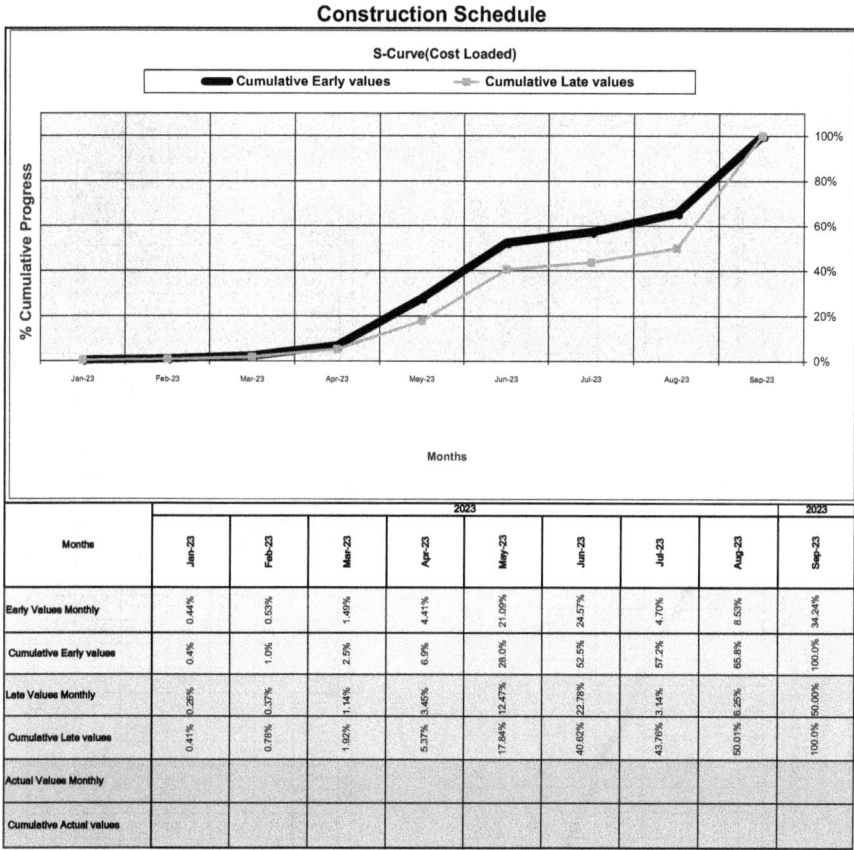

FIGURE 4.32 S-curve (cost loaded-ES-LS).

4.3.6.7.3 Develop Contractor's Quality Control Plan

The contractor's quality control plan (CQCP) is the contractor's everyday tool to insure meeting the performance standards specified in the contract documents. The adequacy and efficient management of CQCP by contractor's personnel have great impact on both the performance of contract and owner's quality assurance surveillance of the contractor's performance.

CQCP is the documentation of contractor's process for delivering the level of construction quality required by the contract. It is a framework for the contractor's process for achieving quality construction. CQCP does not endeavor to repeat or summarize contract requirements. It describes the process which contractor will use to assure compliance with the contract requirements. The quality plan is virtually manual tailor-made for the project and is based on contract requirements.

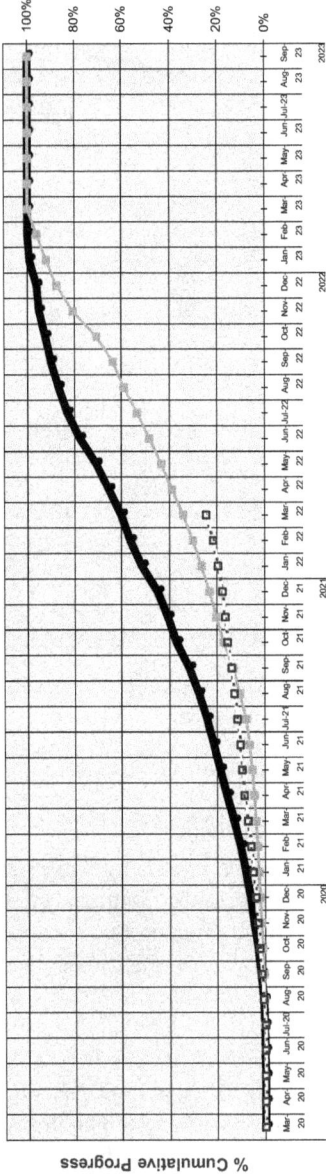

FIGURE 4.33 S-curve (cost loaded-Planned-Actual).

The CQCP is prepared based on project-specific requirements as specified in the contract documents. The plan outlines the procedures to be followed during the construction period to attain the specified quality objectives of the project fully complying with the contractual and regulatory requirements.

In the quality plan, the generic documented procedures are integrated with any necessary additional procedures peculiar to the project in order to attain specified quality objectives. Application of various quality tools, methods, and principles at different stages of construction projects is necessary to make the project qualitative, economical, and meet the owner needs/specification requirements.

Based on contract requirements the contractor prepares his quality control plan and submits the same to the consultant for their approval. Figure 4.34 illustrates logic flow for CQCP. This plan is followed by the contractor to maintain the project quality.

The quality control plan outlines the procedures to be followed during the construction period to attain the specified quality objectives of the project fully complying with the contractual and regulatory requirements. The plan provides the mechanism to achieve the specified quality by identifying the procedures, control, instructions, and tests required during the construction process to meet the owner's objectives.

Table 4.52 illustrates contents of CQCP. However, the contractor has to take into consideration requirements listed under contract documents depending on the nature and complexity of the project.

Appendix A illustrates an outline of a CQCP for a major building construction project. However, the contractor has to take into consideration the requirements listed under the contract documents, depending on the nature and complexity of the project.

4.3.6.8 Develop Management Plans

Apart from construction schedule, S-Curve, and CQCP, the contractor has to submit mainly the following plans to PMC/supervisor (consultant) for their review and approval:

1. Stakeholder Management Plan
2. Resource Management Plan
3. Communication Management Plan
4. Risk Management Plan
5. Contract Management Plan
6. HSE Management Plan

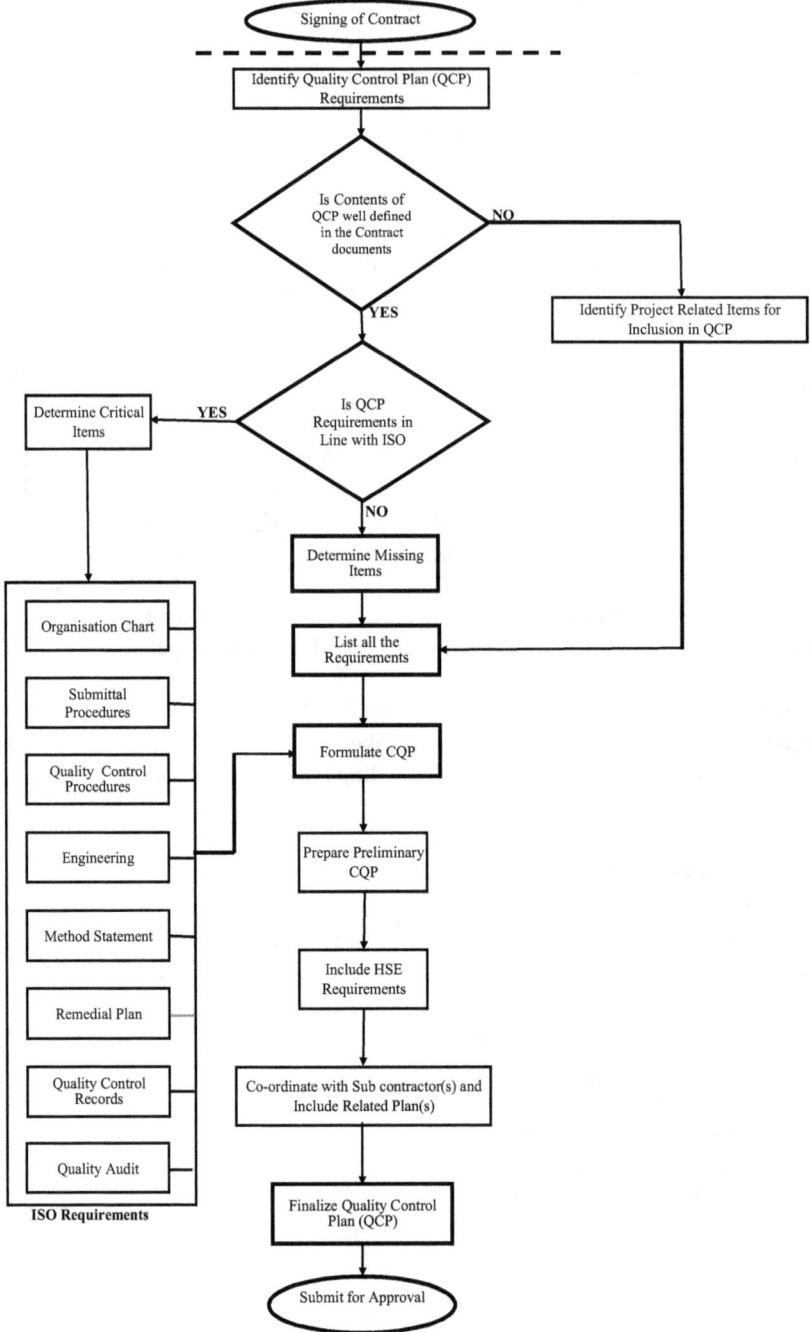

FIGURE 4.34   Logic flow chart for development of contractor's quality control plan.

TABLE 4.52
Contents of Contractor's Quality Control Plan

Section	Topic
1	Introduction
2	Description of Project
3	Quality Control Organization
4	Qualification of QC Staff
5	Responsibilities of QC Personnel
6	Procedure for submittals
	6.1 Submittals for subcontractor(s)
	6.2 Submittals for shop drawings
	6.3 Submittals for materials
	6.4 Modification request
	6.5 Construction program
7	Quality control procedure
	7.1 Procurement
	7.2 Inspection of site activities (checklist)
	7.2.1 Definable feature of work
	7.2.2 Earthworks and site works
	7.2.3 Concrete
	7.2.4 Masonry
	7.2.5 Metal works
	7.2.6 Wood, plastics and composite
	7.2.7 Doors and windows
	7.2.8 Finishes
	7.2.9 Furnishing
	7.2.10 Equipment
	7.2.11 Conveying system
	7.2.12 Mechanical works
	7.2.13 Automation system
	7.2.14 Electrical works
	7.2.15 Landscape
	7.3 Inspection and testing procedure for systems
	7.4 Off-site manufacturing, inspection, and testing
	7.5 Procedure for laboratory testing of material
	7.6 Inspection of material received at site
	7.7 Protection of works
	7.8 Material storage and handling
8	Method statement for various installation activities
9	Project-specific quality procedures
10	Risk management
11	Quality control records
12	Company's quality manual and procedures
13	Periodical testing of construction equipment
14	Quality updating program
15	Quality auditing program
16	Testing, commissioning, and handover
17	Health, safety, and environment

4.3.6.8.1   Develop Stakeholder Management Plan

In order to run a successful project, it is important to address the needs of project stakeholders effectively predicting how the project will be affected and how the stakeholders will be affected. Stakeholder management planning is a process to develop stakeholder engagement plan depending on the roles and responsibilities of the stakeholders and their needs, expectations, and influence on the project. Table 4.48 discussed earlier shows an example matrix for site administration of a construction project. A Distribution Matrix based on the interest and involvement of each stakeholder is prepared and the appropriate documents are sent for their action/information as per the agreed upon requirements.

4.3.6.8.2   Develop Resource Management Plan

Resource management in construction mainly relates to management of following processes:

1. Human resources (Project Teams)
   - Project owner team (Project Manager)
   - Construction Manager (if applicable)
   - Construction Supervisor (Consultant)
2. Construction resources (Contractor)
   - Contractor's Staff (Contractor's core team)
   - Manpower for Construction Works
   - Material for Project
     - Equipment
     - Materials
     - Systems

In most construction projects, contractor is responsible to engage subcontractors, specialist installers, suppliers, arrange for materials, equipment, construction tools, and all type of human resources to complete the project as per contract documents and to the satisfaction of owner/owner's appointed supervision team. Workmanship is one of the most important factors to achieve the quality in construction, therefore it is required that the construction workforce is fully trained and have full knowledge of all the related activities to be performed during the construction process.

Once the contract is awarded, the contractor prepares a detailed plan for all the resources needed to complete the project.

Contract documents normally specify a list of minimum number of core staff to be available at site during construction period. Absence of any of these staff may result in penalty to be imposed on contractor by the owner.

Contractor's human resources mainly consist of two categories. These are

1. Contractor's own staff and workers
2. Subcontractor's staff and workers

The main contractor has to manage all these personnel by

1. Assigning the daily activities
2. Observing their performance and work output
3. Daily attendance
4. Safety during construction process

Figure 4.35 illustrates the contractor's planned manpower chart for the construction project.

Likewise, the contract documents specify that a minimum equipment set is to be available on-site during the construction process to ensure smooth operation of all the construction activities. They are normally listed in the contract documents. These are

- Tower crane
- Mobile crane
- Normal mixture
- Concrete mixing plant
- Dump trucks
- Compressor
- Vibrators
- Water pumps
- Compactors
- Concrete pumps
- Trucks
- Concrete trucks
- Diesel generator sets

Figure 4.36 illustrates the equipment schedule which lists the equipment the contractor has to make available for major construction projects.

It is a general practice in construction projects that the contractor is responsible for procurement of material, equipment, and systems to be installed on the project. The specifications are based on the specifications/contract documents. The contractor also prepares a procurement log based on the project completion schedule. Figure 4.37 illustrates the material management process for construction projects.

4.3.6.8.3 Develop Communication Management Plan

Construction project has involvement of many stakeholders. The project team must provide timely and accurate information to identified stakeholders who will receive communications. Effective communication is one of the most important factors contributing to the success of project. For smooth flow of construction process activities during construction phase, proper communication, and submittal procedure need to be established between all the concerned parties at the beginning of the construction activities.

Table 4.53 illustrates an example guideline to prepare communication matrix for site administration during construction phase, and Table 4.54 illustrates contents of communication management plan.

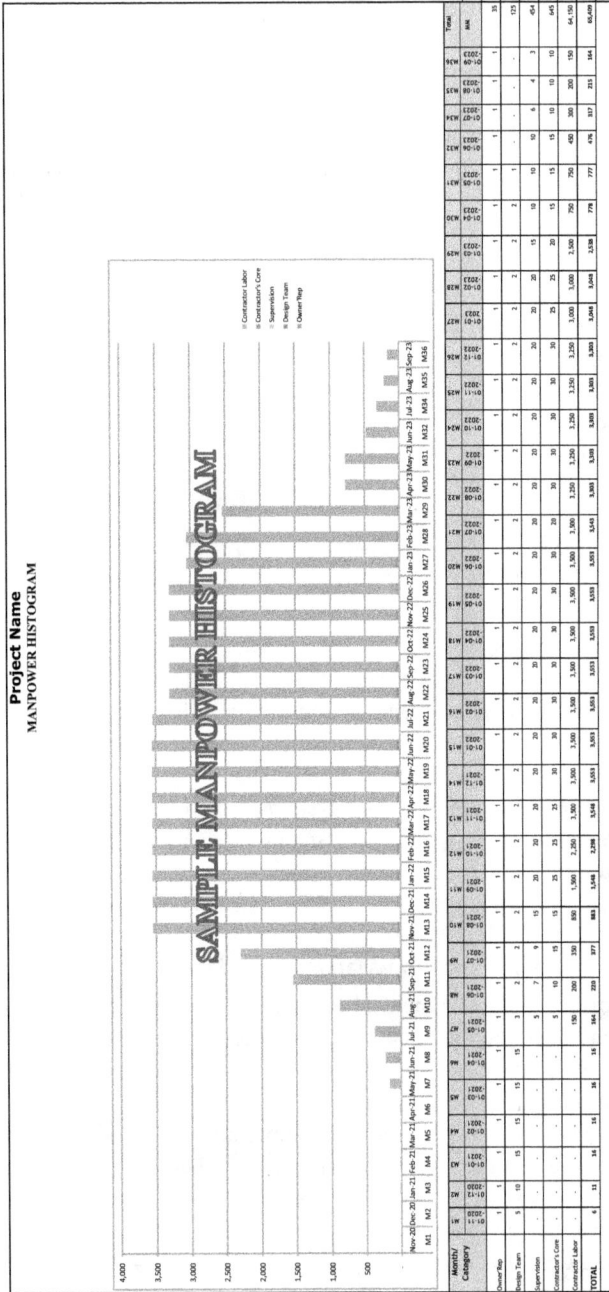

Project Name
MANPOWER HISTOGRAM

SAMPLE MANPOWER HISTOGRAM

FIGURE 4.35   Project manpower histogram.

FIGURE 4.36 Equipment schedule.

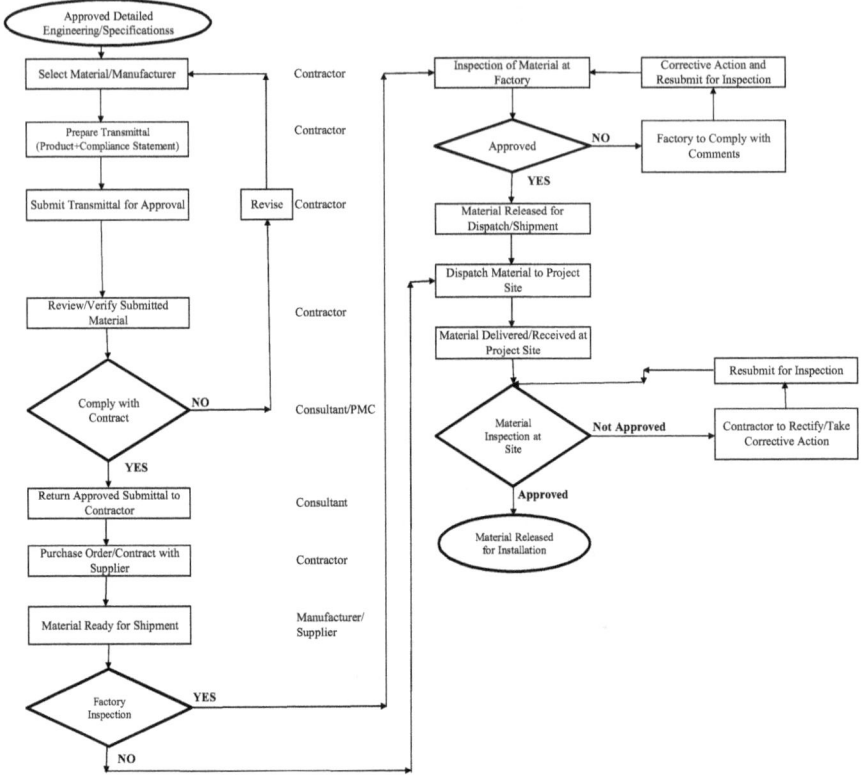

FIGURE 4.37 Material management process for construction project.

The communication between contractor and supervisor takes place through transmittal form. Table 4.55 lists various types of project control documents used during construction projects. Number of copies, distribution, and communication method are specified in the contract documents. These documents are normally transmitted by the contractor using Transmittal Form. Figure 4.38 illustrates sample Transmittal Form.

Correspondence between consultant and contractor is normally though letters or job site instructions. Figure 4.39 is a job site instruction form used by the consultant to communicate with the contractor.

4.3.6.8.4 Develop Risk Management Plan

The probability of occurrence of risk during construction phase is very high compared to design phases. During construction phase, uncertainty comes from various sources as this phase has involvement of various participants. Since the duration of construction phase is longer than earlier phases, the contractor has to also consider occurrence of financial, economical, commercial, political, and natural risks. The contractor has to develop risk management plan. Risk management plan identifies how risk

TABLE 4.53
Guidelines to Prepare Communication Matrix.

Project Name

Name of Construction/Project Manager:						Name of Consultant:		
Contractor Name:						Project Number:		
Serial Number	Type of Document	Originator	Receiver (s)	Purpose	Frequency	Method	Responsible Person for Action	Comments

SAMPLE FORM

TABLE 4.54
Contents of Contractor's Communication Management Plan

Section	Topic
1	Introduction
	1.1 Project Description
2	Stakeholders
	2.1 Project Team Directory
	2.2 Project Organization Chart
	2.3 Roles and Responsibilities
	2.4 Stakeholders Requirement
3	Communication Methods
4	Communication Management Constraints
5	Communication Matrix
6	Distribution of Communication Documents
	5.1 General Correspondence
	5.2 Submittals
	5.3 Status Reports
	5.4 Meetings
	5.5 Management Plans
	5.6 Change Orders
	5.7 Payments
6	Regulatory Requirements
7	Communication Plan Update

TABLE 4.55
List of Project Control Documents

Serial Number	Document Name
I	**Administrative**
	I-1 Material Entry Permit
	I-2 Material Removal Permit
	I-3 Vehicular Entry Permit
	I-4 Site Entry Permit
	I-5 Visitor Entry Permit
	I-6 Municipality Permit
	I-7 Request for Overtime
	I-8 Theft & Damage Report
	I-9 Performance Bonds
	I-10 Advance Payment Guarantee
	I-11 Insurance
	I-12 Accident Report
	I-13 Sample Tag

TABLE 4.55 (Continued)
List of Project Control Documents

Serial Number	Document Name
II	**Contracts Related**
II-1	Notice to Proceed
II-2	Job Site Instruction
II-3	Site Works Instruction
II-4	Attachment to Site Works
II-5	Request for Staff Approval
II-6	Request for Subcontractor Approval
II-7	Request for Vendor Approval
II-8	Material Delivered at Site
II-9	Variation Order
II-10	Attachment to Variation Order
II-11	Baseline Change Request Form
II-12	Extension of Time
II-13	Suspension of Work
II-14	Attendees
II-15	Minutes of Meeting
II-16	Transmittal for Minutes of Meeting
II-17	Submittal Form
III	**Engineering Submittal**
III-1	Master Schedule
III-2	Cost Loaded Schedule
III-3	Engineering Drawings
III-4	Material Approval
III-5	Specification Comparison Statement
III-6	Product Data
III-7	Product Sample
III-8	Work Shop Drawings
III-8	Builders Drawings
III-10	Composite Drawings
III-11	Method Statement
III-12	Request for Information
III-13	Request for Modification
III-14	Variation Order (Proposal)
III-15	Request for Alternative or Substitution
IV	**PCS Reporting Forms**
IV-1	Contractor's Submittal Status Log E-1
IV-2	Contractor's Procurement Log E-2
IV-3	Contractor's Shop Drawing Status Log
IV-4	Daily Progress Report
IV-5	Weekly Progress Report
IV-6	Look Ahead Schedule
IV-7	Monthly Progress Report

(*continued*)

TABLE 4.55 (Continued)
List of Project Control Documents

Serial Number	Document Name	
	IV-8	Progress Photographs
	IV-9	Daily Check List Status
	IV-10	Progress Payment Request
	IV-11	Payment Certificate
	IV-12	Submittal Schedule
	IV-13	Schedule Update Report
V	**Management Plans**	
	V-1	Quality Control Plan
	V-2	Safety Management Plan
	V-3	Environmental Protection Plan
VI	**Quality Control Forms**	
	VI-1	Check List (Request for Inspection)
	VI-2	Check List for Form Work
	VI-3	Notice for Daily Concrete Casting
	VI-4	Check List for Concrete Casting
	VI-5	Quality Control of Concreting
	VI-6	Report on Concrete Casting
	VI-7	Notice for Testing at Lab
	VI-8	Concrete Quality Control Form
	VI-9	Check List for Process Work
	VI-10	Check List for Piping Work
	VI-11	Check List for Instrumentation Work
	VI-12	Check List for Utility Work
	VI-13	Check List for Steel Fabrication Work
	VI-14	Check List for Detection and Protection Works
	VI-15	Check List for Fire Fighting Work
	VI-16	Check List for Architectural Work
	VI-17	Check List for Civil Work
	VI-18	Check List for Mechanical Work
	VI-19	Check List for HVAC Work
	VI-20	Check List for Electrical Work
	VI-21	Check List for Elevator Work
	VI-22	Check List for Low Voltage System Work
	VI-23	Check List for External Work
	VI-24	Check List for Landscape
	VI-25	Remedial Note
	VI-26	Non Conformance/Compliance Report
	VI-27	Material Inspection Report
	VI-28	Safety Violation Notice
	VI-29	Notice of Commencement of New Activity
	VI-30	Removal of Rejected Material
	VI-31	Testing and Commissioning

TABLE 4.55 (Continued)
List of Project Control Documents

Serial Number	Document Name
VII	**Closeout Forms**
	VII-1 As-Built Drawings
	VII-2 Substantial Completion Certificate
	VII-3 Handing Over Certificate
	VII-4 Taking Over Certificate
	VII-5 Manuals
	VII-6 Handing Over of Spare Parts
	VII-7 Defect Liability Certificate

Source: Modified from: Abdul Razzak Rumane. (2013). *Quality Tools for Managing Construction Projects*. Reprinted with permission from Taylor & Francis Group.

associated with the project will be identified, analyzed, managed, and controlled. Risk management is an integral part of project management as the risk is likely to occur at any stage of the project. Therefore, the risk has to be continually monitored and response actions to be taken immediately. The risk management plan outlines how risk activities will be performed, recorded, and monitored throughout the life cycle of the project. It is intended to maximize the positive impact for the benefit of the project and decrease/minimize or eliminate the impact of events adverse to the project. The risk management must commence early in project development stage (study stage) and proceed as the project evolves and more and more information about the project is available. The project plan should

- Define risk management strategy/approach
- Define project objectives, goals related to risk management
- Identify risk owner and team members
- Define risk decisions
- Detail about risk resources
- Include risk management process
  - Methods of risk identification
  - Methods of risk assessment
  - Level of risk
  - Response to risk
  - Management of risk
  - Control of risk
- Process of integrating risk management activities into project scope, schedule, cost, and quality
- Documenting and recording of risks
- Communication procedure for risk reporting
- Update of risk management plan

Project Name		
Consultant Name		
SUBMITTAL TRANSMITTAL FORM		

Contractor Name :		
Contract No. :		
To.	**Resident Engineer**	
Transmittal No.:		**Date :**

Submittal Type:		**Action Requested:**	
ED	Engineering Drawings	1	For Approval
DG	Shop Drawings	2	For Review and Comment
SK	Sketches	3	For Information
PR	Material/Product/System	4	For Construction
MD	Manufacturer's Data	5	For Incorporation Within the Design
SM	Sample	6	For Costing
MM	Minutes of Meeting	7	For Tendering
RP	Reports		
LG	Logs		**SAMPLE FORM**
OT	Others (please specify)		

We are sending herewith the following:

ENCLOSURES

Item	Qty	Ref. No.	Description	Type	Action

Comments:

_____

_____

_____

Issued by:	Received by:
Signature:	Signature:
Date:	Date:

FIGURE 4.38   Transmittal form.

4.3.6.8.5   Develop Contract Management Plan

Contract management in construction projects is an organizational method, process, and procedure to obtain the required construction project/products. It includes the process to acquire construction project complete with all the related material, equipment, system, and services from outside contractors/companies to the satisfaction of the owner/client/end user.

Contract management in construction project involves

Project Name
Consultant Name

JOB SITE INSTRUCTION (JSI)

CONTRACTOR: _____ JSI No. : _____

CONTRACT No.: _____ DATE : _____

The work shall be carried out in accordance with the Contract Documents without change in Contract Sum or Contract Time. Proceeding with the work in accordance with these instructions indicates your acknowledgement that there will be no change in the Contract Sum or Contract Time.

Subject:

_____

SAMPLE FORM

_____
_____
_____
_____

ATTACHMENTS: (List attached documents that support description.)

_____
_____
_____
_____
_____

Signed:	Received by Contractor :
Resident Engineer	Date:

Distribution: ☐ Owner   ☐ Consultant (Supervision)/PMC   ☐ Contractor

FIGURE 4.39 Job site instruction.

- Identification of
  - What are the services in-house available
  - What services to be procured from outside agencies/organizations
  - How to procure (direct contract, competitive bidding)
  - How much to procure
  - How to select a supplier/contractor
  - How to arrive at appropriate price, terms, and conditions

- Signing of contract
- Timely delivery
- Receiving right type of material/system
- Timely execution of work
- Inspection of work to maintain quality of the project
- Completion of project within agreed upon schedule
- Completion of project within agreed upon budget
- Documenting reports and plans

The administration of a contract is the process of formal governance of contract and changes to the contract document. It is concerned with managing contractual relationships between various participants to successfully complete the facility to meet owner's objectives. It includes tasks such as

- Administration of project requirement
- Administration of project team members
- Communication and management reporting
- Execution of contract
- Monitoring contract performance (scope, cost, schedule, quality, risk)
- Inspection and quality
- Variation order process
- Making changes to the contract documents by taking corrective action as needed
- Payment procedures

It is required that the contract administration procedure is clearly defined for the success of the contract and that the parties to the contract understand who does what, when, and how. Following are some typical procedures that should be in place for management of the contract management activities:

1. Contract document maintenance and variation
2. Performance review system
3. Resource management and planning
4. Management reporting
5. Change control procedure
6. Variation order procedure
7. Payment procedure

Table 4.56 lists the contents of the contract management plan.

4.3.6.8.6 Develop HSE Management Plan

The construction industry has been considered to be dangerous for a long time. The nature of work at site always presents some dangers and hazards. There are a relatively high number of injuries and accidents at construction sites. Safety represents an important aspect of construction projects. In construction projects, the requirements

TABLE 4.56
Contents of Contract Management Plan

Serial Number	Topics
1	Contract summary, deliverables and scope of work
2	Type of contract
3	Contract schedule
4	Contract cost
5	Project team members with roles and responsibilities
6	Core staff approval procedure
7	Contract communication matrix/Management reporting
8	Coordination process
9	Liaison with regulatory authorities
10	Engineering drawings submission/approval process
11	Vendor selection process
12	Material/Product/System review/approval process
13	Shop drawing review/approval process
14	Project monitoring and control process
15	Contract change control process
	a) Scope
	b) Material
	c) Method
	d) Schedule
	e) Cost
16	Review of variation/change requests
17	Project holdup areas
18	Quality of performance
19	Inspection and acceptance criteria
20	Risk identification and management
21	Progress payment process
22	Safety management
23	Claims, disputes, conflict, and litigation resolution
24	Contract documents and records
25	Post contract liabilities
26	Contract closeout and independent audit.

Source: Abdul Razzak Rumane. (2013). *Quality Tools for Managing Construction Projects.* Reprinted with permission from Taylor & Francis Group.

to prepare Safety Management Plan (SMP) by the contractor are specified under contract documents. The contractor has to submit the plan for review and approval by supervisor/consultant during mobilization stage of construction phase. Following are the guidelines normally specified in the contract documents which are to be considered by the contractor to establish HSE management plan:

1. Project scope detailing description of project and safety requirements
2. Safety policy statement documenting the contractor's/subcontractor's commitment and emphasis on safety
3. Regulatory requirements about safety
4. Roles and responsibilities of all individuals involved
5. Safety management of different activities
6. Site communication plan detailing how safety information will be shared
7. Hazard identification, risk assessment, and control
8. Emergency evacuation plan
9. Accident reporting system
10. Accident investigation to document root causes and determine corrective and preventive actions
11. Measures for emergency situations
12. Plant, Equipment maintenance and licensing
13. Routine inspections
14. Continuous monitoring and regular assessment
15. Health surveillance
16. System feedback and continuous improvements
17. Safety assurance measures
18. Evaluation of subcontractor's safety capabilities
19. Site neighborhood characteristics and constraints
20. System education and training
21. Safety audit
22. Documentation
23. Records
24. Procedure for project HSE review
25. System update

Following are the main responsibilities of safety engineer/officer:

1. Conducting safety meetings
2. Monitoring on-the-job safety
3. Inspecting the work and identifying hazardous areas
4. Initiating a safety awareness program
5. Ensuring availability of first aid and emergency medical services per local codes and regulations
6. Ensuring that personnel are using protective equipment such as hard hat, safety shoes, protective clothing, life belt, and protective eye coverings
7. Ensuring that the temporary firefighting system is working
8. Ensuring that work areas are free from trash and hazardous material
9. Housekeeping

4.3.6.9 Construction/Execution of Works

The contractor is given few weeks to start the construction works after signing of the contract. A letter from the client/owner is issued to the contractor to begin the project work subject to the conditions of the contract. This letter is known as "Notice to Proceed" letter. (Please refer Figure 4.29) Thereafter the client/owner hand over the project site to start mobilization and the project works.

For smooth implementation of project, proper communication system is established clearly identifying the submission process for correspondence and transmittals. Correspondence between consultant and contractor is normally done through job site instructions; whereas, correspondence between owner, consultant, and contractor is normally done through letters.

The contractor is responsible for executing the contracted works in accordance with the approved material, shop drawings, and specifications as specified in the contract documents.

Prior to starting execution of construction/installation of works, the contractor has to submit material, systems, and shop drawings to the construction supervisor (consultant) for their review and approval.

The detailed procedure for submitting shop drawings, materials, and samples is specified under section "SUBMITTAL" of contract specifications. The contractor has to submit the same to owner/consultant for their review and approval. The consultant reviews the submittal and return the transmittal to the contractor with appropriate action.

4.3.6.9.1   Submittals

During construction phase, there are many documents sent forth and back between owner, construction supervisor (A/E, consultant), and contractor. The originator of these documents is mentioned in the Communication Matrix. Following are the types of documents exchanged among owner, supervisor, and contractor:

- Administrative
- Contract related
- Engineering submittals
- Project monitoring and control
- Quality

Table 4.56 discussed earlier illustrates different types of documents normally used during construction phase to communicate among different stakeholders.

The detailed procedure for submitting materials/products/systems, samples, and shop drawings is specified under section "SUBMITTAL" of contract specifications. The contractor has to submit the same to owner/consultant for their review and approval. The Contract Documents has a special section describing Submittal Procedures (CSI-Format General Requirements) that specifies administrative and procedural requirements for submission of submittals and other documents. Contractor has to comply with the contractual requirements for submittal requirements.

The submittal process, in construction projects, is essential to ensure that contractor's understanding of product specifications, contract drawings, and installation method matches with the designer's intent of product usage and installation method. The submittal process provides the owner the assurance that the contractor is complying with the design concept, and the installed material will function as required by the contract documents. Submittals are documents that are presented by contractor for approval, review, decision, or consideration.

Generally, these submittals fall into three categories. These are:

1. Approval Submittals
2. Review Submittals
3. Information Submittals.

Prior to start of execution/installation of work, contractor has to submit specified material/product/system, shop drawings to the A/E (consultant), construction/project manager as per project specification requirements, for approval, review, or information. The contractor, while preparing the submittal for shop drawing, has to consider the following:

1. Review contract specification
2. Review contract drawings
3. Determine and verify field/site measurements
4. Installation information about the material to be used
5. Installation details relating to the axis or grid of the project
6. Dimensions of the product, equipment to be installed
7. Roughing in requirements
8. Coordination with other trade (disciplines) requirements
9. Clearly marking the changes, deviations to the contract drawings

The consultant reviews the submittal to verify that the proposed product/sample/shop drawing comply with the contract specifications and return the transmittal to the contractor mentioning one of the following actions on the transmittal:

A—Approved
B—Approved as Noted
C—Revise and Resubmit OR Not Approved
D—For Information OR More Information Required

In case of deviation to that of specified items, the contractor has to submit a schedule of such deviation(s) listing all the points that do not conform to the specifications.

Figure 4.40 illustrates Site Transmittal Form for material approval, Figure 4.41 illustrates Specification Comparison Form, and Figure 4.42 illustrates Site Transmittal Form for shop drawing approval.

Contract documents specify number of original (paper print) and copies to be transmitted to A/E (consultant) for review and approval. Figure 4.43a illustrates example submittal process (paper based) and Figure 4.43b illustrates example submittal process (electronic).

Project Name
Consultant Name
SITE TRANSMITTAL
Request for Material Approval

CONTRACT No. :   TRANSMITTAL NO. _____ REV. _____

CONTRACTOR   :

TO   :

WE REQUEST APPROVAL OF THE FOLLOWING MATERIALS/GOODS/PRODUCTS/EQUIPMENT

ITEM NO.	DWG., SPEC. OR BOQ. REF	DESCRIPTION	SUBMITTAL CODE *	ACTION CODE **
		SAMPLE FORM		

DETAILS OF INFORMATION,LITERATURE, CATALOG CUTS, AND THE LIKE ATTACHED ARE:

SAMPLES:
Enclosed   ☐   Submitted under separate cover   ☐   Not applicable   ☐

N.B: We certify that above items have been reviewed in detail & and are correct & in strict performance with the Contract
Drawings & Specification except as otherwise stated.

CONTRACTOR'S REP. : _____ DATE : _____

RECEIVED BY CONSULTANT : _____ DATE : _____

cc: Owner Rep.

Resident Engineer to enter ACTION CODE and REMARKS

R.E.'s REMARKS :

_____
_____
_____
_____ Initials _____ Date _____

Corrections or comments made relative to submittals during this review do not relieve Contractor from compliance with the requirements
of the Drawings and Specifications. This check is only for review of general conformance with the design concept of the project and
general compliance with the information given in the Contract Documents. Contractor is responsible for confirming and correlaing all
quantities and dimensions, selecting fabrication process and techniques of construction; coordinating his work with that of other trades,
and performing his work in a safe and satisfactory manner.

Resident Engineer : _____ DATE : _____

Received by Contractor : _____ DATE : _____

cc: Owner Rep.

* SUBMITTAL CODE:	** ACTION CODE:		
1: Submitted for Approval	A: Approved	C:	Not Approved

FIGURE 4.40   Site transmittal form for material.

Project Name
Consultant Name

SPECIFICATION COMPARISON STATEMENT (SCS)

Contractor: _____   Date : _____

Contract No. _____   A/S No.: _____

| Submittal No. : _____ | Revision: _____ | Transmittal Ref.: _____ |
| Submittal Title: _____ | | Specification Ref : _____ |

S No.	SPECIFICATION REQUIREMENTS	CONTRACTOR'S PROPOSAL	REMARKS
	SAMPLE FORM		

FIGURE 4.41   Specification comparison statement.

4.3.6.9.1.1 Material Submittal

The contractor has to submit the following, as minimum, to the owner/consultant to get their review and approval of materials, products, equipment, and systems. Contractor cannot use these items unless they are approved for use in the project.

Product Data
The contractor has to submit following details:

- Manufacturers technical specifications related to the proposed product
- Installation methods recommend by the manufacturer
- Relevant sheets of manufacturer's catalog(s)
- Confirmation of compliance with recognized international quality standards
- Mill reports (if applicable)
- Performance characteristic and curves (if applicable)
- Manufacturer's standard schematic drawings and diagrams to supplement standard information related to project requirements and configuration of the same to indicate product application for the specified works (if applicable)
- Compatibility certificate (if applicable)
- Single-source liability (this is normally required for systems approval where different manufacturer's items are used)

Compliance Statement
The contractor has to submit a specification comparison statement along with the material transmittal. The compliance statement form is normally included as part

Project Name
Consultant Name
SITE TRANSMITTAL
Request for Shop Drawings Approval

CONTRACT No.:   TRANSMITTAL NO. _____ REV. _____

CONTRACTOR   :

TO   :

WE REQUEST APPROVAL OF THE FOLLOWING ENCLOSED DRAWINGS

ITEM NO.	DWG., SPEC. OR BOQ. REF	DRAWING TITLE	DWG. NOS.	Rev.	SUBMITTAL CODE *	ACTION CODE **
		SAMPLE FORM				

N.B: We certify that above items have been reviewed in detail & and are correct & in strict performance with the Contract Drawings & Specification except as otherwise stated.

CONTRACTOR'S REP. :   _____   DATE : _____

RECEIVED BY CONSULTANT :   _____   DATE : _____

cc: Owner Rep.

Resident Engineer to enter ACTION CODE and REMARKS

R.E.'s REMARKS :

_____
_____
_____
_____
_____
_____
_____ Initials _____ Date _____

Corrections or comments made relative to submittals during this review do not relieve Contractor from compliance with the requirements of the Drawings and Specifications. This check is only for review of general conformance with the design concept of the project and general compliance with the information given in the Contract Documents. Contractor is responsible for confirming and correlating all quantities and dimensions, selecting fabrication process and techniques of construction; coordinating his work with that of other trades, and performing his work in a safe and satisfactory manner.

Resident Engineer :   _____   DATE : _____

Received by Contractor :   _____   DATE : _____

cc: Owner Rep.

cc: Project Manager	** ACTION CODE:		
1: Submitted for Approval	A: Approved	C:	Not Approved
2: Submitted for your Information	B: Approved as Noted	D:	For Information
3:			

FIGURE 4.42   Site transmittal for workshop drawings.

of contract documents. The information provided in the compliance statement helps consultant to review and verify product compliance to the contracted specifications.

In case of any deviation to that of specified item, the contractor has to submit a schedule of such deviation(s) listing all the points not conforming to the specification.

In certain projects owner is involved in approval of materials.

Samples
The contractor has to submit (if required) the sample(s) from the approved material to be used for the work. The samples are mainly required to

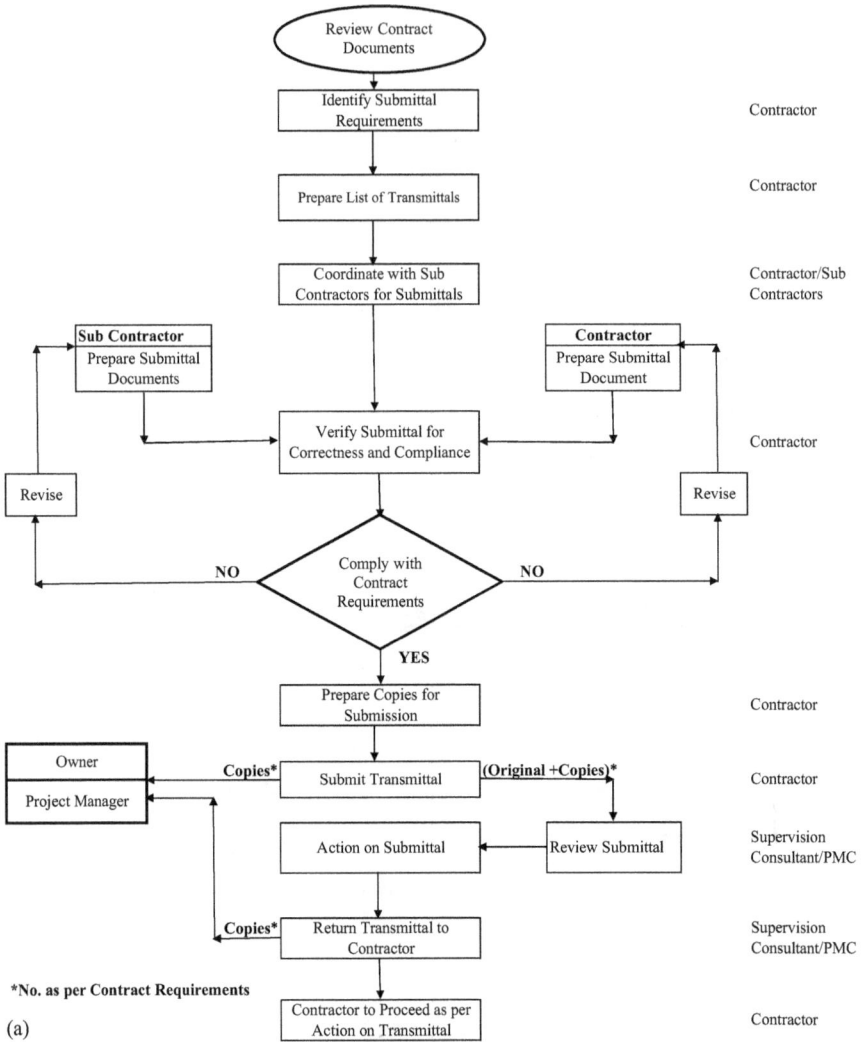

FIGURE 4.43   (a) Submittal process (paper based). (Continued)

- Verify color, texture, and pattern
- Verify that the product is physically identical to the proposed and approved material
- To be used for comparison with products and materials used in the works

At times it may be specified to install the sample(s) in a manner to facilitate review of qualities indicated in the specifications.

Figure 4.44 illustrates procedure for selection of material/product whereas Figure 4.45 illustrates material approval procedure.

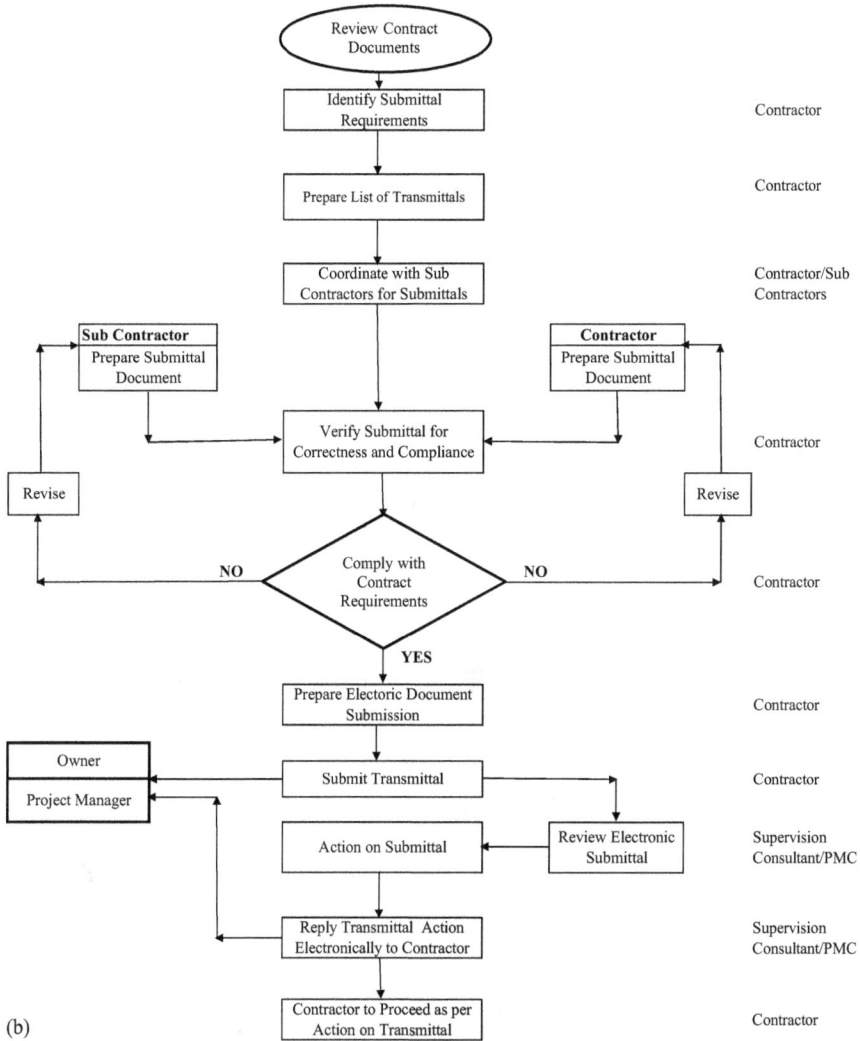

FIGURE 4.43 (Continued) (b) Submittal process (electronic).

4.3.6.9.1.2 Shop Drawings

The contractor is required to prepare shop drawings taking into consideration following as a minimum but not limited to

1. Reference to contract drawings. This help A/E (consultant) to compare and review the shop drawing to that of contract drawing
2. Detail plans and information based on the contract drawings
3. Notes of changes or alterations from the contract documents

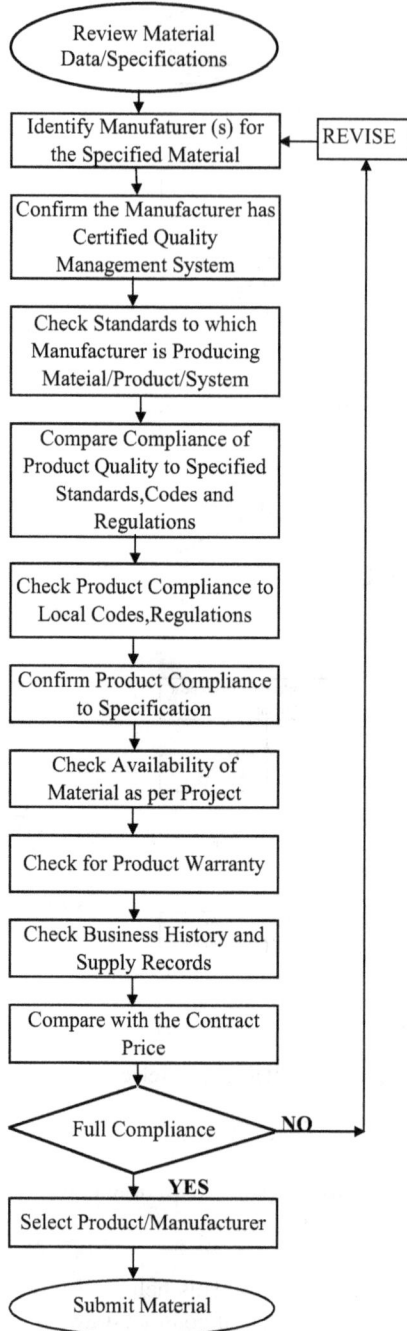

FIGURE 4.44 Material/product selection procedure. (Abdul Razzak Rumane. (2013). *Quality Tools for Managing Construction Projects.* Reprinted with permission from Taylor & Francis Group.)

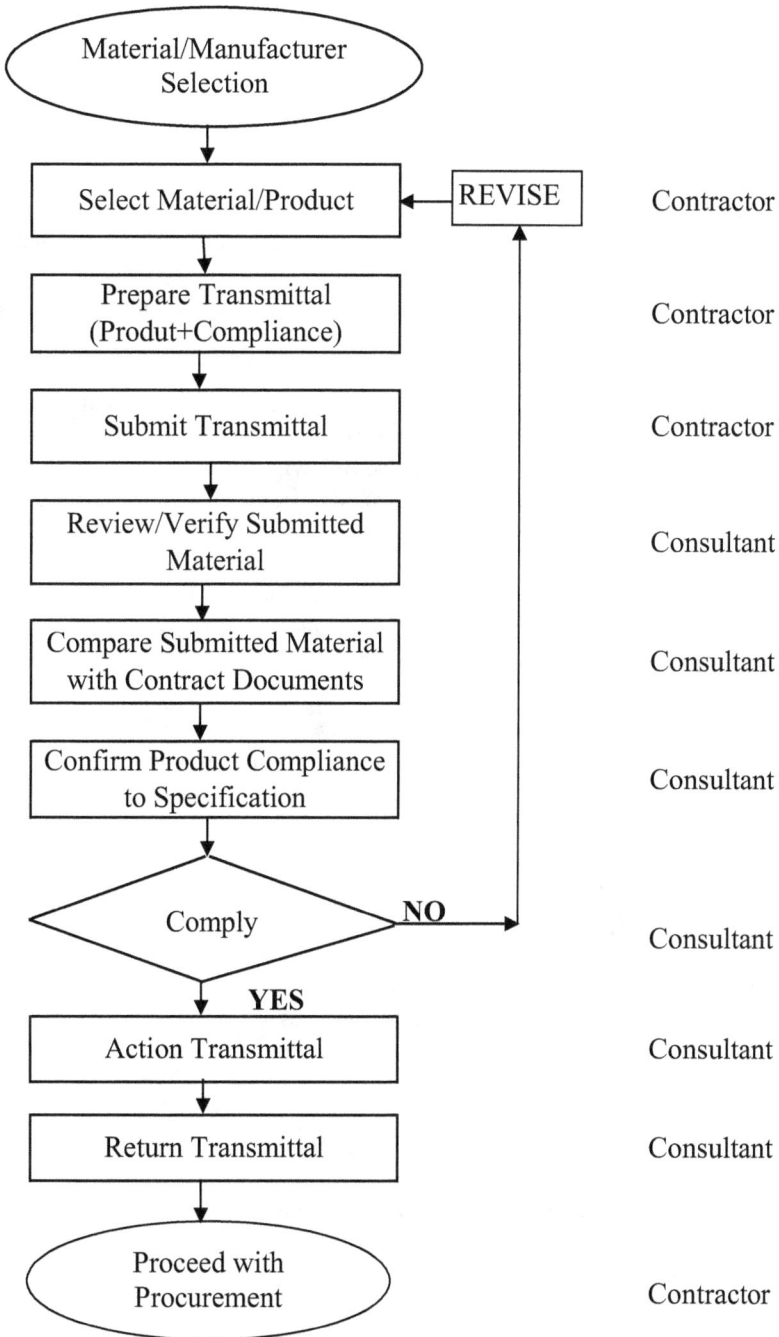

FIGURE 4.45 Material/product/system approval procedure. (Abdul Razzak Rumane. (2013). *Quality Tools for Managing Construction Projects.* Reprinted with permission from Taylor & Francis Group.)

4. Detail information about fabrication or installation of works
5. All dimensions needed to verify at the jobsite
6. Identification of product
7. Installation information about the materials to be used
8. Type of finishes, color, and textures
9. Installation details relating to the axis or grid of the project
10. Roughing in and setting diagram
11. Coordination certification from all other related trades (subcontractors)

The shop drawings are to be drawn accurately to the scale and shall have project-specific information in it. The shop drawings shall not be reproductions of contract drawings.

Based on contract drawings, contractor prepares shop drawings and submits the same to consultant for approval. All the works are executed as per approved shop drawings. The contractor has to consider the following point as the minimum, while developing shop drawings of different trades, to meet the design intents.

4.3.6.9.1.2.1 ARCHITECTURAL WORKS   The architectural works shop drawings mainly cover Masonry, Doors and Windows, Cladding, Partitioning, Reflected ceiling, Stone flooring, Toilet details, Stairs details, and Roofing. A brief requirement to prepare shop drawings for these sections is as follows:

A) Masonry
The shop drawings for masonry works shall include at the minimum

- Area layout
- Guidelines
- Height of masonry
- Type and thickness of blocks used for masonry
- Reinforcement details
- Fixation details
- Openings in the block works for other services
- Sills
- Lintels
- Plastering details, if applicable

B) Doors and Windows
The shop drawings for doors and windows shall include

- Size of doors and windows
- Type of material
- Thickness of frames
- Details of door leaves
- Details of glazing
- Fixing details

- Schedule of Doors and Windows

C) Cladding
The shop drawings for cladding shall include

- Type of cladding material
- Size of panels/tiles and thickness
- Elevations
- Fixation method, anchorage, and supports
- Openings for other services

D) Partitioning
The shop drawings for partitions shall include

- Type of partitioning material
- Size of panels
- Frame size and its installation details
- Partition support system
- Fixation details of panels

E) Reflected Ceiling
The shop drawings for reflected ceiling shall include

- Type of ceiling material
- Size of ceiling panels, thickness of material
- Ceiling level
- Suspension system and framing
- Layout showing jointing layout, faceting, and boundaries between materials
- Location of light fittings, detectors, sprinklers, grills, etc.
- Location of access panel

F) Stone Flooring
The shop drawings for flooring shall include

- Type of flooring material
- Size of flooring tiles
- Layout
- Screeding details
- Jointing details
- Flooring pattern
- Cut out for other services
- Control joints and jointing method
- Anti slip inserts
- Separators
- Thresholds

G) Toilet Details
The shop drawings for toilet details shall include

- Layout
- Plan
- Sections
- Installation details of sanitary wares and fixtures
- Installation details of toilet accessories
- Jointing method
- Material finishes and surface textures
- Installation details of toilet mirror

H) Stairs Details
The shop drawings for stairs details shall include

- Steps details, riser, treads, width, height
- Finishing details
- Handrail details
- Sections
- Plan

I) Roofing-Water Proofing
The shop drawings for roofing details shall include

- Insulation details
- Installation details of insulation
- Installation of water proofing material
- Control joints/expansion joints details
- Tiling details
- Flashings

4.3.6.9.1.2.2 STRUCTURAL WORKS The structural works shop drawings to be prepared according to ACI 315 and shall mainly cover Reinforced Concrete, Formwork, Precast Concrete, and Structural Steel Fabrication. A brief requirement to prepare these shop drawings is as given below:

A) Reinforced Concrete
The reinforced concrete works shop drawings to be prepared according to ACI 315 and shall have following details:

- The size of reinforcement material
- Bar schedule
- Stirrup spacing
- Bar bent diagram

- Arrangement and support of concrete reinforcement
- Dimensional details
- Special reinforcement required for openings through concrete structure
- Type of cement and its strength

B) Formwork

Formwork has great importance from the safety point of view. The shop drawing of formwork shall include

- Details of individual panels
- Position, size, and spacing of adjustable props
- Position, size, and spacing of joints, soldiers, ties, and the like
- Details of formwork for columns, beams, parapet, slabs, and kickers
- Details of construction joints and expansion joints
- Details of retaining walls, core walls, and deep beams showing the position and size of ties, joints, soldiers, and sheeting, together with detailed information on erection and casting sequences and construction joints
- General assembly details including propping, prop bearings, and thorough propping
- All penetrations through concrete
- Full design calculations

C) Precast Concrete

The shop drawings for precast concrete works have great importance as the casting is carried out at precast concrete factory. All the required details need to be shown on these drawings to ensure precast elements are produced without any defects. The shop drawing should provide details of fabrication and installation of precast structural concrete units. It shall indicate member location, plans, elevations, dimensions, shapes, cross sections, openings, and type of reinforcement, including special reinforcement, if any. The following information also has to be indicated in the shop drawings:

- Welded connections by American Welding Society's standards symbol.
- Details of loose and cast-in hardware, insets, connections, joints, and all types of accessories.
- Location and details of anchorage device to be embedded in other construction.
- Comprehensive engineering analysis including fire-resistance analysis.

D) Structural Steel Fabrication

The shop drawings for fabrication of structural steel shall include

- Details of cuts, connections, splices, camber, holes, and other pertinent data
- Welding standards symbols
- Size and length of bolts, indicating details of material and strength of bolts

- Details of steel bars, plates, angles, beams, channels
- Method of erection
- Details of anchorage details, templates, and installation details of bolts

4.3.6.9.1.2.3 ELEVATOR WORKS

- Overall elevation (vertical) for elevator area
- Floor levels
- Hoist way plan, sections, and anchoring details for installation of rails
- Machine room plan and installation details for drive and controller
- Equipment layout in machine room/hoist way
- Cuttings/openings, sleeves required in the concrete slab/walls
- Details and level of cab overhead
- Details and levels about the pit
- Location and level of hall button and hall position indicator
- Openings for landing door, entrance view, and finishes level
- Finishes of cab, door, indicators
- Power supply devices and equipment grounding
- Interface with elevator management system, if any

4.3.6.9.1.2.4 MECHANICAL WORKS (FIRE SUPPRESSION, WATER SUPPLY, AND PLUMBING)

The shop drawings for mechanical works shall include but are not limited to

- Sprinkler layout
- Hose reel details
- Hose reel cabinet size, location, and installation details
- Fire pump location and installation details
- Location of flow switches
- Riser diagram for water supply system
- Size of piping
- Piping route
- Piping levels and slope
- Pipes and sleeves for utility services
- Size of isolating valves and their location
- Equipment plan layout
- Pump room details
- Storage tank details
- Details of toilet accessories and connection details
- Riser diagram for rain water system
- Drainage system piping
- Location of vents and traps
- Location of cleanouts
- Riser diagram for storm water system
- Storm water system
- Electrical power connection details
- Interface with fire alarm and building management system

4.3.6.9.1.2.5   HVAC WORKS

The shop drawing for HVAC works shall include the following but are not limited to

- Location of equipment and their configuration
- Piping size
- Piping route and levels
- Ducting size
- Ducting route and levels
- Insulation details
- Suspension/hanger details
- Equipment layout and plan/plant room details
- Riser diagram for chilled water system
- Installation details of equipment
- Size of diffusers and grills
- Installation details of grills
- Exhaust and ventilation fans layout and details
- Riser diagram for exhaust air system
- Return air opening details
- Equipment schedule
- Electrical connection and power supply details
- Control details
- Sequence of operation
- Schematic diagram for HVAC system
- Schematic diagram for Building Management System by configuring all the equipment and components.

4.3.6.9.1.2.6   ELECTRICAL WORKS   The shop drawings for electrical works shall indicate but are not limited to

- Size and type of conduits, raceways, and exact routing of the same
- Size of cable trays, cable trunking, their installation methods and exact route indicating the level
- Size of wires, and cables
- Small power layout
- Large power layout
- Wiring accessories with circuit references
- Lighting layout with circuit references
- Installation details of light fixtures
- Emergency lighting system
- External lighting layout
- Feeder pillar location and installation details
- Bus duct installation details
- Field installation wiring details for light, power, controls, and signals
- Installation details of lighting control panel
- Installation details of distribution boards, switch boards, and panels

- Location of distribution boards
- Panels, Switch Boards layout in Low Tension (L. T.) Room
- Installation details of Bus Duct
- Schematic diagram showing the configuration of all the equipment
- Load schedules as per actual connected loads
- High voltage panel layout
- Substation layout
- Interface with solar power/alternate energy system
- Voltage drop calculations
- Short-circuit calculations
- Grounding (Earthing) system
- Lightning protection system
- Diesel Generator installation details
- ATS installation details
- Interface with other systems

4.3.6.9.1.2.7 HVAC ELECTRICAL WORKS The shop drawings for HVAC electrical works shall include

- Wiring diagram of individual components and accessories
- MCC and starter panels
- Installation details of MCC panels, starter panels
- Schematic diagram including power and control wiring
- Type and size of conduit and number and size of wires in the conduit
- Type and size of cable
- Details of protection and interlocks
- Details of instrumentation
- Description of sequence of operation of equipment

4.3.6.9.1.2.8 LOW VOLTAGE SYSTEMS The shop drawings for low voltage systems like Building Management System (BMS), Fire Alarm System, Communication System, Public Address System, Audio Visual System, CCTV/Security System, and Access Control System shall consist of

- Size of conduits and raceways, their routing and levels
- Location of components, wiring accessories
- Installation details of components, wiring accessories
- Wiring and cabling details
- Installation details of racks
- Installation details of equipment
- Location and installation details of panels
- Schematic/riser diagram configuring all the components and equipment used in the system
- Interface with other system(s)

4.3.6.9.1.2.9 LANDSCAPE WORKS   The shop drawings for landscape shall include

- Boundary limit
- Excavation area
- Excavation level
- Type of soil in different areas
- Location of plants, trees, shrubs, etc.
- Foundation details for plants, trees, shrubs, etc.
- Location of sidewalk
- Location of driveways
- Foundation for paving
- Grass areas
- Location of services, ducts for utilities
- Location of light poles/bollards
- Foundation for light poles/bollard
- Details of special features
- Details of irrigation system

4.3.6.9.1.2.10 EXTERNAL WORKS (INFRASTRUCTURE)   The shop drawing for external works (infrastructure) shall include

- Width of the road
- Grading details
- Thicknesses of asphalt layers
- Pavement details
- Location of manholes and levels
- Services/utilities pipe routes
- Location of light poles
- Cable routes
- Trench details
- Road marking
- Traffic sign and signals

Immediately, after approval of individual trade shop drawings, the contractor has to submit builder's workshop drawings, composite/coordinated shop drawings taking into consideration following as a minimum. Figure 4.46 illustrates shop drawing preparation and approval procedure.

4.3.6.9.1.3 Builders Workshop Drawings
Builders workshop drawings indicate the openings required in the civil or architectural work for services and other trades. These drawings indicate the size of openings, sleeves, level references with the help of detailed elevation, and plans. Figure 4.47 illustrates builders workshop drawing preparation and approval procedure.

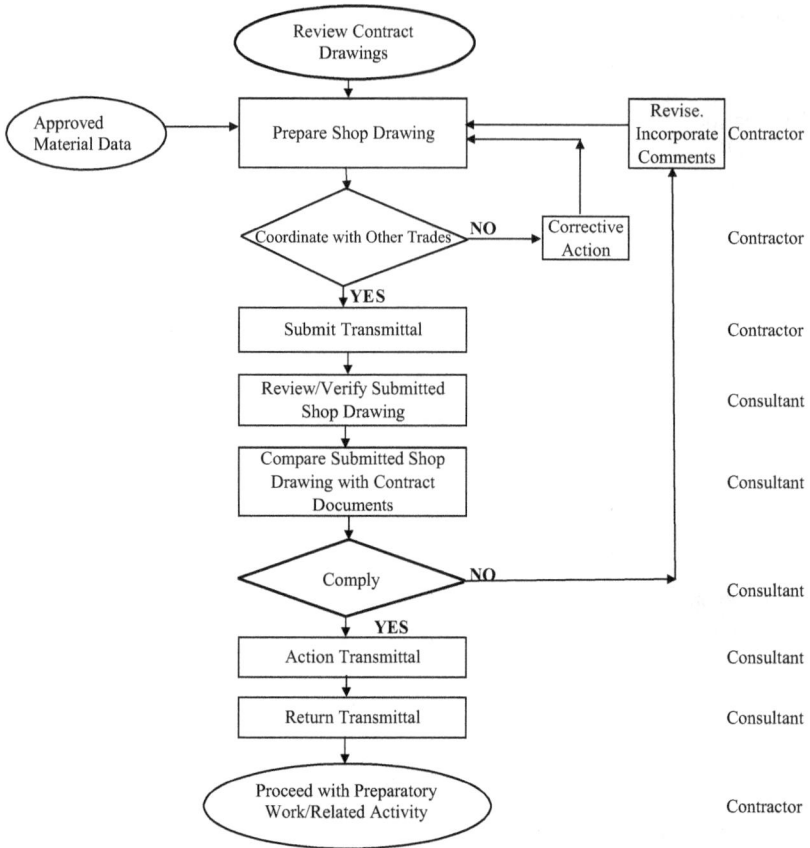

FIGURE 4.46 Shop drawing preparation and approval procedure. (Abdul Razzak Rumane. (2013). *Quality Tools for Managing Construction Projects.* Reprinted with permission from Taylor & Francis Group.)

4.3.6.9.1.4   Composite/Coordination Shop Drawings

The composite drawings indicate the relationship of components shown on the related shop drawings and indicate required installation sequence. Composite drawings shall show the interrelationship of all services with each other and with the surrounding civil and architectural work. Composite drawings shall also show the detailed coordinated cross sections, elevations, reflected plans, and so on resolving all conflicts in levels, alignment, access, space, and so forth. These drawings are to be prepared taking into consideration the actual physical dimensions required for installation within the available space. Figure 4.48 illustrates composite drawing preparation and approval procedure.

4.3.6.9.2   Execute Project Works

The contractor is responsible for executing the contracted works in accordance with contract drawings and specifications as specified in the contract documents. The

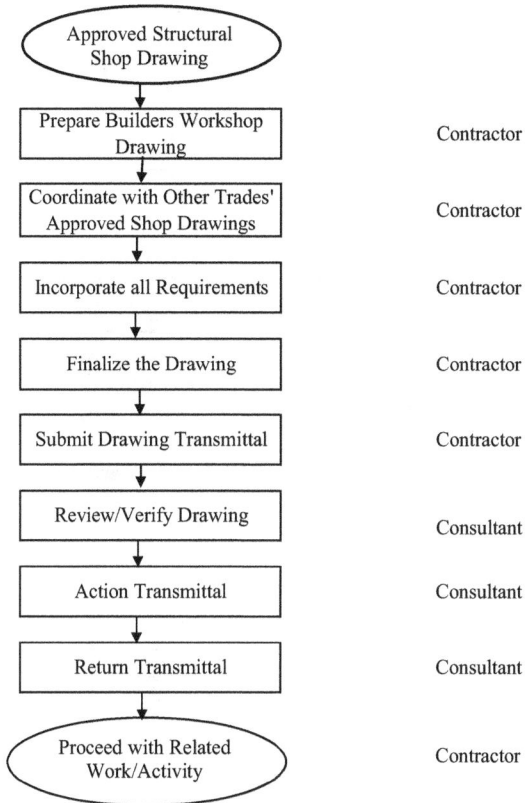

FIGURE 4.47 Builders workshop drawing preparation and approval procedure. (Abdul Razzak Rumane. (2013). *Quality Tools for Managing Construction Projects.* Reprinted with permission from Taylor & Francis Group.)

contractor has to arrange necessary resources to complete the project within the schedule and contracted amount. The contractor has to maintain the executed works until handing over the project to the owner/end user and maintain for additional period if contracted to do so. During construction period, the contractor has to protect executed/installed works to ensure that the works are not damaged. The contractor has to use new and approved material to construct the project/facility.

Construction activities mainly consist of following:

- Site work such as cleaning and excavation of project site
- Construction of foundations including footings and grade beams
- Construction of columns and beams
- Forming, reinforcing, and placing the floor slab
- Laying up masonry walls and partitions
- Installation of roofing system
- Finishes

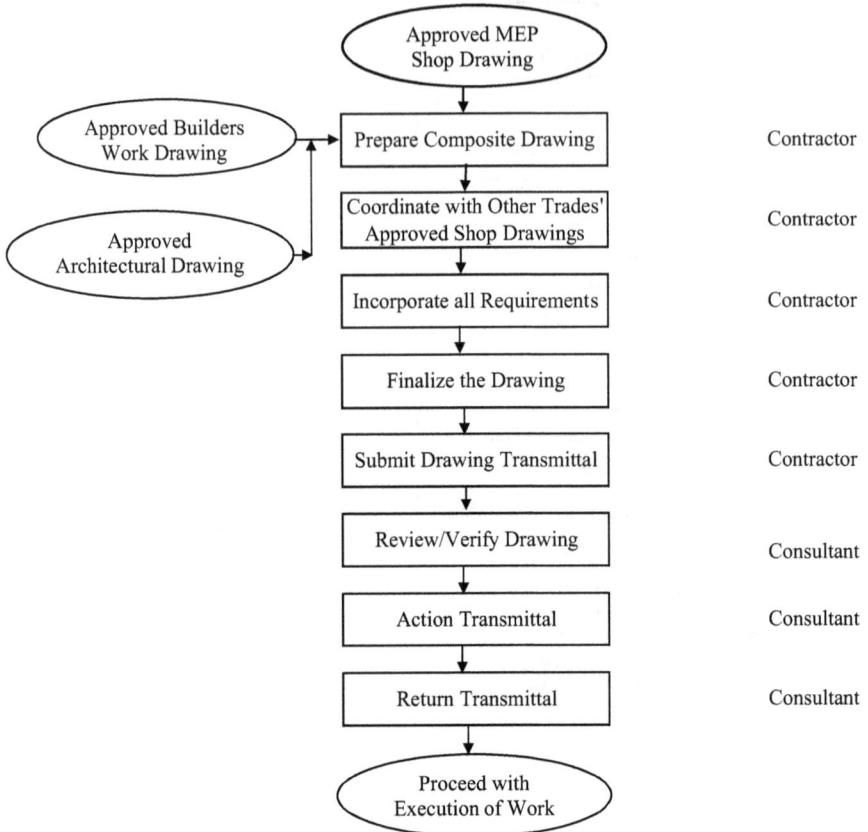

FIGURE 4.48 Composite drawing preparing and approval procedure. (Abdul Razzak Rumane. (2013). *Quality Tools for Managing Construction Projects.* Reprinted with permission from Taylor & Francis Group.)

- Furnishings
- Conveying system
- Installation of fire suppression system
- Installation of water supply, plumbing, and public health system
- Installation of heating, ventilating, and air conditioning system
- Integrated automation system
- Installation of electrical lighting and power system
- Emergency power supply system
- Fire alarm system
- Information and Communication Technology system
- Electronic security and access control system
- Landscape works
- External works

4.3.6.9.2.1   Manage Construction Quality

The construction project quality control process is a part of contract documents which provide details about specific quality practices, resources, and activities relevant to the project. The purpose of quality control during construction is to ensure that the work is accomplished in accordance with the requirements specified in the contract. Inspection of construction works is carried out throughout the construction period either by the construction supervision team (consultant) or appointed inspector agency. Quality is an important aspect of construction projects. The quality of the construction project must meet the requirements specified in the contact documents. Normally the contractor provides on-site inspection and testing facilities at construction site. On a construction site, inspection and testing is carried out at three stages during the construction period to ensure quality compliance.

1. During construction process. This is carried with the checklist request submitted by the contractor for testing of ongoing works before proceeding to the next step.
2. Receipt of subcontractor or purchased material or services. This is performed by a material inspection request submitted by the contractor to the consultant upon receipt of material.
3. Before final delivery or commissioning and handover.

Quality management in construction is a management function. In general, quality assurance and control programs are used to monitor design and construction conformance to established requirements as determined by the contract specifications. Instituting quality management programs reduces costs while producing the specified facility. CQCP discussed under Appendix A is followed throughout construction project. Table 4.57 illustrates contractor's responsibilities to manage construction quality.

4.3.6.9.2.1.1 COST OF QUALITY DURING EXECUTION OF WORK   There are certain hidden costs which may not affect directly the overall cost of the project; however, it may cost the contractor in terms of money and may affect the completion schedule of the project.

Rejection/non-approval of executed/installed works by the supervisor due to noncompliance with specification will cause the contractor loss in terms of

- Material
- Manpower
- Time

The contractor shall have to rework or rectify the work which will need additional resources and will need extra time to do the work as specified.

This may disturb contractor's work schedule and affect execution of other activities. The contractor has to emphasis upon "Zero Defect" policy, particularly for concrete works.

To avoid rejection of works, contractor has to take the following measures:

TABLE 4.57
Contractor's Responsibilities to Manage Construction Quality

AREAS OF QUALITY CONTROL

Sr. No.	ACTIVITY	MAIN CONTRACTOR		SUBCONTRACTORS					
		HEAD OFFICE/ QUALITY MANAGER	PROJECT SITE/ PROJECT MANAGER	STRUCTURAL	INTERIOR	MRCHANICAL (HVAC+PHFF)	ELECTRICAL	LANDSCAPE	EXTERNAL
1	Prepare Quality Control Plan	□	■	□	□	□	□	□	□
2	Construction Schedule	□	■	□	□	□	□	□	□
3	Mobilization	□	■	□	□	□	□	□	□
4	Staff Approval	□	■	□	□	□	□	□	□
5	Prepare Material Submittal		■	■	■	■	■	■	■
6	Submit Material Transmittal		■	□	□	□	□	□	□
7	Prepare Shop Drawings		■	■	■	■	□	■	■
8	Submit Shop Drawing Transmittal		■	□	□	□	□	□	□
9	Material Sample	□	■	■	■	■	■	■	■
10	Receiving Material Inspection		■	■	■	■	■	■	■

No.	Item							
11	Material Testing	■	■	■	■	■	■	■
12	Mock Up	■	■	■	■	■	■	■
13	Site Work Inspection	■	■	■	■	■	■	■
14	Quality of Work	■	■	■	■	■	■	■
15	Prepare Check List	■	■	■	■	■	■	■
16	Submit Check List	□	□	□	□	□	□	■
17	Corrective/ Preventive Action	■	■	■	■	■	■	■
18	Daily Report	□	□	□	□	□	□	■
19	Monthly Progress Report	□	□	□	□	□	□	■
20	Progress Payment	□	□	□	□	□	□	■
21	Site Safety	■	■	■	■	■	■	■
22	Safety Report	□	□	□	□	□	□	■
23	Waste Disposal	■	■	■	■	■	■	■
24	Reply to Job Site Instruction	□	□	□	□	□	□	■
25	Reply to Non-Conformance Report	□	□	□	□	□	□	■
26	Documentation	□	□	□	□	□	□	■
27	Testing and Commissioning	■	■	■	■	■	■	■

(continued)

TABLE 4.57 (Continued)
Contractor's Responsibilities to Manage Construction Quality

| | | MAIN CONTRACTOR | | AREAS OF QUALITY CONTROL SUBCONTRACTORS | | | | | |
		HEAD OFFICE/ QUALITY MANAGER	PROJECT SITE/ PROJECT MANAGER	STRUCTURAL	INTERIOR	MRCHANICAL (HVAC+PHFF)	ELECTRICAL	LANDSCAPE	EXTERNAL
Sr. No.	ACTIVITY								
28	Project Closeout Documents	□	■	□	□	□	□	□	□
29	Punch List		■	■	■	■	■	■	■
30	Request for Issuance of Substantial Completion Letter	□	■	□	□	□	□	□	□

■ **Primary Responsibility**
□ **Advise/Assist**

Source: Abdul Razzak Rumane. (2013). *Quality Tools for Managing Construction Projects*. Reprinted with permission from Taylor & Francis Group.

1. Execution of works as per approved shop drawings using approved material
2. Following approved method of statement or manufacturers recommended method of installation
3. Conduct continuous inspection during construction/installation process
4. Employ properly trained workforce
5. Maintain good workmanship
6. Identify and correct deficiencies before submitting the check list for inspection and approval of work
7. Coordinate requirements of other trades, for example if any opening is required in the concrete beam for crossing of services pipe

Timely completion of project is one of the objectives to be achieved. To avoid delay in completion schedule proper planning and scheduling of construction activities is necessary. Since construction projects have involvement of many participants, it is essential that requirements of all the participants are fully coordinated. This will ensure execution of activities as planned resulting in timely completion of project. Normally the construction budget is fixed at the inception of project, therefore it is required to avoid variations during construction process as it may take time to get approval of additional budget resulting time extension to the project.

Categories of costs related to construction phase can be summarized as follows:

1) Internal Failure Cost
   - Rework
   - Rectification
   - Rejection of checklist
   - Corrective action
2) External Failure Cost
   - Breakdown of System
   - Repairs
   - Maintenance
   - Warranty
3) Appraisal Cost
   - Design review/preparation of shop drawings
   - Preparation of composite/coordination drawings
   - On-site material inspection/test
   - Off-site material inspection/test
   - Mockup
   - Pre-check list inspection
   - Functionality tests
4) Prevention Cost
   - Preventive action
   - Training
   - Work procedures
   - Method statement
   - Calibration of instruments/equipment
   - Planning and managing quality system

PDCA is a well-known tool for continual process improvement. Contractor may use PDCA cycle (Deming Wheel) principle to improve the construction/installation process and avoid rejection of works by the supervision team (consultant).

A) PLAN
- Prepare Shop Drawings
- Submit Materials
- Get Shop Drawings Approved
- Get Materials Approved

B) DO
- Follow Approved Shop Drawings
- Use Approved Material
- Follow Method of Statement
- Follow Manufacturer's Recommendation

C) CHECK
- Check All Conformities against Specification
- Check Work Executed as per Approved Shop Drawings
- Check Approved Material Installed
- Check Functionality of Installed Works/Systems

D) ACT
- Corrective Action
- Prevent Defect from Re-Occurrence

Figure 4.49 illustrates PDCA cycle for execution/installation of works.

4.3.6.10 Monitoring and Controlling

Monitoring and controlling the project is an ongoing process. It starts from the inception of the project and continues till handover of the project. Monitoring and control of construction project is operative during the execution of the project, and its aim is to recognize any obstacles encountered during execution and to apply measures to mitigate these difficulties and to ensure that the goals and objectives on the project are being met. Table 4.58 lists major activities relating to monitoring and controlling processes during construction phase.

Monitoring is collecting, recording, and reporting information concerning any and all aspects of project performance that the project manager or others in the organization need to know. Monitoring of construction project is normally done by collecting and recording the status of various activities and compiling them in the form of progress reports. These are prepared by the consultant, contractor, and distributed to the

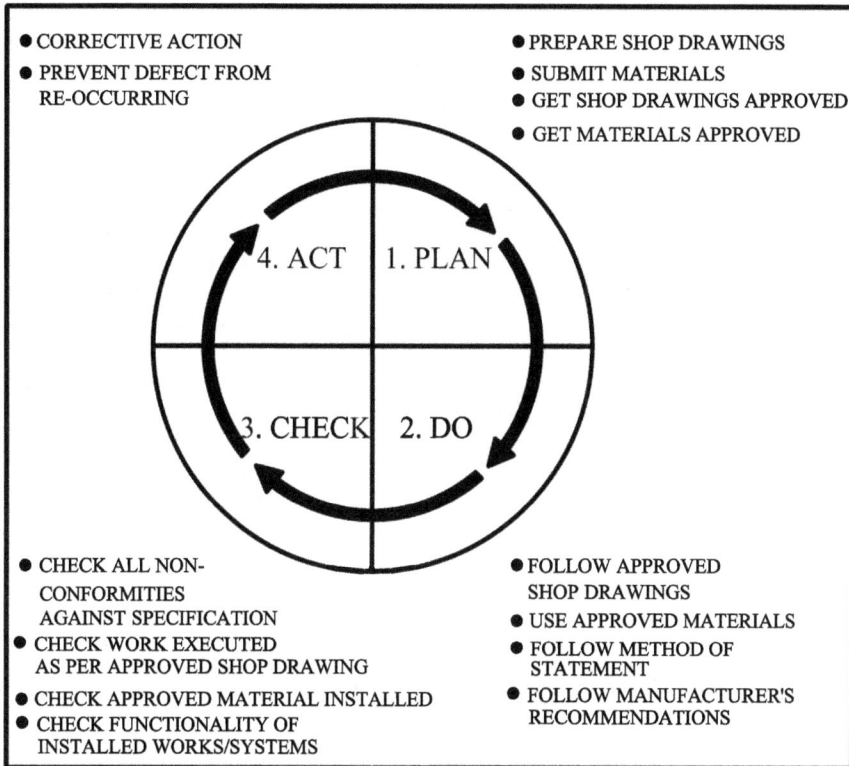

- CORRECTIVE ACTION
- PREVENT DEFECT FROM RE-OCCURRING

- PREPARE SHOP DRAWINGS
- SUBMIT MATERIALS
- GET SHOP DRAWINGS APPROVED
- GET MATERIALS APPROVED

4. ACT | 1. PLAN

3. CHECK | 2. DO

- CHECK ALL NON-CONFORMITIES AGAINST SPECIFICATION
- CHECK WORK EXECUTED AS PER APPROVED SHOP DRAWING
- CHECK APPROVED MATERIAL INSTALLED
- CHECK FUNCTIONALITY OF INSTALLED WORKS/SYSTEMS

- FOLLOW APPROVED SHOP DRAWINGS
- USE APPROVED MATERIALS
- FOLLOW METHOD OF STATEMENT
- FOLLOW MANUFACTURER'S RECOMMENDATIONS

FIGURE 4.49 PDCA cycle (Deming wheel) for execution of works. (Abdul Razzak Rumane (2013), *Quality Tools for Managing Construction Projects.* Reprinted with permission from Taylor & Francis Group.)

concerned members of the project team. Figure 4.50 illustrates logic flowchart for monitoring and controlling process.

Monitoring involves not only tracking time but also budget, quality, resources, and risk. Monitoring construction projects is normally done by compiling status of various activities in the form of progress reports. These are prepared by contractor, supervision team (consultant), construction/project management team. The objectives of project monitoring and control are

1. To report the necessary information in detail and in appropriate form which can be interpreted by management and other concerned personnel to provide the information about how the resources are being used to achieve project objectives

TABLE 4.58
Major Project Activities relating to Monitoring & Controlling Process during Construction Phase

Serial Number	Management Processes	Activities	Elements
1	Integration Management	1.1 Project Start Up	1.1.1 Notice to Process
			1.1.2 Compliance to Contract Documents
			1.1.3 Project Site
			1.1.4 Mobilization
			1.1.5 Forecasted Schedule
			1.1.6 Forecasted Cost
			1.1.7 Issues
		1.2 Change Management System	1.2.1 Design Changes
			1.2.2 Design Errors
			1.2.3 Change Requests
			1.2.4 Scope Change
			1.2.5 Variation Orders
			1.2.6 Site Work Instruction
			1.2.7 Alternate Material
			1.2.8 Specs/Methods
		1.3 Change Analysis	1.3.1 Review, Evaluate Changes
			1.3.2 Approve, Delay, Reject Changes
			1.3.3 Corrective Actions
			1.3.4 Preventive Actions
2	Stakeholder Management	2.1 Project Performance	2.1.1 Progress Reports
			2.1.2 Updates
			2.1.3 Safety Report
			2.1.4 Risk Report
		2.2 Project Updates	2.2.1 Contract Documents
		2.3 Payments	2.3.1 Payment Certificate
		2.4 Change Requests	2.4.1 Site Work Instruction
			2.4.2 Change Orders
		2.5 Issue Log	2.5.1 Anticipated Problems
		2.6 Minutes of Meetings	2.6.1 Progress Meetings
			2.6.2 Other Meetings
3	Scope Management (Contract Documents)	3.1 Validate Scope	3.1.1 Conformance to TOR
			3.1.2 Review of Design Documents
			3.1.3 Conformance to Contract Documents
			3.1.4 Approval of Changes
			3.1.5 Authorities Approval of Deliverables

TABLE 4.58 (Continued)

Major Project Activities relating to Monitoring & Controlling Process during Construction Phase

Serial Number	Management Processes	Activities	Elements
			3.1.6 Stakeholders Approval of Deliverables
			3.1.7 Quality Audit
		3.2 Scope Change Control	3.2.1 Variation Orders
			3.2.2 Change Orders
		3.3 Performance Measures	
4	Schedule Management	4.1 Schedule Monitoring	4.1.1 Project Status
		4.2 Schedule Control	4.2.1 Progress Curve
		4.3 Schedule Changes	4.3.1 Approved Changes
		4.4 Progress Monitoring	4.4.1 Planned VS Actual
		4.5 Submittals Monitoring	4.5.1 Subcontractors
			4.5.2 Material
			4.5.3 Shop Drawings
5	Cost Management	5.1 Cost Control	5.1.1 Work Performance
		5.2 Change Orders	5.1.2 S-Curve
		5.3 Progress Payment	5.1.3 Forecasted Cost
		5.4 Variation Orders	
6	Quality Management	6.1 Control Quality	6.1.1 Quality Metrics
			6.1.2 Quality Check List
			6.1.3 Material Inspection
			6.1.4 Work Inspection
			6.1.5 Rework
			6.1.6 Testing
			6.1.7 Regulatory Compliance
		6.2 Construction Performance	6.2.1 Project Deliverables
			6.2.2 Corrective Action
7	Resource Management	7.1 Conflict Resolution	
		7.2 Performance Analysis	
		7.3 Material Management	
8	Communication Management	8.1 Meetings	8.1.1 Progress Meetings
			8.1.2 Coordination Meetings
			8.1.3 Safety Meetings
			8.1.4 Quality Meetings
		8.2 Submittal Control	8.2.1 Drawings
			8.2.2 Material
		8.3 Documents Control	8.3.1 Correspondence
9	Risk Management	9.1 Monitor & Control Risk	9.1.1 Scope Change Risk

(*continued*)

TABLE 4.58 (Continued)

Major Project Activities relating to Monitoring & Controlling Process during Construction Phase

Serial Number	Management Processes	Activities	Elements
			9.1.2 Schedule Change Risk
			9.1.3 Cost Change Risk
			9.1.4 Mitigate Risk
			9.1.5 Risk Audit
10	Contract Management	10.1 Inspection	
		10.2 Check Lists	
		10.3 Handling of Claims, Disputes	
11	Health, Safety and Environment	11.1 Prevention Measures	11.1.1 Accidents Avoidance/ Mitigation
			11.1.2 Fire Fighting System
			11.1.3 Loss Prevention Measures
		11.2 Application of Codes and Standards	
12	Financial Management	12.1 Financial Control	12.1.1 Payments to Project Team Members
			12.1.2 Payments to Contractor(s)/ Subcontractor(s)
			12.1.3 Material Purchases
			12.1.4 Variation Order Payment
			12.1.5 Insurance and Bonds
		12.2 Cash Flow	
13	Claim Management	13.1 Claim Prevention	13.1.1 Proper Design Review
			13.1.2 Unambiguous Contract Documents Language
			13.1.3 Practical Schedule
			13.1.4 Qualified Contractor(s)
			13.1.5 Competent Project Team Members
			13.1.6 RFI Review Procedure
			13.1.7 Negotiations
			13.1.8 Appropriate Project Delivery System

Source: Abdul Razzak Rumane. (2016). *Handbook of Construction Management: Scope, Schedule, and Cost Control*, Second Edition. Reprinted with permission from Taylor & Francis Group.

```
            ┌────────────────────────┐
            │  Performance and       │
            │  Measurement Baseline  │
            └────────────────────────┘
                        │
                        ▼
        ┌──────────────────────────────┐
        │ Measure Performance (Actual)  │◄────────────────┐
        │ Against Planned Performance   │                 │
        │ Baseline and Metrics          │                 │
        └──────────────────────────────┘                 │
                        │                                  │
                        ▼                                  │
        ┌──────────────────────────────┐                 │
        │      Determine Variance        │                 │
        └──────────────────────────────┘                 │
                        │                                  │
                        ▼                                  │
        ┌──────────────────────────────┐                 │
        │ Identify if Variance Warrants │                 │
        │ any Change Request             │                 │
        └──────────────────────────────┘                 │
                        │                                  │
                        ▼                                  │
        ┌──────────────────────────────┐                 │
        │ Identify the Factors that      │                 │
        │ Caused the Variance            │                 │
        └──────────────────────────────┘                 │
                        │                                  │
                        ▼                                  │
        ┌──────────────────────────────┐                 │
        │ Analyse the Cause(s) for       │                 │
        │ Change                         │                 │
        └──────────────────────────────┘                 │
                        │                                  │
                        ▼                                  │
        ┌──────────────────────────────┐                 │
        │ Recommend Change(s)            │                 │
        │ {Request for Change(s)}        │                 │
        └──────────────────────────────┘                 │
                        │                                  │
                        ▼                                  │
        ┌──────────────────────────────┐                 │
        │ Recommend Preventive and       │                 │
        │ Corrective Action(s)           │                 │
        └──────────────────────────────┘                 │
                        │                                  │
                        ▼                                  │
        ┌──────────────────────────────┐                 │
        │ Perform Integrated Change      │                 │
        │ Control                        │                 │
        └──────────────────────────────┘                 │
                        │                                  │
                        ▼           NO   ┌──────────────┐ │
                    ◇ Change ◇──────────►│ More          │─┘
                        │                │ Information    │
                        │ YES            │ Required       │
                        ▼                └──────────────┘
        ┌──────────────────────────────┐
        │ Perform Quality Control        │
        └──────────────────────────────┘
                        │
                        ▼                  ┌──────────────┐
        ┌──────────────────────────────┐  │ Update Risk   │
        │ Perform Risk Audits            │─►│ Register      │──►
        └──────────────────────────────┘  └──────────────┘
                        │
                        ▼
        ┌──────────────────────────────┐
        │ Manage Reserves                │
        └──────────────────────────────┘
                        │
                        ▼
        ┌──────────────────────────────┐
        │ Update Contract                │
        │ Documents Incorporating        │
        │ Approved Changes               │
        └──────────────────────────────┘
```

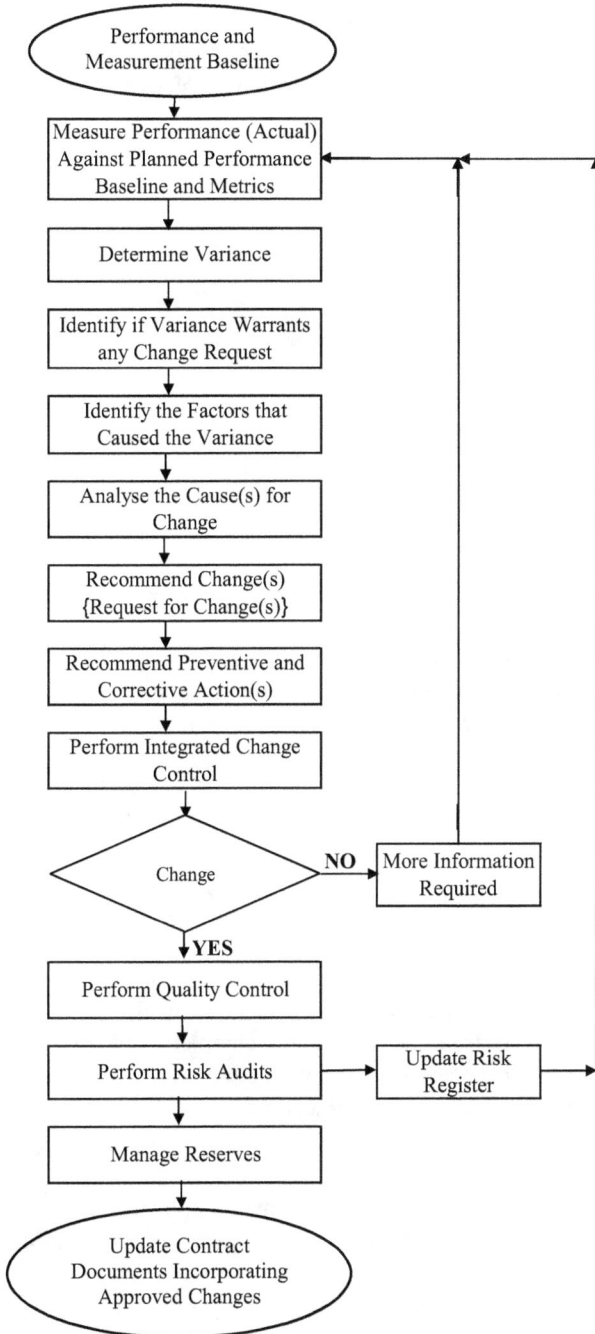

FIGURE 4.50 Logic flowchart for monitoring and controlling process. (Abdul Razzak Rumane. (2013). *Quality Tools for Managing Construction Projects.* Reprinted with permission from Taylor & Francis Group.)

2. To provide an organized and efficient means of measuring, collecting, veri-
    fying, and quantifying data reflecting the progress and status of execution of
    project activities with respect to schedule, cost, resources, procurement, and
    quality
3. To provide an organized, efficient, and accurate means of converting the data
    from the execution process into information
4. To identify and isolate the most important and critical information about
    the project activities to enable decision-making personnel to take corrective
    action for the benefit of the project
5. To forecast and predict about future progress of activities to be performed
    Table 4.59 illustrates monitoring and controlling references for construction
    projects.

Project monitoring and controlling involves a regular comparison of performance
against targets, a search for the cause of deviation, and a commitment to check
adverse variance. It serves two major functions:

1. It ensures regular monitoring of performance
2. It motivates project stakeholders to strive for achieving project objectives

Construction project control is exercised through knowing where to put in the main
effort at a given time and maintaining good communication. There are mainly three
areas where project control is required. These are

1. Quality (Scope)
2. Schedule
3. Budget

All of these areas are to be kept in balance to achieve project objectives. In order to
accomplish the project objectives in construction projects and to achieve successful
project, the supervision consultant/contractor must monitor and control all the major
management processes and handle key attributes effectively and efficiently and track
the progress of the work from the inception to completion of construction and handover of the project. For successful completion of project, monitoring and controlling
of major management processes of a project is discussed below.

4.3.6.10.1 Integration Management

Integration management is coordination and implementation of five Project
Management Process Groups (Initiating, Planning, Executing, Monitoring, and
Controlling) from the time the project is conceived right to closeout stage. Integration
Management involves putting together all the process groups.

TABLE 4.59

Monitoring and Controlling Plan References for Construction Projects

Serial Number	Elements	Contract Reference	Contractor Reference
1	Performance Baseline	Specifications, Drawings	1. Approved Materials 2. Approved Shop Drawings 3. Approved Composite Drawings
		Schedule	1. Contractor's Construction Schedule
		Cost	1. Approved S-Curve
2	Data Collection Methods	Reports	1. Daily Report 2. Weekly Report 3. Monthly Report 4. Safety Report 5. Risk Report 6. Accident Report 7. Check List 8. Risk Report
3	Frequency of Data Collection	1. Daily 2. Weekly 3. Monthly	1. Daily Report 2. Weekly Report 3. Monthly Report
4	Status Information Collection	1. Logs 2. Reports 3. Meetings 4. Check Lists	1. Logs 2. Reports 3. Meetings 4. Check Lists
5	Comparison Between Planned and Actual (Variance)	1. S-Curves 2. Milestones	1. Progress Reports 2. Progress Payment 3. Milestones
6	Analysis	1. Price Analysis	1. Construction Schedule Attachment 2. Progress Payment
7	Corrective Action	1. Comments by Consultant	1. Incorporate Comments
8	Change Order	1. Variation Order	1. Request for Information 2. Request for Modification 3. Request for Variation
9	Document Updates	1. On Regular Basis	1. Reports

Source: Abdul Razzak Rumane. (2013). *Quality Tools for Managing Construction Projects*. Reprinted with permission from Taylor & Francis Group.

Integration Management of construction project includes all the activities performed to effectively control final output of project production (facility), and the input of the process is owner's need for the construction project. To achieve the adequacy of client brief which addresses the numerous complex client/user needs it is required to monitor and control all the related activities.

The supervision consultant/contractor has to monitor and control following activities to ensure that these activities are performed in order to achieve successful project that satisfies owner's/end user's needs.

1. Mobilization requirements
2. Development of project execution scope
   - Review of tender documents and all the requirements for execution of project
   - Review of contract documents
   - Review of construction drawings
   - Review of specifications
3. Sustainability requirements
4. Project execution plan is established
5. All the management plans are developed and approved
6. Construction/execution process and method statement are established
7. Change management—monitor and record all the changes during execution of project

4.3.6.10.2 Stakeholders Management

A stakeholder is anyone who has involvement, interest, or impact in the construction project processes in a positive or negative way. Stakeholders play vital role in determining, formulation, and successful implementation of project processes.

In order to run a successful project, it is important to address the needs of project stakeholders effectively predicting how the project will be affected and how the stakeholders will be affected.

In order to manage stakeholder's expectations in construction projects, following construction-related activities have to be evolved:

1. Develop stakeholders responsibility matrix
2. Develop stakeholders requirements
3. Procedure to distribute performance report
4. Procedure to distribute project update reports
5. Stakeholders' involvement in approvals and reviews
6. Progress payment
7. Meetings

Table 4.60 is an example Stakeholders Responsibility Matrix for construction project.

TABLE 4.60
Stakeholders Responsibilities Matrix during Construction Phase

LEGEND: P=Prepare/Initiate/Responsible, R=Review/Comment, B=Advise/Assist, A=Approve, E=Attend, C=Inform

Serial Number	Activity	Owner/ client	Construction/ Project Manager (if applicable)	Designer/ Consultant	Contractor	Supervisor	Regulatory Authority	Funding Agency	End User/ Facility Manager	Notes/ Comments
1	Selection of Contractor	A	P	B	–	–	–	C	C	
2	Approval of Subcontractor	A	B	B	P	–	–	–	–	
3	Approval of Contractor's Staff	A	B	B	P	–	–	–	–	
4	Execution of Works	C	C	R	P	R	–	–	–	
5	Supervision of Works	C	C	R	P	P	–	–	–	
6	Approval of Material	C	A	R	P	B	–	–	–	
7	Approval of Shop Drawings	C	C	A	P	B	–	–	–	
8	Construction Schedule	C	A	R	P	B	–	–	–	
9	Monitoring Progress	C	P	P	P	B	–	–	–	
10	Monitoring Cost	C	P	P	B	B	–	–	–	
11	Payments	A	R	R	P	B	–	–	–	
12	Request for Information	C	C	R	P	B	–	–	–	
13	Approval of Change	A	B	R	P	B	–	–	–	
14	Quality Plan	C	B	R	P	B	–	–	–	
15	Project Quality	C	R	R	P	P	–	–	–	
16	Meetings	E	E	P	E	E	–	–	–	
17	Safety Plan	C	B	R	P	B	–	–	–	
18	Site Safety	C	C	B	P	P	–	–	–	
19	Testing and Commissioning	C	C	R	P	D	–	–	C	
20	Authorities Approval	C	C	B	P	B	A	–	–	
21	Snag List	C	C	R	P	P	–	–	C	
22	Substantial Completion Certificate	A	R	P	C	–	–	–	C	

To ensure that all the related information and reports are properly distributed to the concerned stakeholder, following are the major activities to be controlled and distribute the information and report to the respective stakeholders:

- Project status/performance
- Changes in scope, schedule, and budget
- Project status
- Project updates
- Project-related issues
- Project payments
- Project conflicts
- Change orders
- Conflicts
- Variation orders
- Anticipated/forecasted problems
- Minutes of meetings

Table 4.61 illustrates consultant's checklist for smooth functioning of project.

4.3.6.10.3   Scope Management

In construction projects, scope management is the process which includes the activities to formulate and define the client's need by establishing project objectives and goals properly addressed in order for the project to have clear direction and controlling what is or is not involved in the project. The project scope documents explain the boundaries of the project, establish project responsibilities for each team member, and sets up procedures for how completed works will be verified and approved. The scope describes the features and functions of the end product or the services to be provided by the project. During project, the scope documentation helps the project team remain focused and on task. The scope statement also provides the project team with guidelines for making decisions about change requests during the project. It is essential that the scope statement should be unambiguous and clearly written to enable all the members of project team understand the project scope to achieve project objectives and goals. Figure 4.51 illustrates scope control process in construction project.

During the construction process, circumstances may come to light that necessitate minor or major changes to the original contract. These changes may occur due to following causes:

1. Differences/errors in contract documents
2. Construction methodology
3. Non availability of specified material
4. Regulatory changes to use certain type of material
5. Technological changes/introduction of new technology

TABLE 4.61
Consultant's Check List for Smooth Functioning of Project

Serial Number	Items to be Checked/Verified	
I	**Project Details**	
	I.1	Scope of work
	I.2	Project objectives
	I.3	Project deliverables
II	**Project Organization**	
	II.1	Organization chart and roles and responsibilities of defined supervision staff
	II.2	Supervision staff deployment matching with project requirements
	II.3	Contractor's staff deployment plan approved as per contract requirements
	II.4	Responsibility matrix prepared and approved by the client and distributed among all project parties
	II-5	Project Directory
III	**Mobilization**	
	III.1	Site permit from Authorities available
	III.2	Project plot boundaries are marked as per the permit
	III.3	Project commencement order issued
	III.4	Copy of permit issued to the contractor
	III.5	Temporary site offices drawings approved
	III.6	Temporary firefighting plan approved by respective authority
	III.7	Copies of Contractor's performance bond, guarantees, insurance policies and licenses available at site
	III.8	Copies of Consultant's performance bond, guarantees, insurance policies and licenses available at site
	III.9	Preconstruction meeting conducted and submittal and approval procedures discussed and agreed
IV	**Project Administration**	
	III-1 Contract Documents	
	IV-1.1	Signed copy of contract between owner and contractor available at site
	IV-1.2	Copies of contract documents available at site
	IV-1.3	Contracted Bill of Quantity (BOQ) is available
	IV-1.4	All volumes of Particular Specifications available
	IV-1.5	Contracted drawings are available
	IV-1.6	Authority approved drawings, duly stamped, available
	IV-1.7	Addendum, if any, to the contract available
	IV-1.8	Replies to tender queries available
	IV-1.9	Copy of signed contract documents and drawings handed over to contractor and has acknowledged the same
	IV-1.10	Log for Codes and Standards available

(continued)

TABLE 4.61 (Continued)
Consultant's Check List for Smooth Functioning of Project

Serial Number	Items to be Checked/Verified
	IV-2 Document Management
	IV-2.1 Document Control System is in place
	IV-2.2 Filing Index is available
	IV-2.3 Material Submittal Log is available
	IV-2.4 Shop Drawing Submittal Log is available
	IV-2.5 Logs for Correspondence between various parties available
	IV-2.6 Log for Check list (Request for Inspection) available
	IV-2.7 Log for JSI (Job Site Instruction) available
	IV-2.8 Log for SWI (Site Work Instruction) available
	IV-2.9 Log for RFI (Request for Information) available
	IV-2.10 Log for VO (Variation Order) available
	IV-2.11 Log for NCR (Non Conformance Report) available
	IV-2.12 Material sample log and place identified
	IV-2.13 Log for Equipment test certificate available
	IV-2.14 Log for Visitor's at site
	IV-2.15 Contractor's staff approval log in place
	IV-2.16 Subcontractor's approval log in place
	IV-2.17 Consultants staff approval in place
	IV-2.18 Overtime request log available
V	**Communication**
	V-1 Communication matrix established and agreed by all the parties
	V-2 Distribution system for transmittals/submittals agreed
VI	**Project Monitoring and Control**
	VI-1 Daily Report log in place
	VI-2 Weekly Report log in place
	VI-3 Monthly Report log in place
	VI-4 Progress Meetings log in place
	VI-5 Minutes of Meetings log in place
	VI-6 Progress Payment log in place
	VI-7 Construction Schedule log in place
VII	**Construction**
	VII-1 Quality Control Plan log in place
	VII-2 Safety Management Plan log in place
	VII-3 Risk Management Plan log in place
	VII-4 Method Statement submittal log in place
	VII-5 Accident and Fire report
	VII-6 Off-site Inspection Visits
	VII-7 Location of Gathering point established

TABLE 4.61 (Continued)
Consultant's Check List for Smooth Functioning of Project

Serial Number	Items to be Checked/Verified	
VIII	**General**	
	VIII-1	Correspondence between site and head office
	VIII-2	Staff related matters
	VIII-3	Copy of Supervision Manual available
	VIII-4	Emergency contact telephones and contact details displayed at site

Source: Abdul Razzak Rumane. (2013). *Quality Tools for Managing Construction Projects*. Reprinted with permission from Taylor & Francis Group.

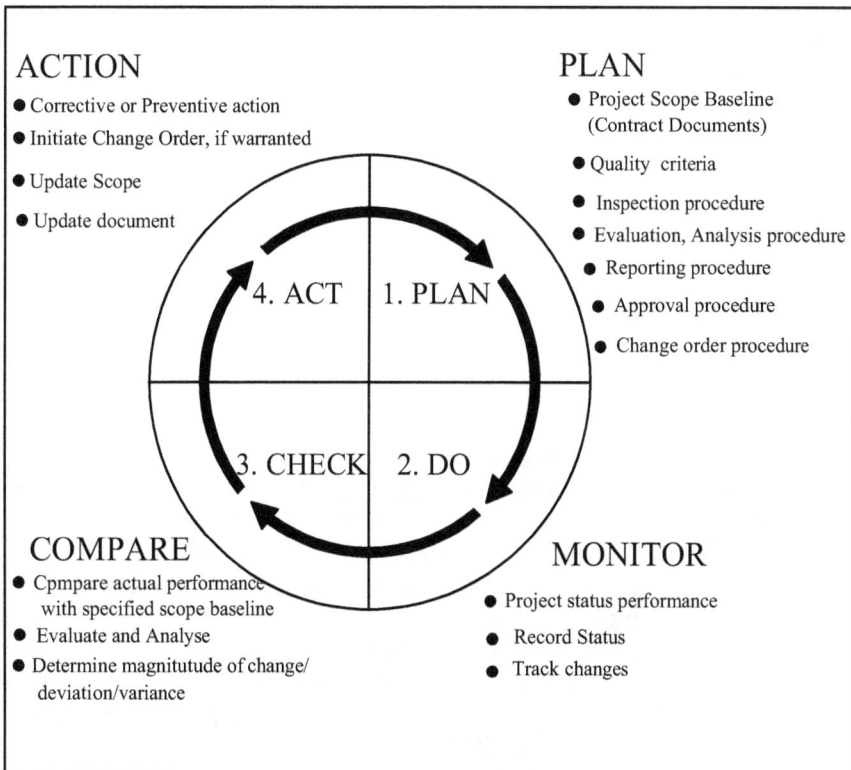

ACTION
- Corrective or Preventive action
- Initiate Change Order, if warranted
- Update Scope
- Update document

PLAN
- Project Scope Baseline (Contract Documents)
- Quality criteria
- Inspection procedure
- Evaluation, Analysis procedure
- Reporting procedure
- Approval procedure
- Change order procedure

4. ACT   1. PLAN

3. CHECK   2. DO

COMPARE
- Cpmpare actual performance with specified scope baseline
- Evaluate and Analyse
- Determine magnitutude of change/deviation/variance

MONITOR
- Project status performance
- Record Status
- Track changes

FIGURE 4.51   Scope control process.

6. Value engineering process
7. Additional work instructed
8. Omission of some works

Table 4.62 lists the causes of changes in construction projects.

These changes are identified as the construction proceeds. These changes or adjustments are beneficial and help build the facility to achieve project objectives. Prompt identification of such requirements helps both the owner and contractor to avoid unnecessary disruption of work and its impact on cost and time. The impacts and consequences of changes in the construction project vary according to

- Type and nature of changes
- Time of occurrence or observance of error or omission
- Change needed for the benefit of the project

A critical change may have negative impact on the project baseline(s). It is important to establish a reliable change control system to manage changes to the project scope. The changes should be managed to maximize the benefits to the project, minimize the negative impacts and effects on the project. The change control system should include

- Submission procedure
- Evaluation and review procedure
- Impact on project
- Approval
- Reporting
- Baseline(s) updates
- Document update

The changes that arise during construction can be either initiated by the owner or by the contractor or even by any of the stakeholders to the construction project.

Identification of discrepancies/errors and changes in the specified scope are common in construction projects. Prompt identification of such requirements helps both the owner and contractor to avoid unnecessary disruption of work and its impact on cost and time. Contractor uses Request for Information (RFI) form to request technical information from the supervision team. These queries are normally resolved by the concerned supervision engineer. However, it is likely that the matter has to be referred to the designer as RFI has many other considerations to be taken care of which may be beyond the capacity of supervision team member to resolve. Normally there is a defined period to respond RFI. Such queries may result in variation to the contract documents. It is in the interest of both owner and contractor to resolve RFI expeditiously to avoid its effect on construction schedule.

TABLE 4.62
Causes of Changes in Construction Projects

Serial Number	Causes	
I	**Owner**	
	I-1	Delay in making the site available on time
	I-2	Change of plans
	I-3	Financial problems/payment delays
	I-4	Change of schedule
	I-5	Addition of work
	I-6	Omission of work
	I-7	Project objectives are not well defined
	I-8	Different site conditions
	I-9	Value engineering
II	**Designer (Consultant)**	
	II-1	Inadequate specifications
		Design errors
		Omissions
	II-2	Scope of work not well defined
	II-3	Conflict between contract documents
	II-4	Coordination among different trades and services
	II-5	Design changes/modifications
	II-6	Introduction of latest technology
III	**Contractor**	
	III-1	Process/methodology
	III-2	Substitution of material
	III-3	Non availability of specified material
	III-4	Charges payable to outside party due to cancellation of certain items/ products
	III-5	Delay in approval
	III-6	Contractor's financial difficulties
	III-7	Unavailability of manpower
	III-8	Unavailability of equipment
	III-9	Material not meeting the specifications
	III-10	Workmanship not to the mark
IV	**Miscellaneous**	
	IV-1	New regulations
	IV-2	Safety considerations
	IV-3	Weather conditions
	IV-4	Unforeseen circumstances
	IV-5	Inflation
	IV-6	Fluctuation in exchange rate
	IV-7	Government policies

Source: Abdul Razzak Rumane. (2013). *Quality Tools for Managing Construction Projects.* Reprinted with permission from Taylor & Francis Group.

Figure 4.52 illustrates RFI form which contractor submits to the consultant to clarify differences/errors observed in the contract documents, change in construction methodology, change in the specified material, and so on.

Figure 4.53 illustrates Process to Resolve Scope Change (Contractor Initiated).

Figure 4.54 illustrates a Variation Order Request Form contractor submits to the owner/consultant for approval of change(s) in the contract.

Figure 4.55 illustrates Process to Resolve Request for Variation.

4.3.6.10.4   Schedule Management

Monitor and control schedule is the process to determine the current status of the schedule, identify the influencing factors that causes the schedule changes, determine that the schedule has changed, and manage the changes in the approved project schedule baseline by updating and taking appropriate actions, if necessary, to minimize deviation from the approved schedule.

Monitoring is collecting, recording, and reporting information concerning project performance. Monitoring involves measurement of current status of the project accomplishment and performance.

Current schedule (as-built schedule) is reflection of current situation of all activities/tasks, milestones, sequencing, resources, duration, constraints, and project update.

Control process is established for managing the current schedule. Controlling is using the actual data collected through monitoring and comparing the same to the planned performance to bring actual performance to planned performance by correcting the variances or implementing approved changes. Analysis of variance between the baseline and current schedule dates and duration provides necessary information for management and stakeholders' decision.

Monitoring construction projects is normally done by compiling status of various activities in the form of progress reports. These are prepared by the contractor, supervision team (consultant), and construction/project management team. The objectives of project monitoring and control are

1. To report the necessary information in detail and in appropriate form which can be interpreted by management and other concerned personnel to provide them with the information about how the resources are being used to achieve project objectives.
2. Provide an organized and efficient means of measuring, collecting, verifying, and quantifying data reflecting the progress and status of execution of project activities with respect to schedule, cost, resources, procurement, and quality.
3. Provide an organized, efficient, and accurate means of converting the data from the execution process into information.

Project Name

Consultant Name

REQUEST FOR INFORMATION (TECHNICAL)

CONTRACT NO. : _____ R.F. I. NO. _____

CONTRACTOR. : _____ DATE : _____

To: Resident Engineer _____

REF:

SUBJECT:

REQUEST FOR INFORMATION (Technical)

This form is used by the contractor to request information and is normally sent to the A/E who responds on the same form.

SAMPLE FORM

CONTRACTOR: _____

DISTRIBUTION: Employer ☐   Engineer ☐   R.E. ☐

RESPONSE BY R.E.:

Signature of R. E. _____   Date _____

RESPONSE RECEIVED:
FOR CONTRACTOR: _____   DATE: _____

DISTRIBUTION: Employer ☐   Engineer ☐   R.E. ☐

FIGURE 4.52   Request for information.

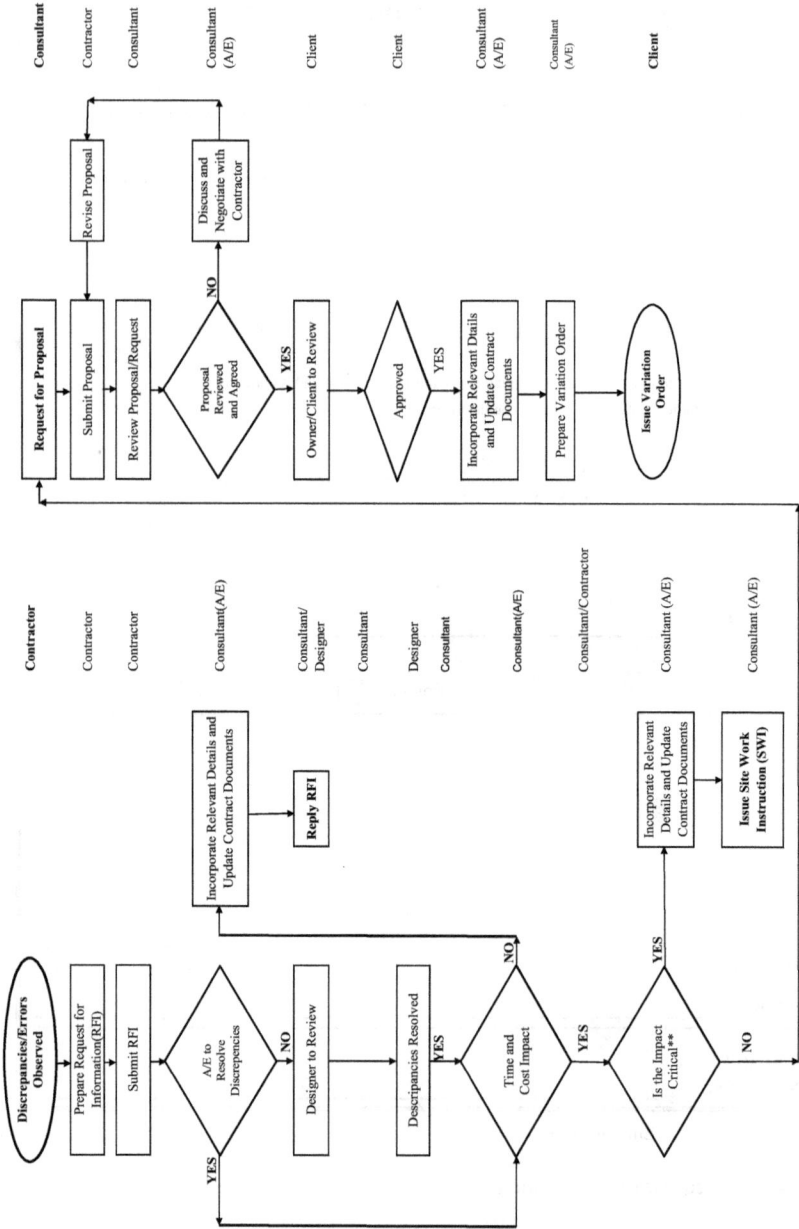

FIGURE 4.53 Process to resolve scope change (contractor initiated).

Project Name
Consultant Name

REQUEST FOR VARIATION

CONTRACT NO. : _____ NO. : _____

CONTRACTOR: _____ DATE : _____

TO :

PROPOSED VARIATION:

☐ PRODUCT ☐ METHOD OF FABRICATION ☐ METHOD OF INSTALLATION

SPECIFIED PRODUCT

PROPOSED PRODUCT _____

SPEC. SECTION # _____ PAGE # _____ ARTICLE # _____

DRG REF _____ DRG # _____ REV# _____

SPECIFIED MANUFACTURER

PROPOSED MANUFACTURER _____

BRIEF PRODUCT DESCRIPTION _____

REASON FOR PROPOSED VARIATION

CHANGE IN DESIGN	REQUIRED BY AUTHORITIES	SITE CONDITIONS	SWI

COST AND TIME EFFECT

COST NO ☐ YES ☐ AMOUNT --------(ADDITION)

TIME NO ☐ YES ☐ DAYS ----------

ATTACHMENTS:
1 Schedule of additions/ommissions
2 Bill Summary
3 Rate Analysis
4 Measurements

Technical and cost comparison sheets must be attached with this request, other wise it will not be reviewed. Contractor shall fill and submit two formsto the OWNER.
Front sheet only shall be returned to Contractor with OWNER action.

WE (THE MAIN CONTRACTOR) CERTIFIES AND UNDERTAKES THAT:

CONTRACTOR'S REP _____ DATE/TIME _____

RECEIVED BY A/E _____ DATE/TIME _____

REVIEW AND ACTION BY OWNER

☐ Approved ☐ Not Approved ☐ Approved as Noted ☐ Incomplete Data Resubmit

COMMENTS

APPROVED SUBJECT TO COMPLIANCE WITH CONTRACT DOCUMENTS

Authorised Signature _____ DATE/TIME: _____

THE APPROVAL OF ANY VARIATION REQUEST SHALL BE SOLELY AT THE DIRECTION OF THE OWNER AND
SUCH APPROVAL SHALL IN NO WAY RELIEVE THE CONTRACTOR OF ANY OF HIS LIABILITIES AND OBLIGATIONS UNDER THE CONTRACT.

RECEIVED BY CONTRACTOR: _____ DATE/TIME: _____

cc: OWNER ☐ EMPLOYER ☐ R.E. ☐ ☐

FIGURE 4.54 Request for variation.

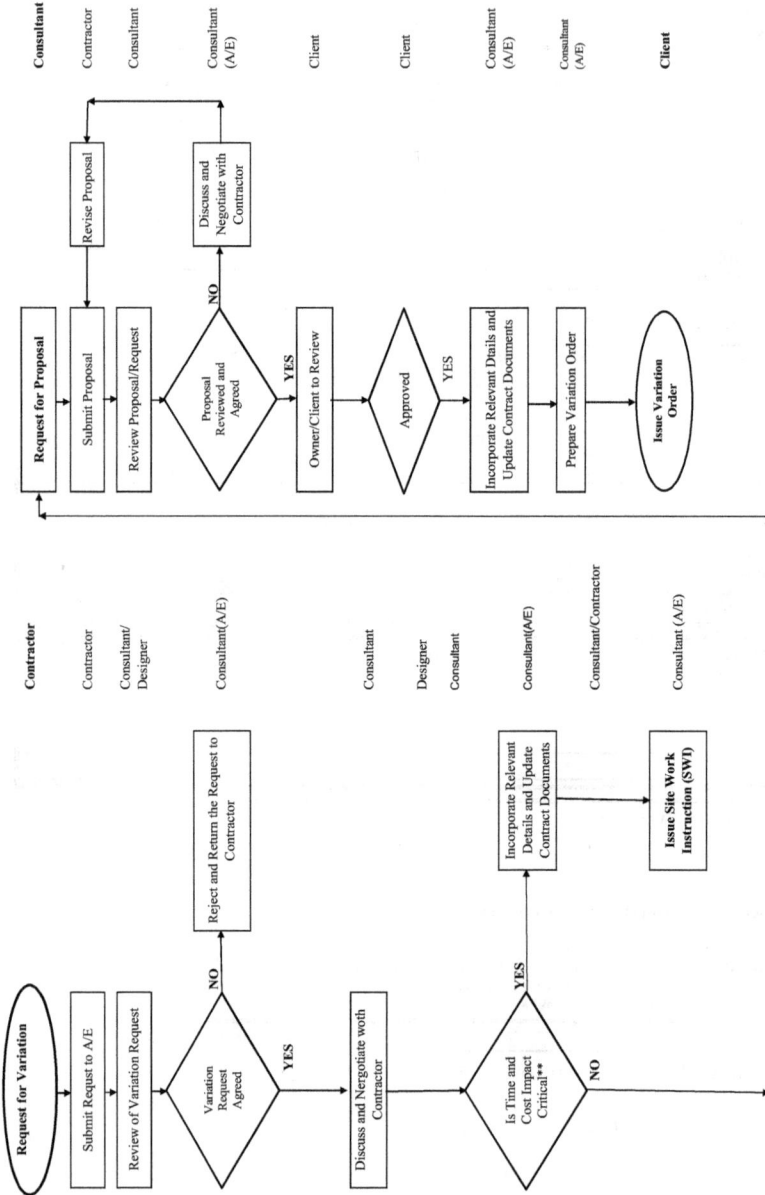

FIGURE 4.55　Process to resolve request for variation.

4. To identify and isolate the most important and critical information about the project activities to enable decision-making personnel to take corrective action for the benefit of the project.
5. Forecast and predict the future progress of activities to be performed.

Following information is required to prepare current (as-built) schedule and compare with the baseline schedule:

- Percentage completion of each activity based on approved checklist
- Actual start/finish dates for the completed activities
- Activities scheduled to start but not yet started
- Activities scheduled to complete but under progress
- Remaining duration to complete each activity
- Percentage completed activities
- Percentage partially completed activities
- Percentage not yet started
- Material, equipment yet to be received
- Available resources
- Regulatory approvals
- Milestones not yet reached
- Logic and duration revision to keep the schedule unchanged
- Problems and issues
- Risks
- Change orders

After analyzing the current status with the actual, the schedule performance report is prepared consisting of following information:

- Project status: where the schedule stands at current situation
- Project progress: plan versus actual
- Forecasting: prediction of future status and progress tend

Figure 4.56 illustrates Schedule Monitoring and Controlling Process.

4.3.6.10.4.1 Monitoring of Work Progress

The work progress is normally monitored through daily and monthly progress reports. Monthly progress report consists of progress photographs to document physical progress of work. These photographs are used to compare compliance with the planned activities and actual performance. Figure 4.57 illustrates Traditional Monitoring System.

With the advent of technology, it is possible to monitor and evaluate construction activities using cameras and related software technologies. In this process, digital images are captured through use of cameras. These photographs are processed using photo modeler software and developing 3D model view of the digital picture captured from the site. The captured as-built data is compared to the planned activities by

Baseline Schedule	MONITORING	COMPARING	
	Collect current status	Compare Actual vs Baseline	
	Measure current status		
	Record current status		
	Report current status		

CONTROLLING
Analyze performance
Evaluate performance
Determine variance
Evaluate Impact on the Project

Updated Schedule	ACTION
	Adjust
	Preventive action
	Corrective action
	Develop forecast
	Modify schedule,if approved
	Update
	Report

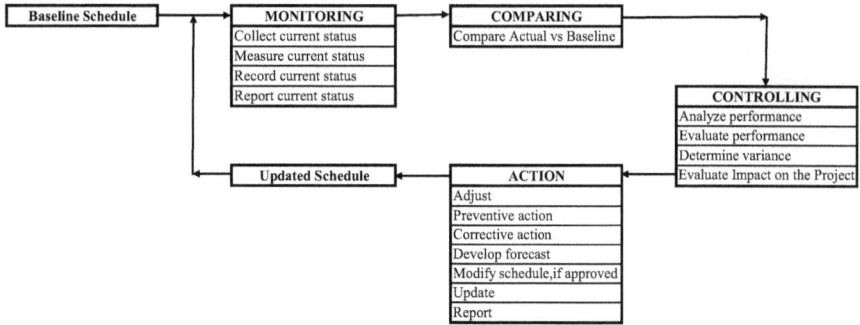

FIGURE 4.56 Schedule monitoring and controlling process. (Abdul Razzak Rumane. (2016). *Handbook of Construction Management: Scope, Schedule, and Cost Control*. Reprinted with permission from Taylor & Francis Group.)

interfacing through Integrated Information Modeling System. The uses of the system are to

- improve the accuracy of information
- avoid delays in getting the information
- improve communication among all parties
- improve effective control of the project
- improve document recording
- help reduce claims

Figure 4.58 illustrates schematic for Digitized Monitoring System.

4.3.6.10.4.2   Logs

The approved CCS is the performance baseline for construction projects which is achieved by collecting information through different methods. Work progress is monitored through various types of logs, S-Curves, reports, and meetings.

There are various types of logs used in construction projects to monitor and control construction activities. The main logs used in a construction project are as follows:

1. Subcontractors Submittal and Approval Log
2. Submittal Status Log
3. Shop Drawings and Materials Logs—E1
4. Procurement Log—E2
5. Equipment Log
6. Manpower Logs

These logs provide necessary information about the status of subcontractors, materials, shop drawings, procurement, and availability of contractor's resources and help determine its effects on project schedule and project completion.

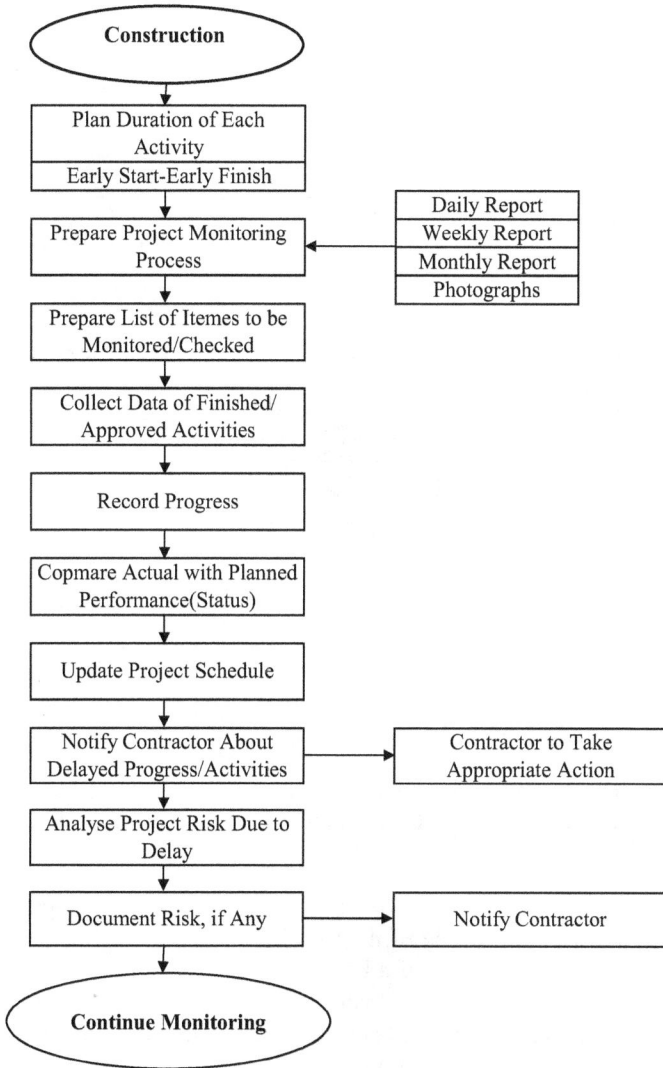

FIGURE 4.57   Traditional monitoring system.

4.3.6.10.4.3   Progress Reports

Apart from different types of logs and submittals, progress curves, time control charts, contractor's progress is monitored through various types of reports and meetings. These are

1. Daily report
2. Weekly report
3. Monthly repot

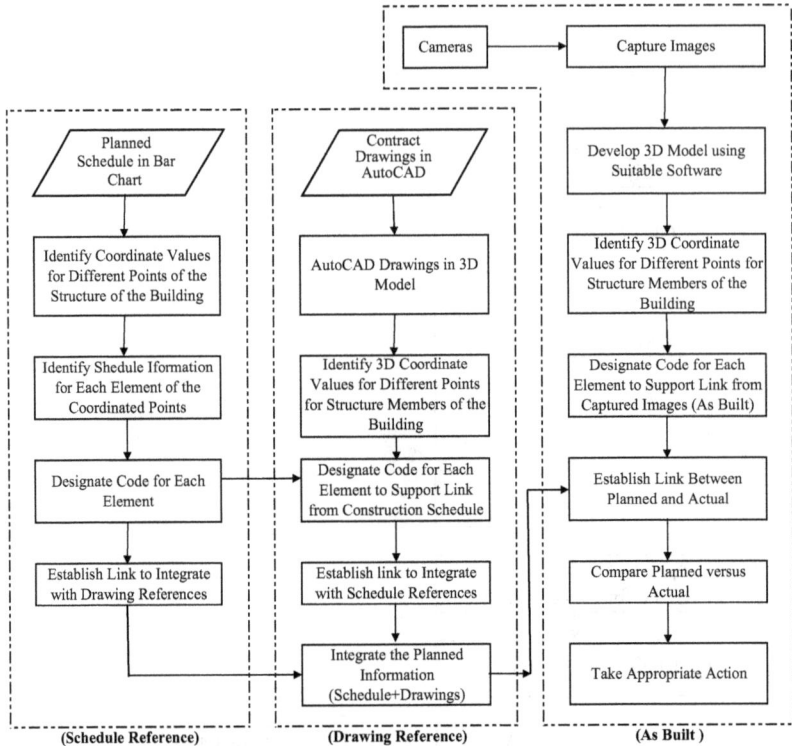

FIGURE 4.58 Digitized progress monitoring. (Abdul Razzak Rumane. (2013). *Quality Tools for Managing Construction Projects.* Reprinted with permission from Taylor & Francis Group.)

Contractor's daily progress is monitored through daily progress report submitted by the contractor on the morning of the working day following the day which the report relates. It gives the status of all the resources available at site for that particular day. It shows the details of contractor's staff and manpower, contractor's plant and equipment, and material received at site. Details of the subcontractor's work and resources are also included in the report.

4.3.6.10.4.4 Progress Meetings

Progress meetings are conducted at an agreed upon interval to review the progress of works and discuss the problems, if any, for smooth progress of construction activities. The contractor submits pre-meeting submittal to project manager/consultant normally two days in advance of the scheduled meeting date. The submittal consists of

1. List of completed activities
2. List of current activities
3. Two weeks look ahead
4. Critical activities
5. Materials submittal log

6. Shop drawings submittal log
7. Procurement log

Apart from the issues related to the progress of works and programs, site safety and quality related matters are also discussed in these meetings. These meetings are normally attended by owner's representative, designer/consultant staff, contractor's representative, and subcontractor's responsible personnel. Normally following meeting are conducted to assess the work progress:

1. Coordination Meetings
   Coordination meetings are held from time to time to resolve coordination matters among various trades.
2. Quality Meetings
   Quality meetings are conducted to discuss quality issues at site and how to improve the construction process to avoid/reduce rejection and rework.
3. Safety Meetings
   Safety meetings are also held to discuss related health, site safety, and environmental matters.

Frequency of conducting meetings is agreed between all the parties at the beginning of construction phase. Normally the CM/R.E. prepares the agenda for the meeting and circulates to all the participants. Contractor informs R.E. in advance about the points contractor would like to discuss, which are included in the agenda. The minutes of meetings are recorded and circulated among all the attendees and others per the approved responsibility/site communication matrix.

4.3.6.10.5   Cost Management

Cost management is the process involving planning, cost estimating, budgeting, and cost controlling to ensure that the project is successfully completed within approved budget.

Cost management in a construction project is planning and managing the cost of facility throughout the project life cycle.

4.3.6.10.5.1   Control Cost

Monitoring and control of project payment is essential with the budgeted amount. This is done through monitoring cash flow with the help of S-curves and progress curves which give exact status of payment and also identifies if it is exceeding the budget. Uninterrupted cash flow is one of the most important elements in the overall success of the project. Figure 4.59 illustrates Planned and Actual Cost S-Curve.

4.3.6.10.6   Quality Management

Quality management is an organization-wide approach to understand customer needs and deliver the solutions to fulfill and satisfy the customer. Quality management is managing and implementation of quality system to achieve customer satisfaction at

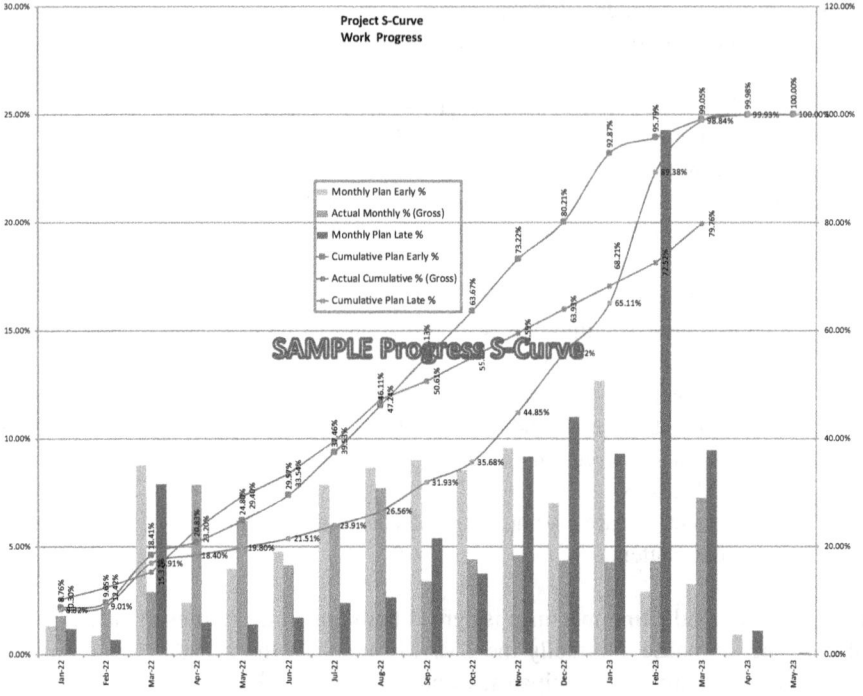

FIGURE 4.59   S-curve (work progress).

the lowest overall cost to the organization while continuing to improve the process. Quality system is a framework for quality management. It embraces the organization structure, policies, procedures, and processes needed to implement quality management system.

Quality management in construction projects is different to that of manufacturing.

Quality in construction projects is not only the quality of products and equipment used in the construction, it is the total management approach to complete the facility as per the scope of works to customer/owner satisfaction to be completed within specified schedule and within the budget to meet owner's defined purpose. Quality management in construction addresses both the management of project and the product of the project and all the components of the product. It also involves incorporation of changes or improvements, if needed. Construction project quality is fulfillment of owner's needs as per defined scope of works within a budget and specified schedule to satisfy owner's/user's requirements.

Quality management system in construction projects mainly consists of

- Quality management planning
- Quality assurance
- Quality control

4.3.6.10.6.1   Develop Quality Management Plan

The quality management plan for construction projects is part of the overall project documentation addressing and describing the procedures to manage construction quality and project deliverable. The quality management plan identifies following key components:

- Details of the quality standards and codes to be complied
- Project objectives, project scope of work
- Stakeholders' quality requirements
- Regulatory requirements
- Quality matrix for different stages
- Design criteria
- Design procedures
- Detailed construction drawings
- Detailed work procedure
- Well-defined specification for all the materials, products, components, and equipment to be used to construct the facility
- Manpower and other resources to be used for the project
- Inspection and testing procedures
- Quality assurance activities
- Quality control activities
- Defect prevention, corrective action, and rework procedure
- Project completion schedule
- Cost of the project
- Documentation and reporting procedure

4.3.6.10.6.2   Perform Quality Assurance

Quality assurance in construction projects covers all activities performed by design team, contractor, and quality controller/auditor (supervision staff) to meet owner's objectives as specified and to ensure and guarantee that the project/facility is fully functional to the satisfaction of owner/end user. Auditing is part of the quality assurance function.

Quality assurance is the activity for providing evidence to establish confidence among all concerned that quality related activities are being performed effectively. All these planned or systematic actions are necessary to provide adequate confidence that a product or service will satisfy given requirements for quality.

Quality assurance covers all activities from design, development, production/construction, installation, servicing to documentation, and also includes regulations of the quality of raw materials; assemblies, products, and components; services related to production; and management, production, and inspection processes. Following are major activities to be performed for quality assurance of the construction project:

- Confirm that owners needs and requirements are included in the scope of works (TOR)
- Review and confirm design compliance to Terms of Reference (TOR)

- Executed works comply with the specified standards and codes
- Conformance to regulatory requirements
- Works executed as per approved shop drawings
- Installation of approved material, equipment on the project
- Method of installation as per approved method statement or manufacturer's recommendation
- Coordination among all the trades
- Continuous inspection during construction/installation process
- Identify and correct the deficiencies
- Timely submission and review of transmittals

4.3.6.10.6.3   Control Quality

Quality control in construction projects is performed at every stage through use of various control charts, diagrams, checklists and so on and can be defined as:

- Checking and review of project design
- Checking and review of bidding/tendering documents
- Analysis of contractor's bids
- Checking of executed/installed works to confirm that works have been performed/executed as specified, using specified/approved materials, installation methods, and specified references, codes, standards to meet intended use
- Controlling budget
- Planning, monitoring, and controlling project schedule

The construction project quality control process is a part of contract documents which provide details about specific quality practices, resources, and activities relevant to the project. The purpose of quality control during construction is to ensure that the work is accomplished in accordance with the requirements specified in the contract. Inspection of construction works is carried out throughout the construction period either by the construction supervision team (consultant) or appointed inspector agency. Quality is an important aspect of construction project. The quality of construction project must meet the requirements specified in the contract documents. Normally contractor provides on-site inspection and testing facilities at construction site. On a construction site, inspection and testing is carried out at three stages during the construction period to ensure quality compliance.

1. During construction process. This is carried with the checklist request submitted by the contractor for testing of ongoing works before proceeding to next step.
2. Receipt of subcontractor or purchased material or services. This is performed by a material inspection request submitted by the contractor to the consultant upon receipt of material.
3. Before final delivery or commissioning and handover

Quality management in construction is a management function. In general, quality assurance and control programs are used to monitor design and construction conformance to established requirements as determined by the contract specifications. Instituting quality management programs reduces costs while producing the specified facility. CQCP developed as per the contents discussed in Table 4.52 under Section 4.3.6.7.3 is to be followed throughout construction project. Table 4.63 illustrates responsibilities for site quality control.

4.3.6.10.7 Resource Management

The success of a construction project depends largely on availability, performance, and utilization of resources. In construction projects, the resources are linked with duration of project, and each activity is allocated a specific resource to be available at the specific time. The construction resource mainly consists of

1. Construction workforce
   i. Contractor's own staff and workers
   ii. Subcontractor's staff and workers
2. Construction equipment, machinery
3. Construction material, equipment, and systems to be installed on the project

In most construction projects, contractor is responsible to engage all types of human resources to complete the project, subcontractors, specialist installers, suppliers, arrange equipment, construction tools, and materials as per contract documents, and to the satisfaction of owner/owner's appointed supervision team. Workmanship is one of the most important factors to achieve the quality in construction; therefore, it is required that the construction workforce is fully trained and have full knowledge of all the related activities to be performed during the construction process.

4.3.6.10.7.1 Construction Workforce

Once the contract is awarded, the contractor prepares a detailed plan for all the resources he needs to complete the project. Contract documents normally specify a list of minimum number of core staff to be available at site during construction period. Absence of any of these staff may result in penalty to be imposed on contractor by the owner.

A typical list of contractor's minimum core staff needed during construction period for execution of work of a major building construction project is discussed under Section 4.3.6.1.3.

Contractor's human resources mainly consists of two categories

1. Contractor's own staff and workers
2. Subcontractor's staff and workers

The human resources required to complete the projects are based on a resource loading program. It is necessary that all the construction resources are coordinated and brought together at the right time in order to complete on time and within budget. The main contractor has to manage all these personnel by

TABLE 4.63
Responsibility for Site Quality Control

	OWNER	SUPERVISOR/ CONSULTANT	CONTRACTOR				
			LINEAR RESPONSIBILITY CHART				
Sr.No.	DESCRIPTION	OWNER/ PROJECT MANAGER	CONSULTANT/ DESIGNER	CONTRACTOR MANAGER	QUALITY INCHARGE	QUALITY ENGINEERS	SITE ENGINEERS	SAFETY OFFICER
---	---	---	---	---	---	---	---	---
1	Specify Quality Standards	□	■					
2	Prepare Quality Control Plan			□	■	□		
3	Control Distribution of Plans and Specifications			□	■	□		
4	Submittals			■	□		■	
5	Prepare Procurement Documents			□			■	
6	Prepare Construction Method Procedures			□	□	□	■	
7	Inspect Work in Progress		■	■		□	■	
8	Accept Work in Progress		■					

9	Stop Work in Progress
10	Inspect Materials Upon Receipt
11	Monitor and Evaluate Quality of Works
12	Maintain Quality Records
13	Determine Disposition of Nonconforming Items
14	Investigate Failures
15	Site Safety
16	Testing and Commissioning
17	Acceptance of Completed Works

■ **Primary Responsibility**
□ Advise/Assist

Source: Abdul Razzak Rumane (2017). *Quality Management in Construction Projects*, Second Edition. Reprinted with permission from Taylor & Francis Group Company.

1. Assigning the daily activities
2. Observing their performance and work output (productivity)
3. Daily attendance
4. Safety during construction process

4.3.6.10.8 Communication Management

For smooth implementation of project, proper communication system is established clearly identifying the submission process for correspondence, submittals, minutes of meeting, and reports.

4.3.6.10.8.1 Correspondence

Normally all the correspondence between contractor and consultant is through Submittal Transmittal Form (please refer Figure 4.38). Correspondence between consultant and contractor is normally done through job site instructions whereas correspondence between owner, consultant, and contractor is normally done through letters. Figure 4.39 discussed earlier is a sample job site instruction form used by the consultant to communicate with the contractor.

4.3.6.10.8.2 Submittals

Contract documents specify the number of copies to be submitted to various stakeholders. Please refer Figure 4.43 already discussed for submittal process. Table 4.55 discussed earlier lists different types of forms (list of control documents) used in construction projects.

4.3.6.10.8.3 Meetings

There are various types of meetings conducted during execution of project. These meetings are conducted at an agreed upon frequency. The meetings during construction phase are held for specific reasons. For example,

1. Kickoff meeting is held to acquaint project team members and discuss project objectives, procedures, and other contract information
2. Progress meetings to review work progress
3. Coordination meetings to coordinate among different disciplines and resolve the issues
4. Quality meetings to discuss on-site quality issues and improvements to the construction process
5. Safety meetings to discuss site safety and environmental issues

Apart from the above discussed meeting, the contractor, consultant, owner can call for a meeting to discuss any project relevant issue and information.

Prior to conducting any meeting, an agenda is circulated to stakeholders and team members to attend the meeting. Figure 4.30 discussed earlier illustrate Sample Agenda Format for Meeting. The proceedings of the meeting are recorded and minutes of meeting is circulated among all attendees and others per the approved form for minutes of meeting.

4.3.6.10.8.4   Reports

During construction phase, following reports are submitted by the contractor:

- Progress Reports such as
  - Daily report
  - Weekly report
  - Monthly progress report
- Safety report
- Risk report

4.3.6.10.9   Risk Management

Risk management is a set of coordinated activities to direct and control the risk. Risk management is the process of identifying, analyzing, assessing, prioritizing different kinds of risks, responding to any risk by planning risk mitigation, implementing mitigation plan, and controlling the risks. It is a process of thinking systematically about the possible risks, problems, or disasters before they happen and setting up the procedure that will avoid the risk, or minimize the impact, or cope with the impact. The objectives of project risk management are to increase the probability and impacts of positive events and decrease the probability and impacts of events adverse to the project objectives.

Risk management is application specific. In some circumstances, it can therefore be necessary to supplement the vocabulary in this guide. Where terms related to the management of risk are used in a standard, it is imperative that their intended meanings within the context of the standard are not misinterpreted, misrepresented, or misused. ISO 31000 focuses on best practice principles for implementing, maintaining, and improving a framework for risk management. Table 4.64 lists the clauses of ISO 31000:2018.

Risk management is an ongoing process that continues throughout the lifecycle of a project. Managing risk is based on Principles, Framework, and Process outlined under clause 4, 5, and 6 of ISO 31000 Risk Management System. These activities have coordinated relationship. Figure 4.60 shows the relationship between these activities.

As per ISO 31000, there are eight principles of risk management. The purpose of risk management principles is for value creation and protection. Figure 4.61 illustrates eight principles of risk management.

These principles provide guidance on the characteristics of effective and efficient risk management, communicating its value, and explaining its intention and purpose. The principles are foundation for risk management and should be considered for establishing the risk management framework and processes.

Table 4.65 illustrates potential risks on scope, schedule, and cost during construction phase and its effects and mitigation action.

Risk monitoring and controlling is a systematic process of tracking identified risks, monitoring residual risks, identifying new risks, execution of risk response plan, and evaluating the effectiveness of implementation of actions against established levels of risk in the area of scope, time, cost, and quality throughout the life cycle of the project. It involves timely implementation of risk response to identified risk to ensure

TABLE 4.64
ISO 31000:2018 Sections (Clauses)

Section (Clause) No.	Relevant Clause in 31000:2018	Description
1	**Scope**	
2	**Normative References**	
3	**Terms and References**	
	3.1	Risk
	3.2	Risk management
	3.3	Stakeholder
	3.4	Risk source
	3.5	Event
	3.6	Sequence
	3.7	Likelihood
	3.8	Control
4	**Principles**	
	4.1	Integrated
	4.2	Structured and Comprehensive
	4.3	Customized
	4.4	Inclusive
	4.5	Dynamic
	4.6	Best available information
	4.7	Human and cultural factors
	4.8	Continual improvement
5	**Framework**	
	5.1	General
	5.2	Leadership and commitment
	5.3	Integration
	5.4	Design
	5.4.1	Understanding the organization and its context
	5.4.2	Articulating risk management commitment
	5.4.3	Assisting organizational roles, authorities, responsibilities and accountabilities
	5.4.4	Allocating resources
	5.4.5	Establishing communication and consultation
	5.5	Implementation
	5.6	Evaluation
	5.7	Improvement
	5.7.1	Adapting
	5.7.2	Continually improving

TABLE 4.64 (Continued)
ISO 31000:2018 Sections (Clauses)

Section (Clause) No.	Relevant Clause in 31000:2018	Description
6	**Process**	
	6.1	General
	6.2	Communication and consultation
	6.3	Scope, context, and criteria
		6.3.1 General
		6.3.2 Defining the scope
		6.3.3 External and internal context
		6.3.4 Defining risk criteria
	6.4	Risk assessment
		6.4.1 General
		6.4.2 Risk identification
		6.4.3 Risk analysis
		6.4.4 Risk evaluation
	6.5	Risk treatment
		6.5.1 General
		6.5.2 Selection of risk treatment options
		6.5.3 Preparing and implementing risk treatment plans
	6.6	Monitoring and review
	6.7	Recording and reporting

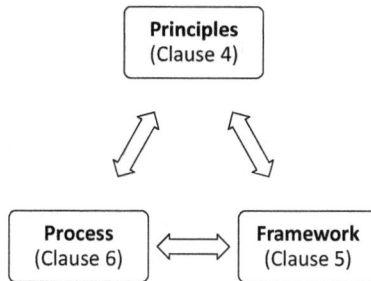

FIGURE 4.60 Risk management clauses relationship.

the best outcome for a risk to a project. Figure 4.62 illustrates a typical flowchart for risk monitoring process.

4.3.6.10.10 Contract Management

Contract management during construction phase is an organizational method, process, and procedure to manage all contract agreements involved between the owner,

ISO 31000 Risk Management Principles Clause Model

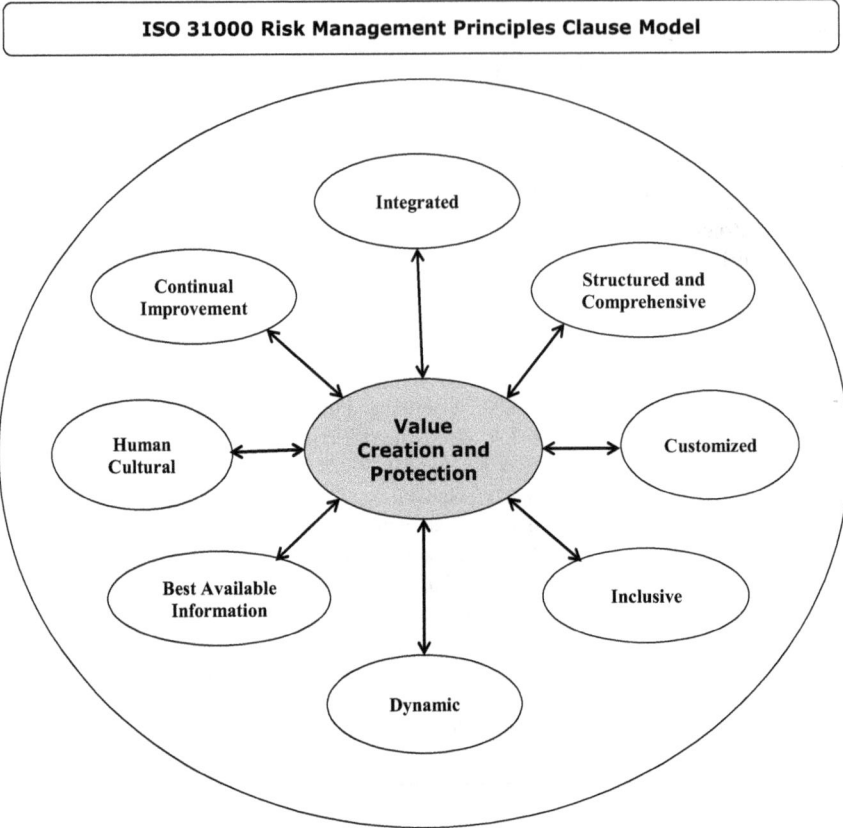

FIGURE 4.61 Risk management principles clause model. (Abdul Razzak Rumane. (2022). *Risk Management Applications Used to Sustain Quality in Projects*, CRC Press, Florida. Reprinted with permission from Taylor & Francis Group.)

contractor, subcontractor, manufacturers, and suppliers. During construction phase, contracts are managed mainly by following parties who are directly involved in the execution of project:

- Consultant/Construction (Project) Manager
- Contractor

Apart from these two parties, subcontractor and vendors also have their contract management system.

Contract management process starts once the contract is signed. The consultant is responsible for managing the contract on behalf of owner. The consultant monitor scope, schedule, cost, and quality of the construction to ensure contract conditions are met. The contractor is responsible to ensure that all project works are executed

TABLE 4.65
Potential Risks on Scope, Schedule, and Cost during Construction Phase and its Effects and Mitigation Action

Sr. No.	Potential Risk	Probable Effects	Control Measures/Mitigation Action
1.0 Scope			
1.1	Scope/design changes	• Project schedule • Project cost • Claim	• Compress schedule • Resolve change order issues in order not to delay the project
1.2	Different site conditions to the information provided	• Change in scope of work • Delay in project	• Contractor to investigate site conditions prior to starting the relevant activity
1.3	Inadequate site investigation data	• Additional work • Scope change	• Contractor to investigate site conditions prior to starting the relevant activity
1.4	Conflict in contract documents	• Project delay	• Amicably resolve the issue
1.5	Incomplete design	• Project scope • Project schedule • Project cost	• Raise Request for Information (RFI) • Resolve issue in accordance with contract documents
1.6	Incomplete scope of work	• Project scope • Project schedule • Project cost	• Raise Request for Information (RFI) • Resolve issue in accordance with contract documents
1.7	Design changes	• Project scope • Project schedule • Project cost	• Follow contract documents for change order
1.8	Design mistakes	• Project scope • Project schedule • Project cost	• Raise Request for Information (RFI) • Resolve issue in accordance with contract documents
1.9	Errors & omissions in contract documents	• Project scope • Project schedule • Project cost	• Raise Request for Information (RFI) • Resolve issue in accordance with contract documents
1.10	Incomplete specifications	• Project scope • Project schedule • Project cost	• Raise Request for Information (RFI) • Resolve issue in accordance with contract documents
1.11	Conflict with different trades	• Project delay	• Coordinate with all trades while preparing coordination and composite drawings

(*continued*)

TABLE 4.65 (Continued)
Potential Risks on Scope, Schedule, and Cost during Construction Phase and its Effects and Mitigation Action

Sr. No.	Potential Risk	Probable Effects	Control Measures/Mitigation Action
1.12	Inappropriate construction method	• Project delay • Claim	• Raise Request for Information (RFI) and correct the method statement
1.13	Quality of material	• Project delay	• Locate suppliers having proven record of supplying quality product
2.0 Schedule			
2.1	Incompetent subcontractor	• Project delay • Project quality	• Contractor has to monitor the workmanship and work progress.
2.2	Delay in transfer of site	• Project delay	• Contractor to adjust the construction schedule
2.3	Delay in mobilization	• Project delay	• Adjust construction schedule accordingly
2.4	Project schedule	• Project completion	• Compress duration of activities
2.5	Inappropriate schedule/plan	• Project delay	• Contractor to prepare schedule taking into consideration site conditions all the required parameters
2.6	Delay in changer order negotiations	• Project schedule	• Request owner/supervisor/project manager to expedite the negotiations and resolve the issue
2.7	Resource availability(material)	• Project delay	• Contractor to make extensive search
2.8	Resource (labor) low productivity	• Project quality • Project delay	• Contractor to engage competent and skilled labors
2.9	Equipment/plant productivity	• Project delay	• Contractor hire/purchase equipment to meet project productivity requirements
2.10	Insufficient skilled workforce	• Project duration	• Contractor arrange workforce from alternate sources
2.11	Failure/delay of machinery and equipment	• Project delay	• Contractor to plan procurement well in advance
2.12	Failure/delay of material delivery	• Project delay	• Contractor to plan procurement well in advance
2.13	Delay in approval of submittals	• Project delay	• Notify owner/project manager
2.14	Delays in payment	• Project delay • Claim	• Contractor to have contingency plans • Owner to pay as per contract

TABLE 4.65 (Continued)
Potential Risks on Scope, Schedule, and Cost during Construction Phase and its Effects and Mitigation Action

Sr. No.	Potential Risk	Probable Effects	Control Measures/Mitigation Action
2.15	Statutory/regulatory delay	• Project delay	• Regular follow up by the contractor, owner with the regulatory agency
3. Cost			
3.1	Low bid project cost	• Project quality	• Contractor to try competitive material, improve method statement and higher production rate from its manpower
3.2	Variation in construction material price	• Project quality • Project cost	• Contractor to negotiate with supplier manufacturer for best price. Contractor to request for change order if applicable as per contract
3.3	Damage to equipment	• Schedule	• Regularly maintain the equipment. Take immediate action to repair damage equipment
3.4	Damage to stored material	• Project delay • Material quality	• Contractor to follow proper storage system
3.5	Structure collapse	• Injuries • Project delays	• Contractor to ensure that formwork and scaffolding is properly installed
3.6	Leakage of hazardous material	• Safety hazards	• Contractor to take necessary protect to avoid leakage. Store in safe area
4.0 General			
4.1	Failure of team members not performing as expected	• Project quality • Project delay	• Select competent candidate. Provide training.
4.2	Change in laws and regulations	• Scope/specification changes • variation order	• Contractor to inform owner/consultant and raise Request for Information (RFI)
4.3	Access to worksite	• Extra/additional time to access site	• Access road to be planned in coordination with adjacent area and local authority
4.4	Theft at site	• Project delay	• Contractor to monitor access to site. Record entry/exit to the site. Provide fencing around project site.

(continued)

TABLE 4.65 (Continued)

Potential Risks on Scope, Schedule, and Cost during Construction Phase and its Effects and Mitigation Action

Sr. No.	Potential Risk	Probable Effects	Control Measures/Mitigation Action
4.5	Fire at site	• Project delay	• Contractor to install temporary fire fighting system. Inflammable material to be stored in safe and secured place with necessary safety measures
4.6	Injuries	• Project delay	• Contractor to keep first aid provision at site. Take immediate action to provide medical aid.
4.7	New technology	• Scope change • Schedule • Cost	• Owner/contractor to mutually agree for changes in the contract for better performance of project

within the agreed upon time and cost in accordance with the contract conditions and specification.

For successful contract management, the contractor as well as consultant/CM/PM has to consider following points while executing the project:

1. Use of RFI to get clarification on some aspects of the project. There are two parts in RFI. These are
   i.   "Question" by the contractor
   ii. "Answer" by the owner (consultant)
2. Execution of project works using specified and approved materials, equipment, and systems
3. Developing project execution plan considering realistic duration for each activity
4. Execution of contracted works in a timely manner in accordance with agreed upon schedule
5. Dealing variations to the specified product, method, work in accordance with related specification, contract clauses, and by providing substantiation and justifications that have resulted proposing alternative or substitute material
6. Managing errors, omissions, and additions strictly in accordance with contract terms and avoiding any delays to the project
7. Conducting meetings to monitor progress and clarify prevailing project issues
8. Cooperating with all team members to fulfill their contractual obligations
9. Resolving disputes in an amicable way by adopting cooperative approach

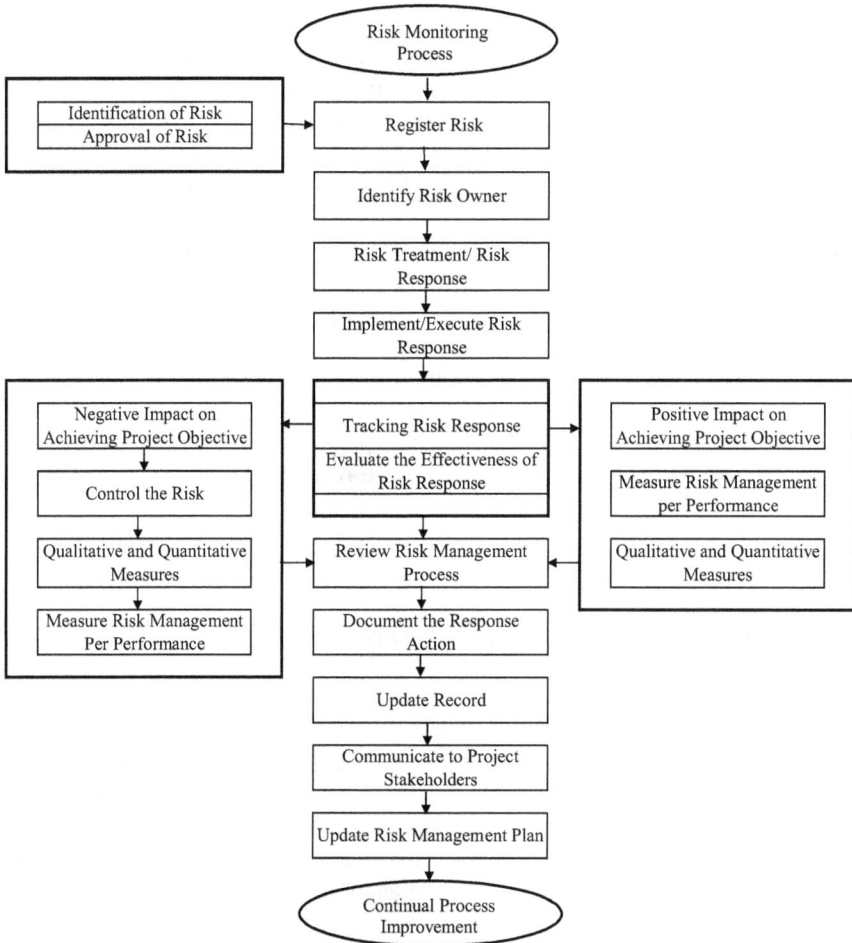

FIGURE 4.62 Typical flowchart for risk monitoring process.(Abdul Razzak Rumane. (2022). *Risk Management Applications Used to Sustain Quality in Projects*, CRC Press, Florida. Reprinted with permission from Taylor & Francis Group.)

10. Providing resources to ensure timely availability of competent workforce as per resource schedule
11. Taking action on all the transmittals within agreed upon period
12. Timely reply to all correspondences, queries
13. Communicating issues and problems well in advance
14. Not to ignore problems/issues with the hope that they might go away
15. Arranging payment of progress payment as per contractual entitlement within stipulated time
16. Maintaining list of claims on monthly basis
17. Settling claims in an accordance with contract terms
18. Maintaining proper logs and records

4.3.6.10.11   HSE Management

HSE (Health, Safety, and Environment) management system plays an important role in the construction industry. HSE management system plays an important role in projects right from project inception through substantial completion of the project. Worker's health and safety and the safety of worksite/workplace are important factors and are prime responsibility of everyone in HSE management. Safety is the corner-stone of a successful project.

ISO 14000 is a series of international standards that have been developed to incorporate environmental aspects into business operations and product standards. ISO 14001 is a specific standard in the series for a management system that incorporates a set of interrelated elements designed to minimize harmful effects on the environment due to the activities performed by an organization, and to achieve continual improvement of its environmental performance. ISO 14000 outlines how to put an effective environment system in place. ISO 14001 incorporates quality management system philosophy, terminology, and requirement structure similar to that of ISO 9001 and provides system compatibility.

Safety management system is a systematic and coordinated approach to managing safety, including the necessary organizational structures, procedures, and account-abilities to optimally manage safety.

There are three core aspects of safety management system. These are

1. Systematic—HSE management activities are in accordance with predetermined HSE policies, strategic objectives, regulatory compliance, and organizational responsibilities to prevent accidents, safety hazards which cause or have potential to cause personal injury, fatalities or property damage, and protection of natural resources due to emission of petroleum hazards.
2. Proactive—An approach that emphasizes safety hazard identification, investigation, and analysis, and risk control and mitigation before events that affect safety occur.
3. Explicit—All safety management activities are documented, communicated, and implemented.

The following are general safety guidelines the contractor may follow to avoid accidents/injuries:

1. Ensure safe access.
2. Ensure that ladders are in good condition and properly secured.
3. Provide barricades around openings, holes, and platforms.
4. Choose the right system of scaffolding for the job to be performed.
5. Check lifting gears for capacity, condition of wire rope, slings, hooks, eyebolts, shackles, proprietary lighting equipment, and spreader beam.
6. Ensure that the crane is
   a. On firm, level ground and outriggers are fully extended.
   b. Operating with a safe working load.

      c. Working with minimum load swing.

      d. Operating with the right type of chain.

      e. Working with the load kept clear of personnel.

7. The hoist is certified by a third party, and the certificate is valid.

8. Wear all protective equipment and garments necessary to be safe on the job.

9. Wear and use eye protection coverings.

10. Use safety shoes.

11. Use hard hats.

12. Use safety belts.

13. Use respiratory masks whenever required.

14. Use the right size of hand tools.

15. Before using any plant or mechanized equipment, check that it is certified for safe operation.

16. Before using electric tools:

      a. Check that it is properly earthed.

      b. Ensure that cable, plugs, or connectors are in sound condition and properly wired up.

17. Use proper guards while using power tools such as circular saws, portable grinders, and bench grinders.

18. Use safe loading/unloading techniques.

19. All material stored at site should be stacked properly to ensure that it is stable and secured against sliding or collapse.

20. Work areas and means of access should be maintained in a safe and orderly condition.

21. Mark access and escape routes.

22. Keep passageways and accessways free from materials, supplies, and obstructions all the time.

23. Mark all the hazardous areas.

24. Prohibit storage of flammable and combustible material.

25. "No smoking" sign to be displayed.

26. Install sirens and alarm bells at site.

27. Ensure that the temporary firefighting system is working all the time.

28. All formwork, shoring, and bracing should be designed, fabricated, erected, supported, braced, and maintained so that it will safely support all vertical and lateral loads that might be applied, until such loads can be supported by the structure.

29. While placing concrete:

      a. Make use of protective clothing and equipment.

      b. Use appropriate gloves.

      c. Use rubber boots.

      d. Wear protective goggles.

      e. Take necessary precautions while using concrete skips or concrete.

30. Display safety signs such as DANGER, CAUTION, and WARNING on all live electrical panels.

31. Post safety, warning signs, and notices of weekly project safety record.
32. Conduct "Safety Awareness" programs.

4.3.6.10.12   Finance Management

In construction projects, maximum amount is expended during construction phase. During this phase

1. Owner has to make payments to
    a. Main contractor
    b. Supervisor (consultant
    c. Construction/project manager, if applicable
    d. Specialist consultant
    e. Specialist contractor
    f. Any other party such as direct appointments
    g. Owner-supplied items, if any
2. Main contractor has to make payments to
    a. Subcontractors
    b. Suppliers (material procurement)
    c. Designer, if any design work is involved
    d. Workforce
    e. Rent (equipment rent, rental vehicles)
3. Subcontractor has to make payment to
    a. Suppliers
    b. Specialist

Owner's payment is mainly related to progress payment claimed and approved by the consultant, advance payments (if any as per contract) to the contractor, monthly fees to the consultant's construction/project manager.

Contractor and subcontractor's payments are linked to the approved executed works. The management of finance for project is done through project cash flow forecast.

4.3.6.10.12.1   Cash Flow

Cash flow = Cash in–Cash out

Forecasting of cash flow is important for smooth functioning of contract to ensure that an appropriate level of funding is in place and suitable draw-down facilities are available. Cash flow prediction is normally made with the help of S-curve. The S-curve stands for 'Standard' curve but the name has also taken into account the shape of letter "S".

The contractor prepares "S" based on expected work progress on monthly basis. Forecasting of cash flow is important to

- Ensure that sufficient funds are available to meet the demand during construction to fulfill the commitment/obligations
- Sufficient funds are available when it is required (timing)
- Ensure maximum utilization of funds

In order to prepare the "S" Curve, accurate information of following elements is required:

- Total expenses
- Total income
- Timing of payment

The resource loaded construction schedule can be used to calculate cash flow forecast. "S" curve for owner is different than that of contractor.

The owner has to take into consideration following points while preparing cash flow:

- Errors and omissions
- Effects of change order or variation
- Effect of inflation
- Effect of international exchange rate
- Change in the sequence of work schedule to expedite or to mitigate delay
- Payment toward material at-site or off-site
- Provisional sum in the contract

Similarly, while preparing the cash flow, the contractor has to consider following points:

- Delay in execution of work
- Productivity less than estimated
- Delay in receipt of payment
- Advance payment, if any, for material
- Different site conditions
- Stoppage of work due to bad weather
- Re-sequencing of work
- Delay in material delivery at site resulting payment cannot be claimed
- Higher material price than estimated
- Fluctuation in international exchange rate

Project accounting system is required to manage the project needs and expenses and provide the financial information to all the interested stakeholders. Table 4.66 lists different logs maintained by finance department.

4.3.6.10.13   Claim Management

Claim management is the process to mitigate the effects of the claims that occur during the construction process and resolve quickly and effectively. If the claims are not managed effectively, it can lead to disputes ending in litigation. Even under most ideal circumstances, contract documents cannot provide full information; therefore, claims do occur in the construction projects. In construction projects, claim is defined as seeking adjustment or consideration by one party against other party with respect to

TABLE 4.66
Logs by Finance Department

Section	Log
1	Incoming and outgoing correspondence
2	Progress (interim) payment
3	Subcontractor payment
4	Material purchase
5	Equipment purchase
6	Procurement (general)
7	Procurement (consumables)
8	Letter of credit
9	Freight/transportation charges
10	Custom clearance
11	Equipment rent
12	Vehicle rent
13	Insurance, bonds, guarantees
14	Regulatory, license fee
15	Staff salaries
16	Labor salaries
17	Office rent
18	Camp rent
19	Cash in hand

- Extension of time
- Scope
- Method
- Payment

There are mainly three types of claims. These are

1. Contractual Claims: Claims that fall within the specific clause of the contract.
2. Extra-Contractual Claims: Claims that result from the breach of contract.
3. Ex-Gratia Claims: Claims that the contractor believes the rights on moral ground.

Table 4.67 lists major causes of claims in construction projects.

All the claims are to be resolved and the contract to be closed. The method of resolution depends on the type of claim, size of the claim, severity of the claim, and effects and consequences of the claim on the project. The submitted claims are to be resolved in a justifiable manner. Following are the method used to resolve the claim:

- Negotiation
- Mediation

TABLE 4.67
Major Causes of Construction Claims

Serial Number		Causes
I		**Owner Responsible**
	I-1	Delay in issuance of notice to proceed
	I-2	Delay in making the site available on time
	I-3	Different site conditions
	I-4	Project objectives are not well defined
	I-5	Inadequate specifications
		a) Design errors
		b) Omissions
	I-6	Scope of work not well defined
	I-7	Conflict between contract documents
	I-8	Change/modification of design
	I-9	Change in schedule
	I-10	Addition of work
	I-11	Omission of work
	I-12	Delay in approval of subcontractor
	I-13	Delay in approval of materials
	I-14	Delay in approval of shop drawings
	I-15	Delay in response to contractor's queries
	I-16	Delay in payment to contractor
	I-17	Lack of coordination among different contractor directly under the control of owner
	I-18	Interference and change by the owner
	I-19	Delay in owner-supplied material
	I-20	Acceleration
II		**Contractor Responsible**
	II-1	Delay to meet milestone dates
	II-2	Non-compliance with specifications
	II-3	Changes in specified process/methodology
	II-4	Substitution of material
	II-5	Non-compliance to regulatory requirements
	II-6	Charges payable to outside party due to cancellation of certain items/products
	II-7	Material not meeting the specifications
	II-8	Workmanship not to the mark
	II-9	Suspension of work
	II-8	Termination of work
III		**Miscellaneous**
	III-1	New regulations
	III-2	Weather conditions
	III-3	Unforeseen circumstances

Source: Abdul Razzak Rumane. (2013). *Quality Tools for Managing Construction Projects*. Reprinted with permission from Taylor & Francis Group.

- Arbitration
- Litigation

In all these cases a comprehensive analysis is necessary to come to an amicable solution. Figure 4.63 illustrates Claim Management Process.

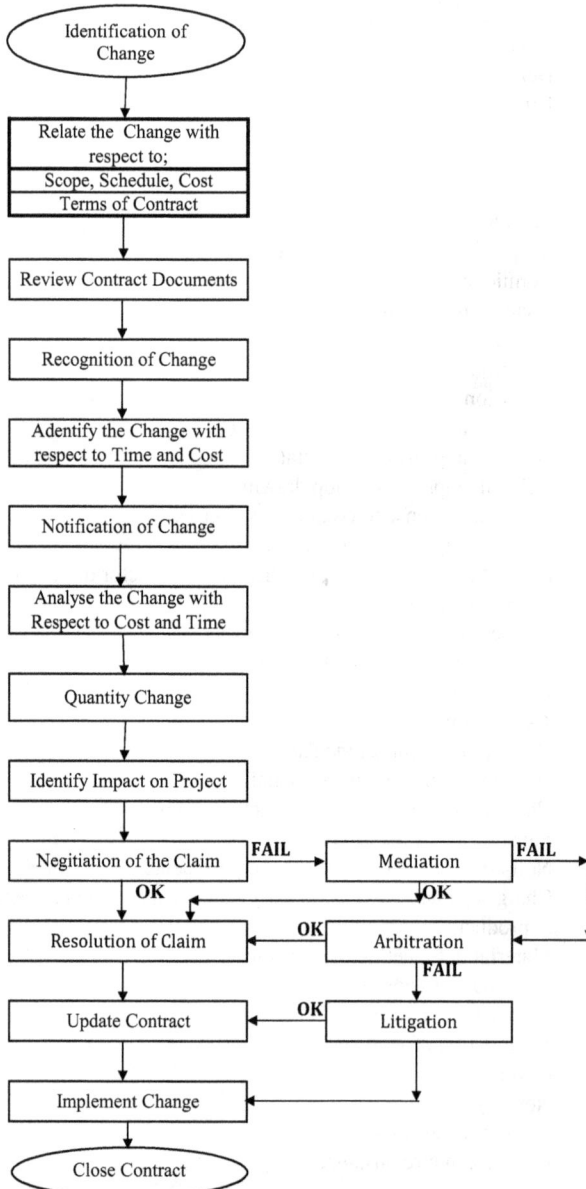

FIGURE 4.63 Claim management process. (Abdul Razzak Rumane. (2016). *Handbook of Construction Management: Scope, Schedule, and Cost Control.* Reprinted with permission from Taylor & Francis Group.)

4.3.6.11 Inspection of Executed Works/Systems

The inspection of construction works is performed throughout the execution of project. Inspection is an ongoing activity to physically check the installed works. Checklists are submitted by the contractor to the consultant who inspects the executed works/installations. If the work is not carried out as specified then it is rejected and the contractor has to rework or rectify the same to ensure compliance with the specifications. During construction all the physical and mechanical activities are accomplished on the site. The contractor carries final inspection of the works to ensure full compliance with the contract documents.

During the construction process the contractor has to submit the checklists to the consultant to inspect the works. Submission of check or RFI is an ongoing activity during the construction process to ensure proper quality control of construction. Concrete work is one of the most important components of building construction. The concrete work has to be inspected and checked at all the stages to avoid rejection or rework. Necessary care has to be taken right from the control of design mix of the concrete till the casting is complete and cured. The contractor has to submit checklist at different stages of concrete work and has to make certain tests, specified in the contract, during casting of concrete. In order to ensure that structure concrete works are executed without any defects or rejection and achieve concrete strength as specified, proper sequencing of works is important. Figure 4.64 illustrates work sequence for formwork.

Figure 4.65 illustrates checklist for general works to be inspected by the consultant. Apart from above-mentioned checklists, the contractor submits following checklist at different stages of concreting process:

1. Checklist for Quality Control of Form Work
2. Notice for Daily Concrete Casting
3. Checklist for Concrete Casting

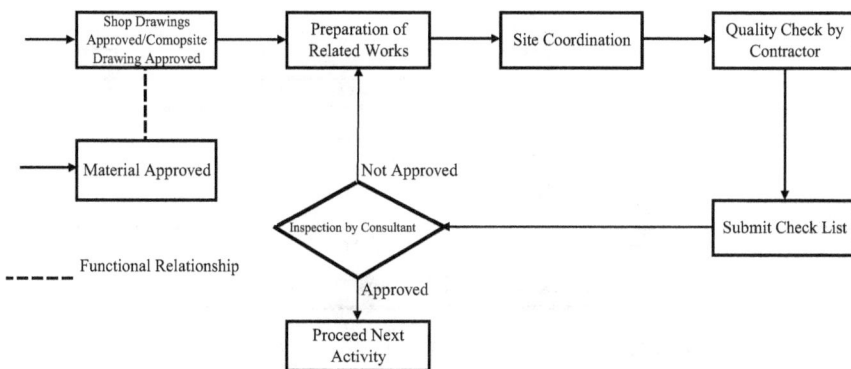

FIGURE 4.64 It is claim management process. Sequence of execution of works. (Abdul Razzak Rumane (2017), *Quality Management in Construction Projects*, Second Edition. Reprinted with permission from Taylor & Francis Group Company.)

Owner Name
Project Name
Consultant Name
CHECK LIST

CONTRACTOR   : _____   CHECK LIST No. : [_____]

CONTRACT No. : _____   QC Code No. [_____]

TO   : **Resident Engineer**

CCS ACTIVITY NO : _____   SPECIFICATION DIVISION : _____   SECTION : _____

     [ ] Process Work   [ ] Civil Works   [ ] Information and Communication

AREA :   [ ] Piping Work   [ ] Electrical Works   [ ] Low Voltage System

     [ ] Mechanical Works   [ ] Instrumentation Works   [ ] Other (Specify)

Please inspect the following :-

Location : _____

Work : _____

Sketch(es) attached { } No.

The work to be inspected has been coordinated with all related subcontractors.

Estimated Quantity of Work : _____   Date & Time Inspection Required : _____

Contractor Signature : _____   Date & Time _____

Received By : _____   Date & Time _____

C.C.: Owner Rep: _____   Date & Time _____

(All request must be submitted at least 24 hours prior to the required inspection)

Reply: The above is Approved/Not approved for the following :-

_____

_____

_____

_____

_____

Inspected by _____ Date & Time _____ Resident Engineer _____ Date & Time _____

Received by Contractor _____ Date & Time _____

C.C.: Owner Rep: _____ Date & Time _____

FIGURE 4.65 Checklist

   4. Checklist for Quality Control of Concreting
   5. Report on Concrete Casting
   6. Notice for Testing at Lab
   7. Concrete Quality Control Form

4.3.6.12   Validate Executed Works

The inspection of construction works is performed throughout the execution of project. Inspection is an ongoing activity to physically check the installed works. Checklists are submitted by the contractor to the consultant to inspect the executed works/installations. Table 4.68 lists the items to be checked by the consultant and probable reasons for rejection of executed works.

TABLE 4.68
Reasons for Rejection of Executed Works

Sr. No.	Description of Works	Probable Reasons for Rejection
Shoring		
1	Shoring	No adequate support for shoring. Vertical plumb level not proper. No proper bracing. Not enough depth. Anchor test not approved
2	Dewatering	Water level not under control
Earth Works		
1	Excavation	Not as per specified level. Surface is not even. Excavated material not removed from the site.
2	Backfilling	Compaction is not proper. Backfilling thickness is not as specified. Soil is rubbish and loose. Soil test (strength) failed
Concrete Sub Structure		
1	Blinding concrete	Thickness not as specified. Concrete strength not as specified
2	Termite Control	Uneven spray.
3	Reinforcement of steel	Reinforcement arrangement not as specified. Support not proper. Water stops are not provided. Construction joints are not provided.
4	Concrete casting	Concrete strength not as specified.
5	Shuttering for beams and columns	Shuttering dimensions not as specified. Shuttering is not strong enough. Shutters are not vertical. Shuttering height not proper.
Concrete Super Structure		
1	Reinforcement of steel for beams and columns	Reinforcement arrangement not as specified. Support not proper.
2	Shuttering for columns and beams	Shuttering dimensions not as specified. Shuttering is not strong enough. Shutters are not vertical. Shuttering height not proper.

(*continued*)

TABLE 4.68 (Continued)
Reasons for Rejection of Executed Works

Sr. No.	Description of Works	Probable Reasons for Rejection
3	Concrete Casting of beams and columns.	Concrete strength not as specified.
4	Formwork for Slab	Props spacing not correct. Props are not sturdy. Formwork surface is not clean.
5	Reinforcement for slab	Reinforcement arrangement not as specified. Reinforcement is not properly placed and secured. Spacers are not provided. Minimum concrete cover not provided. Construction joints not provided.
6	Concrete casting of slab	Concrete strength not as specified. Casting level is not proper.
7	Precast Panels	Panels are not fixed properly. Load test on panel not performed before erection.
Masonry		
1	Block work	Block alignment is not proper. Joints are not aligned. Guidelines are missing. Anchor beads are not provided. Reinforcement mesh is not provided. Mortar is not as specified.
2	Concrete Unit Masonry	Block alignment is not proper. Joints are not aligned. Reinforcement mesh is not provided. Mortar is not as specified.
Partitioning		
1	Installation of Frames	Stud spacing not correct. Fixation method is not as specified. Insulation is not provided.
2	Installation of Panels	Alignment not proper. Joints are not proper. Panels are not painted.
Metal Work		
1	Structural Steel Work	Anchorage and fixing not proper. Method of erection not as specified. Fire protection is not applied. Finishing is different than specified.
2	Installation of Cat Ladders	Fixation is not proper. Alignment is not done. Finishing is not as specified.
3	Installation of Balustrade	Fixation is not proper. Alignment is not done. Finishing is not as specified.
4	Installation of Space Frame	Fixation is not proper. Alignment is not done. Finishing is not as specified.
5	Installation of Handrails and Railings	Fixation is not proper. Alignment is not done. Finishing is not as specified.

TABLE 4.68 (Continued)
Reasons for Rejection of Executed Works

Sr. No.	Description of Works	Probable Reasons for Rejection
Thermal and Moisture Protection		
1	Applying Waterproofing Membrane	Number of layers not as specified. There is no overlap between the rolls. Application not done properly. No. of coats as per specs not applied. There is no skirting. Leakage test failed.
2	Installation of Insulation	Fixation is not properly. Insulation thickness is not as specified.
Doors and Windows		
1	Aluminum Windows and Doors	Dimensions are not as specified. Accessories and hardware are different than specified. Windows are not air/water tight and not sealed properly. Doors are not opening properly.
3	Glazing	Color and type is different than specified. Fixing method is not proper. Cracks are observed in some of the units.
4	Steel Doors and Windows	Alignment not proper. Hardware and accessories not as specified. Finishing is not proper.
6	Wooden Frames	Dimensions are different. Finishing is not as specified.
7	Wooden Doors	Alignment not proper. Finishing is not as specified. Fixation hardware is not as specified.
8	Curtain Wall	Pattern is not as specified. Glazing is different than specified. Drainage system is not provided. Expansion joints not provided. Wind pressure and water test failed.
9	Hatch Doors	Door is not aligned. Finishing is not matching as per architectural requirements. Fixing is not proper.
11	Louvers	Alignment and levels not proper. Accessories not as specified.
Internal Finishes		
1	Plastering	Cracks in the plaster. Voids are observed. Specified accessories not used. Hollow sound observed. Curing is not enough.
2	Painting	No. of layers not as specified. Color and texture is not proper.
3	Cladding-Ceramic	Alignment, angles and joints are not proper. Grouting is not done properly. Color and texture pattern is not matching
4	Cladding-Marble	Fixation is not proper. Alignment, angles and joints are not proper. Lines are not matching.
5	Ceramic Tiling	Alignment, angles and joints are not proper. Grouting is not done properly. Color and texture pattern is not matching
6	Stone Flooring	Fixation is not proper. Alignment, angles and joints are not proper.

(*continued*)

TABLE 4.68 (Continued)
Reasons for Rejection of Executed Works

Sr. No.	Description of Works	Probable Reasons for Rejection
7	Acoustic Ceiling	Suspension system is not as specified. Alignment, levels, and joints are not proper. Services openings are not provided. Ceiling height not matching with approved level.
8	Demountable Partitions	Fixation is not proper. Alignment is not done properly.
Furnishing		
1	Carpeting	Joints are not smooth. Carpet fixed without adhesive. Edges are not trim properly. Color and texture not matching.
2	Blinds	Stacking/Rolling not proper. Fixing is not secured.
3	Furniture	Location is not as per approved layout. Dimensions are not correct. Finishing is nor proper. Fixation is not secured properly. Accessories are not matching with the furniture.
External Finishes		
1	Painting	No. of layers not as specified. Color and texture is not proper.
2	Brickwork	Brick alignment is not proper. Joints are not aligned. Grouting is not proper.
3	Stone	Fixation is not proper. Alignment, angles and joints are not proper. Color and texture not matching.
4	Cladding-Aluminum	Alignment is not proper. Cladding is not leveled and aligned. Fixation is not secured. Joints are without sealant.
5	Cladding-Granite	Fixation is not proper. Alignment, angles and joints are not proper. Color and texture not matching.
6	Curtain Wall	Accessories are not properly fixed. Leveling and alignment not done. Drainage system is not provided. Wind pressure and water tests failed.
7	Glazing	Glass type is different than specified and approved. Fixation is not secured. Leveling and alignment not done properly. Joints have gap. Frame finishing is not as per approved sample.
Equipment		
1	Installation of Maintenance Equipment	Brackets are not fixed at specified location.
2	Installation of Kitchen Equipment	Equipment are not fixed as per approved layout.
3	Installation of Parking Control	Barrier is not fixed at specified location. Control panel is not terminated.

TABLE 4.68 (Continued)
Reasons for Rejection of Executed Works

Sr. No.	Description of Works	Probable Reasons for Rejection
Roof		
1	Parapet Wall	Parapet level is not correct. Finishing is not as specified.
2	Thermal Insulation	Type of insulation and thickness is different than specified.
3	Waterproofing	No. of layers are less than specified. Application is not smooth. There is no overlap between the sheets. Skirting is not provided. Water test failed.
4	Roof Tiles	Thickness of tiles is less than specified. Tiles are not aligned and leveled. Joints are not proper.
5	Installation of Drains	Drains are not located as per approved shop drawings. Fixation is not proper.
6	Installation of Gutters	Type of gutters is different than specified. Gutters are not aligned and leveled.
Elevator System		
1	Installation of Rails	Fixation is not proper.
2	Installation of Door Frames	Frame is not aligned properly.
3	Installation of Cabin	Cabin is not leveled.
4	Cabin Finishes	Cabin finishes are incomplete.
5	Installation of Wire Rope	Rope joints are weak.
6	Installation of Drive Machine	Drive is leveled properly. Equipment earthing is not done.
7	Installation of Controller	Location of controller is not proper. Controller is not terminated. Equipment earthing is not provided. Identification label is not provided.
Mechanical Works		
A. Fire Fighting System		
1	Installation of Piping System	Pipes are dirty need cleaning and painting.
2	Installation of Sprinklers	Sprinkler spacing not as per authorities approved shop drawing.
3	Installation of Foam System	Foam agent concentration level not proper.
4	Installation of Pumps	Eccentric reducer need to fixed at the suction side of the pump. Pumps are not leveled. Cable termination not complete. Equipment earthing not done.

(continued)

TABLE 4.68 (Continued)
Reasons for Rejection of Executed Works

Sr. No.	Description of Works	Probable Reasons for Rejection
5	Installation of Hose Reels/ Cabinet	Cabinet door not closing properly. Color of cabinet to be as per authorities requirement.
6	Installation of Fire Hydrant	Fire hydrant to be fixed properly at specified location.
7	External Fire Hose Cabinet	External fire hose cabinet need hose rack.

B. Water Supply

1	Installing of Piping System	Pipe supports required to keep the pipe in straight position.
2	Installation of Pumps	Pumps are not leveled. Cable termination not complete. Equipment earthing not done.
3	Filter Units	Filter installed without bypass line and tab point for pressure gauge.
4	Toilet Accessories	Towell rod not fixed.
5	Installation of Insulation	Insulation has cuts. Insulation is not covered with canvas.
6	Water Heaters	Water heater drain not installed.
7	Hand Dryers	Installation height is more than specified. Termination is not proper.
8	Water Tank	Float switch for pump not installed.

C. Drainage System

1	Installation of Pipes below grade	Drainage pipe to be tightened properly. Slope not proper.
2	Installation of Pipes above grade	Slope not proper. More support are required.
3	Installation of Manholes	Manhole level is not correct.
4	Installation of Clean Out, Floor Drains	Floor drains and clean out are not flush with the floor finish.
5	Installation of Sump Pumps	Height of float switch need adjustment.
6	Installation of Gratings	Grating frame not flush with floor finish level.

D. Irrigation System

1	Installation of Piping System	Pipe crossing under road need sleeves.

TABLE 4.68 (Continued)
Reasons for Rejection of Executed Works

Sr. No.	Description of Works	Probable Reasons for Rejection
3	Installation of Pumps	Pumps installed without suction strainer.
4	Installation of Controls	Controllers are not weatherproof. Installation height is not proper. Fixing method is not proper.

HVAC Works

1	Installation of Piping	Hanger supports are not properly fixed. Invert level is not correct. Piping is rusty, not clean and not painted.
2	Installation of Ducting	Duct level not correct. Duct metal not clean. Suspension system not properly fixed. No proper sealant around duct joints.
3	Installation of Insulation	Insulation has cuts. Straps not fixed at specified interval.
4	Installation of Dampers, Grills and Diffusers	Material is squeezed. Color is faded. Material not matching with architectural requirements.
5	Installation of Cladding	Cladding surface is not even. Cladding is not overlapped and sealed properly.
6	Installation of Fans	Fans fixing is not secured
7	Installation of Fan Coil Units (FCU)	FCU level not correct FCU fixed without spring isolators. No slope for drain pipe.
8	Installation of Air Handling Units(AHU)	No flexible connector between chilled water piping and AHU.AHU body is damaged. Body paint is peeled off.
9	Installation of Pumps	Pumps are not leveled. Cable termination not complete. Equipment earthing not done.
10	Installation of Chillers	Rubber isolators not installed. Anchorage not fixed properly.
11	Installation of Cooling Towers	Vibration springs not installed. Damage in cooling tower surface. Spray nozzle damaged
12	Installation of Thermostat and Controls	Location not as specified. Thermostat body damaged.
13	Installation of Starters	Location to be near the equipment. Installation height not as specified.
14	Building Management System(BMS)	All the components not installed. Termination not complete. Wiring not dressed and bunched properly.

(*continued*)

TABLE 4.68 (Continued)
Reasons for Rejection of Executed Works

Sr. No.	Description of Works	Probable Reasons for Rejection
Electrical Works		
1	Conduiting – Raceways	Method of installation not proper. Conduit run is not parallel .Minimum clearance from other services less than specified. Supports are not secured, straps or clamp not provided. Spacing, spacing between embedded conduit and concrete aggregate is not as specified. Fastening of conduit with reinforcement steel is not secured. Location of sleeves not as per app roved shop drawings. Location of boxes is not as specified and is not coordinated.
2	Cable Tray – Trunking	Cable trays are not aligned properly. Cable tray run is not parallel and no proper bends. Supports are not as per approved shop drawings. Minimum spacing between any cable tray and other services not maintained. Supports are not secured.
3	Floor Boxes	Location is not coordinated. Level of floor boxes is not proper
4	Wiring	No. of wires in the conduit are not as per approved shop drawing. Circuit wiring for switches not properly done. Termination method is not proper.
5	Cabling	Distance between two cables on cable tray is less than twice the diameter of larges cable diameter. Cable tie is not provided circuit identification is missing
6	Installation of Bus Duct	Supports are not fixed properly. connection between the ducts is not proper. Level of installation is not as specified. Spacing between parallel is less than specified. Supports are not secured.
7	Installation of Wiring Devices/ Accessories	Installation height is not as specified,, coordination with architectural requirements not done to match with the finishes.
8	Installation of Light Fittings	Installation methods is not correct. Proper size of hangers and support not provided. Location and levels are not matching with architectural requirements.
9	Grounding	Welding method for joining tapes is not thermoweld type. Connection methods between tape and clamp to be as specified. Provide clamp at specified distance.
10	Distribution Switch Boards	Installation height of boards not as specified. Termination of wires/cables not done properly. Shrouding for cable not provided. Wires and cables do not have continuity. Wires in distribution board is not properly dressed and bunched.
11	Installation of Diesel Generator Set	D.G Set not aligned properly. Exhaust system and piping not properly installed. Cables not terminated. Panels not installed. Day tank not installed. Pump system not properly fixed.

TABLE 4.68 (Continued)
Reasons for Rejection of Executed Works

Sr. No.	Description of Works	Probable Reasons for Rejection
Fire Alarm System		
4	Installation of Detectors, Bells, Pull Stations, Interface Modules	Location is not as specified. Height of installation not matching with shop drawing and architectural requirements.
5	Installation of Repeater Panel	Installation height not as specified, method of installation not proper. Termination of cables is not done.
6	Installation of Mimic Panel	Installation height not as specified. Method of installation not proper. Termination of cables is not done.
7	Installation of Fire Alarm Panel	Installation height not as specified. Method of installation not proper. Termination of cables is not done.
Telephone/Communication System		
1	Installation of Racks	Method of installation not as per approved shop drawing. Termination of cables not done properly. Racks are not leveled. Cables are not labeled.
2	Installation of Switches	Switches are not properly installed. Switched are not installed in sequence. Cabling is not complete. Identification label not provided for cables
Public Address System		
5	Installation of Speakers	Fixing of speakers is not secured. Height of installation is not as specified. Speakers installed in false ceiling do not have proper support.
6	Installation of Racks	Method of installation not as per approved shop drawing. Termination of cables not done properly. Racks are not leveled. Cables are not labeled.
7	Installation of Equipment	Equipment are not properly installed. Cabling is not proper.
Audio Visual System		
5	Installation of Speakers	Fixing of speakers is not secured. Height of installation is not as specified. Speakers installed in false ceiling do not have proper support.
6	Installation of Monitors/ Screens	Method of fixing nor proper. Installation height is not as per approved shop drawings.
7	Installation of Racks	Method of installation not as per approved shop drawing. Termination of cables not done properly. Racks are not leveled. Cables are not labeled.
8	Installation of Equipment	Equipment are not properly installed. Cabling is not proper.

(*continued*)

TABLE 4.68 (Continued)
Reasons for Rejection of Executed Works

Sr. No.	Description of Works	Probable Reasons for Rejection
CCTV/Security System		
3	Installation of Cameras	Fixing of cameras not proper. Installation height is not as specified. Termination of cameras not proper.
5	Installation of Panels	Panels not properly fixed. Cabling not complete.
7	Installation Monitors/ Screens	Method of fixing nor proper. Installation height is not as per approved shop drawings.
Access Control		
3	Installation of RFID proximity readers, Finger print readers	Readers not installed properly.
4	Installation of Magnetic locks, Release buttons, Door contacts	Magnetic locks not fixed properly. Release buttons not installed at specified height and location.
5	Installation of Panel	Panels not properly fixed. Cabling not complete.
6	Installation of Server	Cable termination not proper. Cables to be properly dressed. Identification labels not provided.
Systems Integration		
1	Installation of Switches	Switches not stacked as per sequence. Identification labels not provided
2	Installation of Servers	Cable termination not proper. Cables to be properly dressed. Identification labels not provided.
External Works		
1	Site Works	Compaction failure. Grading level not proper. Slope is not provided.
2	Asphalt Work	Asphalt levels not proper. Asphalt material not as specified.
3	Pavement Works	Pavement limits not as per approved layout. Curb stones not laid properly.
4	Piping Works	Pipe not laid at specified depth. No protection on pipes. Pipe joints not done well. Leakage observed.
5	Electrical Works	Cable not buries properly. Location of light poles not correct. Light poles are not vertically installed.
6	Manholes	Location of manholes not correct. Manhole level not correct.
7	Road Marking	Marking color not as specified.

Source: Abdul Razzak Rumane. (2017). *Quality Management in Construction Projects*, Second Edition. Reprinted with permission from Taylor & Francis Group.

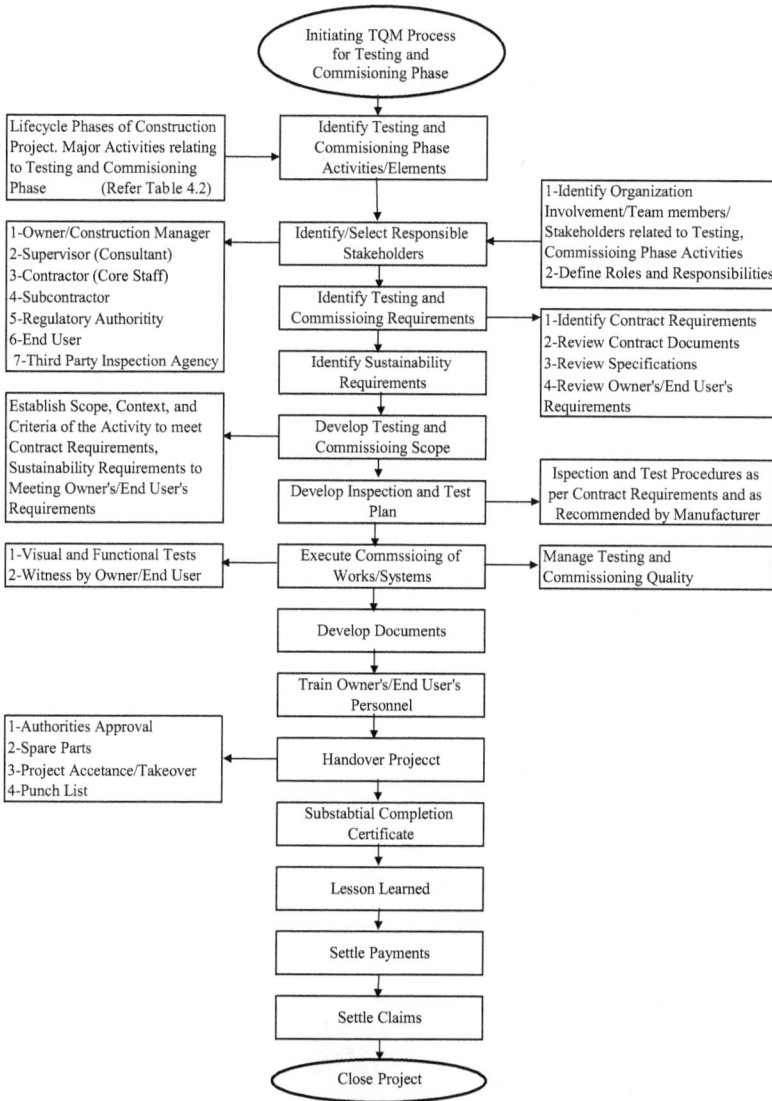

FIGURE 4.66 Typical flowchart for total quality management applications in quality of testing and commissioning phase activity.

If the work is not carried out as specified, then it is rejected, and the contractor has to rework or rectify the same to ensure compliance with the specifications. During construction all the physical and mechanical activities are accomplished on the site. The contractor carries out a final inspection of the works to ensure full compliance with the contract documents.

Table 4.69 summarizes the TQM concept for construction phase activities.

TABLE 4.69
Total Quality Management Concept for Construction Phase Activities

Serial Number	Phase Activity	Stakeholders/ Team Members	Customer Specifications/ Requirements	System for Managing/Tool
1	Identification of Stakeholders/ Project Team Members	1. Project owner 2. Project manager 3. Supervision consultant 4. Contractor/ subcontractor	1. Contract documents 2. Owner's need and requirements	Please refer Section 4.3.6.1 and Table 4.47, Table 4.48 and Table 4.49
2	Mobilization	1. Project Manager 2. Supervision consultant 3. Contractor 4. Subcontractor	1. Contract requirements 2. Owner's requirements 3. Regulatory requirements	Please refer Section 4.3.6.2
3	Development of Project Site Facilities	1. Project Manager 2. Supervision consultant 3. Contractor 4. Subcontractor	1. Contract requirements 2. Regulatory requirements	Please refer Sections 4.3.6.3
4	Identification of Project Execution/ Installation Requirements	1. Project Manager 2. Supervision consultant 3. Contractor 4. Subcontractor	1. Contract requirements	Please refer Section 4.3.6.4
5	Identification of Sustainability Requirements	1. Project Manager 2. Supervision consultant 3. Contractor 4. Subcontractor	1. Contract requirements 2. Regulatory requirements	Please refer Section 4.3.6.5
6	Establish Project Execution/ Installation Scope	1. Contractor 2. Supervision consultant 3. Subcontractor	1. Contract requirements	Please refer Section 4.3.6.6
7	Project Planning and Scheduling	1. Contractor 2. Supervision consultant 3. Subcontractor 4. Planning manager 5. Quality manager	1. Contract requirements	Please refer Section 4.3.6.7 and Table 4.50, Table 4.51 and Figure 4.31
8	Develop Management Plans	1. Contractor 2. Supervision consultant 3. Subcontractor	1. Contract requirements	Please refer Section 4.3.6.8

TABLE 4.69 (Continued)
Total Quality Management Concept for Construction Phase Activities

Serial Number	Phase Activity	Stakeholders/ Team Members	Customer Specifications/ Requirements	System for Managing/Tool
9	Construction/ Execution of Works	1. Contractor 2. Subcontractor	1. Approved material, shop drawings, and method statements. 2. Contract specifications	Please refer Section 4.3.6.9
10	Monitoring and Control	1. Project manager (owner) 2. Supervision consultant 3. Contractor 4. Planning manager	Approved schedule, S-curve, and management plans	Please refer Section 4.3.6.10
11	Execution of Executed Works/ Systems	1. Contractor 2. Subcontractor 3. Supervision consultant	1. As per approved material, shop drawings, and method statement	Please refer Section 4.3.6.11 and Figure 4.64 and Figure 4.65
12	Validate Executed Works	1. Supervision consultant 2. Owner 3. Third-party inspection agency (if applicable)	1. Contract requirements	Please refer Section 4.3.6.12 and Table 4.68

4.3.7 TQM in Testing, Commissioning, and Handover Phase

Testing, commissioning, and handover is the last phase of the construction project lifecycle. This phase involves testing of electromechanical systems, commissioning of the project, obtaining authorities' approval, training of user's personnel, handing over of technical manuals, documents, and as-built drawings to the owner/owner's representative. During this period the project is transferred/handed over to the owner/ end user for their use and substantial completion certificate is issued to the contractor. Figure 4.66 illustrates typical flowchart for TQM applications for the testing and commissioning, and handover phase activities.

4.3.7.1 Identify Stakeholders

Following stakeholders are involved during testing, commissioning, and hand-over phase:

1. Owner
   - Owner's representative/project manager
2. Construction supervisor
   - Construction supervisor (consultant)
   - Specialist contractor
3. Contractor
   - Main contractor
   - Subcontractor
   - Testing and commissioning specialist
3. Regulatory authorities
4. End user
5. Third-party inspecting agency

4.3.7.1.1 Select Team Members

It is essential to select team members who have experience in testing and commissioning of major projects. The team members can be from the same supervision team, which was involved during execution of the project, if they have experience to carry out testing and commissioning. In most cases the manufacturer's representative is involved in testing the supplied equipment/systems. The owner may engage specialist firm(s) to perform start up activities and commission the project.

4.3.7.1.2 Develop Responsibility Matrix

Table 4.70 illustrates the contribution of various participants during testing, commissioning, and handover phase and Table 4.71 lists responsibilities of consultant during closeout phase.

TABLE 4.70

Responsibilities of Various Participants during Testing, Commissioning, and Handover Phase

| Phase | Responsibilities | | |
	Owner	Consultant/Supervisor	Contractor
Testing, Commissioning, and Handover	• Acceptance of project • Takeover • Substantial completion certificate • Training • Payments	• Witness tests • Witness commissioning • Check close out requirements • Recommend take over • Recommend issuance of Substantial completion certificate	• Testing • Commissioning • Authorities' approvals • Documents • Training • Handover

TABLE 4.71
Typical Responsibilities of Supervision Consultant during Project Testing, Commissioning, and Handover Phase

Serial Number	Responsibilities
1	Ensure that all the equipment, systems are functioning and operative
2	Ensure that performance tests are carried out on all the equipment, systems and equipment, systems are performing as per intended/design requirements
3	Ensure that Job Site Instruction (JSI), Non-Conformance Report (NCR) are closed
4	Ensure that site is cleaned, and all the temporary facilities and utilities are removed
5	Ensure as-built drawings handed over to the client/end user
6	Ensure that operation and maintenance manuals handed over to the client
7	Ensure that record books are handed over to the client
8	Ensure that guarantees, warrantees, bonds are handed over to the client
9	Ensure that test reports, test certificates, inspection reports handed over to the client
10	Ensure that spare parts are handed over to the client
11	Ensure that snag (punch) list prepared and handed over to the client
12	Ensure that training for client/end user personnel completed
13	Ensure that substantial completion certificate issued and maintenance period commissioned
14	Ensure that all the dues of suppliers, subcontractors, contractor paid
15	Ensure that retention money is released
16	Ensure that supervision completion certificate from the owner is obtained
17	Lesson learned documented

Source   Modified from; Abdul Razzak Rumane. (2013). *Quality Tools for Managing Construction Projects*. Reprinted with permission from Taylor & Francis Group.

4.3.7.2   Identify Testing, Commissioning, and Handover Requirements

Testing and start up (commissioning) requirements are specified in the contract documents. It is essential to inspect and test all the installed/executed works prior to hand over the project to the owner/end user. Generally, all works are checked and inspected on regular basis while the construction is in progress; however, there are certain inspection and tests to be carried out by the contractor in presence of owner/ consultant. These are especially for rotating equipment, systems, conveying systems, electrical works, low voltage systems, information and technology related products, emergency power supply system, and electrically operated equipment which are energized after connection of permanent power supply. Testing of all these equipment, systems start after completion of installation works. By this time facility is connected to permanent electrical power supply, and all the equipments are energized.

Testing and commissioning is to be carried out on installed equipment, machinery, systems to ensure that they are safe and meet the intended requirements of the project to the satisfaction of owner/end user. The testing is normally undertaken to prove the quality and workmanship of the installation. It is also known as static testing. Upon completion of static testing, dynamic testing can be undertaken, this is "commissioning." Commissioning is carried out to prove that the systems operate and perform to the design intent and specification.

Commissioning is the orderly sequence of testing, adjusting. and balancing the equipment, system, and bringing the equipment, systems, and subsystems into operation and starts when the construction and installation of works are complete. Commissioning is normally carried out by contractor or specialist in presence of consultant and owner/owner's representative and user's operation and maintenance personnel to ascertain proper functioning of the systems to the specified standards.

4.3.7.3 Identify Sustainability Requirements

Following items to be checked for compliance with requirements

1. Energy-efficient equipment, systems
2. Daylighting system
3. Sensor-controlled lighting system
4. Performance testing to meet owner's requirements and usage

4.3.7.4 Develop Testing, Commissioning Scope

The contract documents specify the testing and commission works to be performed by contractor, subcontractor, and specialist supplier of equipment/systems. Following are the main scope of works to be carried out during this phase:

1. Testing of all equipment
2. Commissioning of all equipment
3. Testing of all systems
4. Commissioning of all systems
5. Performance testing of all equipment, systems
6. Conformance of all systems meeting sustainability requirements
7. Obtaining authorities' approvals
8. Submission of as-built drawings
9. Submission of technical manuals and documents
10. Submission of record books
11. Submission of warranties and guarantees
12. Training of owner/user's personnel
13. Handover of spare parts
14. Handover of the project to owner/end user
15. Preparation of punch list
16. Issuance of substantial certificate
17. Lessons learned

Table 4.72 lists the items to be tested and commissioned prior to handing over of the project.

4.3.7.5 Develop Inspection and Testing Plan

Figure 4.67 illustrates logic flowchart for development of inspection and testing plan.

TABLE 4.72
Major Items for Testing and Commissioning of Equipment

Serial Number	Discipline	Items
1.0	Elevator	1. Power Supply 2. Speed 3. Capacity to Carry Design Load 4. Number of Stops 5. Emergency Landing 6. Emergency Call System 7. Elevator Management System 8. Interface with Fire Alarm System
2.0	Mechanical	
2.1	Mechanical (Fire Suppression)	1. Sprinklers 2. Piping 3. Fire Pumps 4. Power Supply and Controls 5. Emergency Power Supply for Fire Pumps 6. Hydrants 7. Hose Reels 8. Water Storage Facility 9. Gaseous Protection System for Communication Rooms 10. Fire Protection System for Diesel Generator Room 11. Interface with Fire Alarm System 12. Interface with BMS
2.2	Mechanical (Public Health)	1. Piping 2. Pipe Flushing and Cleaning 3. Pumps 4. Boilers 5. Hot Water System 6. Water Supply and purity 7. Fixtures 8. Power Supply for Equipment 9. Controls 10. Drainage System 11. Irrigation System
3.0	HVAC	1. Pipe Cleaning and Flushing 2. Chemical Treatment 3. Pumps 4. Duct Work 5. Air Handling Unit 6. Heat Recovery Unit 7. Split Units 8. Chillers 9. Cooling Towers

(continued)

TABLE 4.72 (Continued)
Major Items for Testing and Commissioning of Equipment

Serial Number	Discipline	Items
		10. Heating System (Controls, Piping, Pumps)
		11. Fans (Ventilation, Exhaust)
		12. Humidifiers
		13. Starters
		14. Variable Frequency Drive
		15. Motor Control Centers (MCC Panels)
		16. Chiller Control Panels
		17. Building Management System (BMS)
		18. Interface with Fire Alarm System
		19. Thermostat
		20. Air Balancing
4.0	Electrical	
4.1	Electrical (Power)	1. Lighting Illumination Levels
		2. Working of Photo Cells and Controls
		3. Wiring Devices (Power Sockets)
		4. Lighting Control Panels
		5. Electrical Distribution Boards
		6. Electrical Bus Duct System
		7. Main Switch Boards/Sub Main Switch Boards
		8. Min Low Tension Panels
		9. Isolators
		10. Emergency Switch Boards
		11. Motor Control Centers (MCC Panels)
		12. Audio Visual Alarm Panel
		13. Diesel Generator
		14. Automatic Transfer Switch (ATS)
		15. UPS (Uninterrupted Power Supply)
		16. Earthing (Grounding) System
		17. Lightning Protection System
		18. Surge Protection System
		19. Power Supply to Equipment (HVAC, Mechanical, Elevators, Others)
		20. IP Rating of Out Door Switches, Isolators, Switch Boards
		21. Interface with BMS
		22. Emergency Power System
		23. Exhaust Emission of Generator
4.2	Electrical (Low Voltage)	1. Fire Alarm System
		2. Communication System
		3. CCTV System
		4. Access Control System
		5. Public Address System
		6. Audio Visual System
		7. Master Satellite Antenna System

TABLE 4.72 (Continued)
Major Items for Testing and Commissioning of Equipment

Serial Number	Discipline	Items
5	External Works	1. Lighting Poles
		2. Boundary Wall Lighting
		3. Lighting Bollards
		4. Irrigation System
		5. Electrical Distribution Boards
6	General	1. Power Supply for Gate Barriers
		2. Automatic Gates
		3. Rolling Shutters
		4. Window Cleaning System
		5. Gas Detection System
		6. Water/Fluid Leak Detection System
		7. Waste Treatment System

Source: Abdul Razzak Rumane. (2013). *Quality Tools for Managing Construction Projects*. Reprinted with permission from Taylor & Francis Group.

4.3.7.6 Execute Commissioning of Works/Systems

The testing is mainly carried out on electromechanical works/systems, electrically operated equipment/systems, and rotating equipment which is energized after connection of permanent power supply to the facility. These include the following:

1. Pumps (all types)
2. Piping works (pressure test)
3. Compressors
4. Generators
5. Coolers
6. Conveying system
7. Supervisory control system
8. Water supply, plumbing, and public health system
9. Fire suppression system
10. HVAC system
11. Integrated automation system (building automation system)
12. Electrical switchgear
13. Electrical lighting and power system
14. Instrumentation and control system
15. Grounding (earthing) and lightning protection system
16. Fire alarm system.
17. Information and communication system.
18. Electronic security and access control system
19. Public address system
20. Parking control system
21. Material handling equipment

```
        ⬭ Review Contract
          Documents
              │
              ▼
    ┌──────────────────────┐
    │ Determine Scope of Work │
    └──────────────────────┘
              │
              ▼
    ┌──────────────────────┐
    │  Identify Inspection/  │
    │   Testing/Start Up     │
    │     Requirements       │
    └──────────────────────┘
              │
              ▼
    ┌──────────────────────┐
    │   Prepare Check List   │
    └──────────────────────┘
              │
              ▼
    ┌──────────────────────┐
    │ Coordinate with All other │
    │        Trades          │
    └──────────────────────┘
              │
              ▼
    ┌──────────────────────┐                    ┌──────────────────┐
    │   Submit Check List    │◄───────────────── │ Corrective Action │
    └──────────────────────┘                    └──────────────────┘
              │                                          ▲
              ▼                                          │
    ┌──────────────────────┐
    │ Inform Concerned Authorities │
    │ to Witness Inspection/Test   │
    └──────────────────────┘
              │
              ▼
    ┌──────────────────────┐
    │  Perform Inspection/Test │
    └──────────────────────┘
              │
              ▼
           ◇ Approved ◇  ──NO──────────────────►
              │ YES
              ▼
    ┌──────────────────────┐
    │ Record Results on Check List │
    └──────────────────────┘
              │
              ▼
    ┌──────────────────────┐
    │ Record Performance Results │
    └──────────────────────┘
              │
              ▼
        ⬭ Proceed with Next
            Activity
```

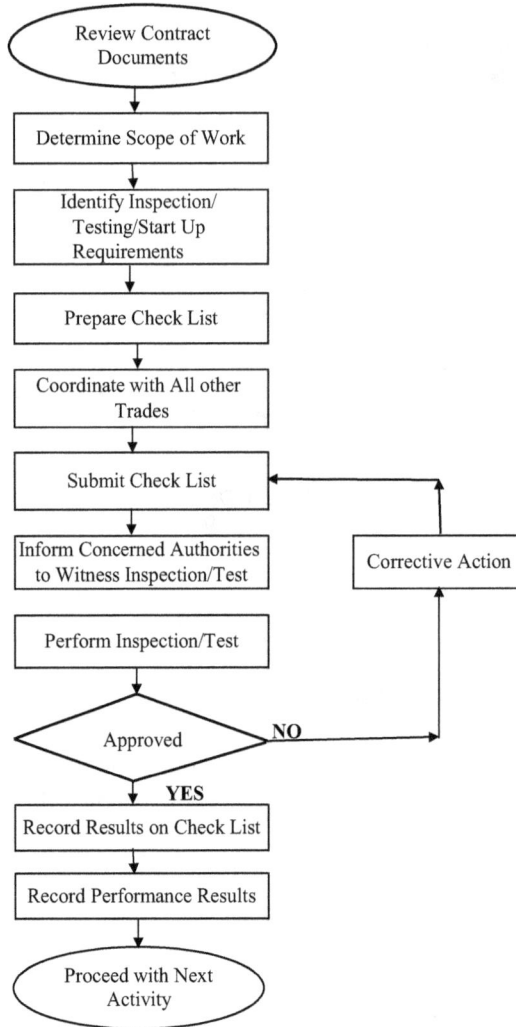

FIGURE 4.67 Logic flowchart for development of inspection and test plan. (Abdul Razzak Rumane. (2016). *Handbook of Construction Management.* Reprinted with permission from Taylor & Francis Group.)

22. Emergency power supply system
23. Electrically operated equipment

The testing of these works/systems is essential to ensure that each individual work/system is fully functional and operates as specified. The tests are normally coordinated and scheduled with specialist contractors, local inspection authorities, third-party inspection authorities, and manufacturer's representatives. Sometimes owner's representative may accompany the consultant to witness these tests.

Test procedures are submitted by the contractor along with the request for final inspection. Standard forms, charts, and checklists are used to record the testing results.

4.3.7.7 Manage Testing, Commissioning Quality

The contract document identifies testing and commissioning procedure to be followed on the installed equipment.

In order to follow the specified procedure and smooth functioning of testing and commissioning, the contractor/consultant has to plan quality (planning of design work), perform quality assurance, and control quality for managing testing and commissioning activities. This will mainly consist of following:

1. Plan Quality
    - Identify testing and commissioning requirements
    - Identify codes and standards to be followed for testing and commissioning
    - Identify contract specifications to be followed for testing and commissioning
    - Identify regulatory requirements to be followed for testing and commissioning
    - Identify (List) the items to be tested
    - Identify (List) the items to be commissioned
    - Establish testing and commissioning criteria
    - Identify manufacturer recommendation procedure
    - Establish detail procedure for testing of equipment
    - Ensure mechanical completion of the plant/equipment
    - Identify pre-commissioning requirement
    - Establish schedule for testing and commissioning
    - Establish sequence of commissioning of equipment
    - Identify and pre-qualify testing and commissioning agency, if mandated by the contract
    - Identify stakeholders to witness the testing and commissioning
    - Calibration/Certification of testing equipment to be used for testing and commissioning
    - Acceptance criteria
    - Establish corrective action plan

2. Perform Quality Assurance
    - Witness by the stakeholders
    - Ensure inspection of all installed equipment is finished
    - Manufacturer's recommended method statement
    - Ensure operating and maintenance manual requirements are followed
    - Sequence of testing and commissioning
    - Testing test points
    - Approval of third-party inspection agency
    - Approval of specialist commissioning agency
    - Records of testing and commissioning
    - Calibration certificate of testing equipment used for testing and commissioning
    - Ensure availability of "Punch List"

3. Quality Control
    - Check line by line as per flowchart
    - Test points (location)

- Flow test (water)
- Control system
- Storage capacity of tank
- Installation safety
- Environmental compatibility

4.3.7.8   Develop Documents

Table 4.73 illustrates project closeout documents to be submitted for project closeout.

4.3.7.8.1   As-Built Drawing

Most contracts require contractor to maintain a set of record drawings. These drawings are marked to indicate the actual installation where the installation varies appreciably from installation shown in the original contract. Revisions and changes to the original contract drawings are almost certain for any construction project. All such revisions and changes are required to be shown on the record drawings. As-built drawings are prepared by incorporating all the modifications, revisions, and changes made during the construction. These drawings are used by the user/operator after taking over the

TABLE 4.73
Project Closeout Documents

Description	Testing and Commissioning	As-Built Drawings	Operation and Maintenance	Guarantees	Warranties	Government Authorities	Record Documents	Test Certificates	Samples	Spare Parts	Punch Lists	Final Cleaning	Training	Taking Over Certificate	Remarks
ARCHITECTURAL WORKS															
CIVIL WORKS															
MECHANICAL WORKS															
HVAC WORKS															
ELECTRICAL WORKS (LIGHT & POWER)															
ELECTRICAL WORKS (LOW VOLTAGE)															
FINISHES															
EXTERNAL WORKS															

SAMPLE FORM

project for their reference purpose. It is the contractual requirements that the contractor handover as-built drawings along with record drawings, record specifications, and record product data to the owner/user before handing over of the project and issuance of substantial completion certificate. In certain projects contractor has to submit field records on excavation and underground utility services detailing their location and levels.

4.3.7.8.2 Technical Manuals and Documents

Technical manuals, design and performance specifications, test certificates, warranties and guarantees of the installed equipment are required to be handed over to the owner as part of contractual conditions.

Systems and equipment manuals submitted by the contractor to the owner/end user generally consist of

- Source Information
- Operating Procedures
- Manufacturer's Maintenance Documentation
- Maintenance Procedure
- Maintenance and Service Schedules
- Spare Parts List and Source Information
- Maintenance Service Contract
- Warranties and Guarantees

Procedure for submission of all these documents is specified in the contract document.

4.3.7.8.3 Record Books

- Manufacturing record book
- Project record book
- Engineering record book
- Construction record book

4.3.7.8.4 Warranties and Guarantees

Contractor has to submit warranties and guarantees in accordance with the contract documents. Normally the guarantee for waterproofing works varies from 15 to 20 years. Similarly, the warranty for diesel generator is set as five years.

4.3.7.9 Train Owner's/End User's Personnel

Normally training of user's personnel is part of contract terms. The owner/user's commissioning, operating and maintenance personnel are trained and briefed before commissioning starts in order to familiarize the owner/user's personnel about the installation works and also to ensure that the project is put into operation rapidly, safely, and effectively without any interruption. Timings and details of training vary widely from project to project. Training must be completed well in advance of the requirement to make the operating teams fully competent to be deployed at the right time during commissioning. This needs to be planned from project inception, so that the roles and activities of the commissioning and operating staff are integrated into a coherent team to maximize their effectiveness.

4.3.7.10 Handover Project

Once the contractor considers that the construction and installation of works have been completed as per the scope of contract and final tests have been performed and all the necessary obligations have been fulfilled, the contractor submits a written request to the owner/consultant for handing over of the project and for issuance of substantial completion certificate. This is done after testing and commissioning is carried out, and it is established that project can be put in operation or owner can occupy the same. In most construction projects, there is a provision for partial handover of the project.

4.3.7.10.1 Authority's Approval

Necessary regulatory approvals from the respective concerned authorities are obtained so that owner can occupy the facility and start using/operating the same. In certain countries all such approvals are needed before electrical power supply is connected to the facility. It is also required that the building/facility is certified by the related fire department authority/agency that it is safe for occupancy.

4.3.7.10.2 Handover of Spare Parts

Most contract documents include the list of spare parts, tools, and extra materials to be delivered to the owner/end user during the close out stage of the project. The contractor has to properly label these spare parts and tools clearly indicating the manufacturer's name and model number if applicable. Figure 4.68 illustrates spare parts handing over form used by the contractor.

4.3.7.10.3 Accept/Takeover of Project

Normally a final walk-through inspection of the project is carried out by the committee which consists of owner's representative, design and supervision personnel, and contractor to decide the acceptance of works and that the project is complete enough to be put to use and operational. If there are any minor items that remain to be finished, then such list is attached with the certificate of substantial completion for conditional acceptance of the project. Issuance of substantial completion certifies acceptance of works. If the remaining works are of minor nature, then the contractor has to submit a written commitment that he shall complete said works within the agreed upon period. A memorandum of understanding is signed between the owner and the contractor that the remaining works will be completed within an agreed upon period.

The contractor starts handing over all completed works/systems which are fully functional, and the owner has agreed to take over the same. A handing over certificate is prepared and signed by all the concerned parties. Figure 4.69 illustrates sample handing over certificate.

Project Name
Project Name
HANDING OVER OF SPARE PARTS

CONTRACTOR: _____   CERTIFICATE No. :

SUBCONTRACTOR:_____   DATE

SPECIFICATION NO : _____   DIVISION _____   SECTION : _____

DRAWING No. _____   BOQ Ref._____

AREA : ☐   Process Works   ☐   Electrical Works   ☐   Mechanical Works

☐   Piping Works   ☐   Instrumentation Works   ☐   Low Voltage System

---

Following Spare Parts have been handed over to the owner/end user

Description of Spare Parts

Sr.No.	Description	BOQ Reference	Spec. Ref.	Manufacturer	Specified Qty	Delivered Qty

SAMPLE FORM

(Attach additional sheet,if required)

SIGNED BY:

OWNER/END USER: _____ _____   CONTRACTOR: _____

CONSULTANT: _____   SUBCONTRACTOR:_____

FIGURE 4.68 Handing over of spare parts.

4.3.7.10.4 Punch List

Owner/consultant inspects the works and informs contractor of unfulfilled contract requirements. A punch list (snag list) is prepared by the consultant listing all the items still requiring completion or correction. The list is handed over to the contractor for rework/correction of the works mentioned in the punch list. Contractor resubmit

FIGURE 4.69 Handing over certificate.

inspection request after completing or correcting previously notified works. A final snag list is prepared if there are still some items which need corrective action/completion by the contractor, then such remaining works are to be completed within the agreed period to the satisfaction of the owner/consultant. Table 4.74 is sample form for preparation of punch list.

TABLE 4.74
Punch List

OWNER NAME:
PROJECT NAME:
PUNCH LIST:
Punch List Number: Date:
Area   Type of Work
Zone

Serial Number	Item	Remark
1	Piping	
2	Storage Tank	
3	Pumps	
4	Vessel	
5	Fire Alarm Detectors	
6	Sprinklers	
7	Communication System Devices	*Sample ListSample List*
8		
9		
10		
11		
12		
13		
14		
15		
16		
17		
18		
19	Any Other Item	

4.3.7.11   Issue Substantial Completion Certificate

A substantial certificate is issued to the contractor once it is established that the contractor has completed works in accordance with the contract documents and to the satisfaction of the owner. Contractor has to submit all the required certificates and other documents to the owner before issuance of the certificate.

The certificate of substantial completion is issued to the contractor and the facility is taken over by the owner/end user. By this stage, owner/end user already takes possession of the facility and operation and maintenance of the facility commences. The project is declared complete and is considered as the end of the construction project life cycle.

The defect liability period starts after issuance of substantial certificate.

During this period, the contractor has to complete the punch list items and also to rectify the defects identified in the project/facility.

4.3.7.12 Lesson Learned

Construction/project manager, consultant, and contractor have to prepare lesson learned and document the same for future references to improve the processes and organizational performance. This includes:

- Reasons for delay
- Cost overrun
- Reasons for rejection/rework
- Preventive/corrective actions
- Causes for claims

4.3.7.13 Settle Payments

The owner has to settle all the dues toward the consultant, contractor, and other parties involved. Similarly, the contractor has to settle due payments to their subcontractors and suppliers.

4.3.7.14 Settle Claims

The entire project-related claims to be amicably settled as per contract conditions to close the project.

4.3.7.15 Close Contract

Table 4.75 illustrates a list of activities that need to be considered for project closeout.

Table 4.76 summarizes the TQM concept for testing and commissioning phase activities.

TABLE 4.75
Project Close out Check List

Sr. No.	Description	Yes/No
Project Execution		
1	Contracted works completed	
2	Site work instructions completed	
3	Job site instructions completed	
4	Remedial notes completed	
5	Non-compliance reports completed	
6	All services connected	
7	All the contracted works inspected and approved	
8	Testing and commissioning carried out and approved	
9	Any snags	
10	Is project fully functional?	
11	All other deliverable completed	
12	Spare parts delivered	
13	Is waste material disposed?	
14	Whether safety measures for use of hazardous material established?	
15	Whether the project is safe for use/occupation?	
Project Documentation		
16	Authorities approval obtained	
17	Record drawings submitted	
18	Record documents submitted	
19	As-Built drawings submitted	
20	Technical manuals submitted	
21	Operation and maintenance manuals submitted	
22	Equipment/material warrantees/guarantees submitted	
23	Test results/test certificates submitted	
Training		
24	Training to owner/end user's personnel imparted	
Payments		
25	All payments to subcontractors/specialist suppliers released	
26	Bank guarantees received	
27	Final payment released to main contractor	
Handing over/Taking over		
28	Project handed over/taken over	
29	Operation/maintenance team taken over	
30	Excess project material handed over/taken over	
31	Facility manager in action	

TABLE 4.76

Total Quality Management Concept for Testing and Commissioning Phase Activities

Serial Number	Phase Activity	Stakeholders/ Team Members	Customer Specifications/ Requirements	System for Managing/Tool
1	Identification of Stakeholders/ Project Team Members	1. Project owner/ project manager 2. Supervision consultant 3. Contractor 4. Subcontractor 5. Third-party testing agency	1. Contract documents	Please refer Table 4.70, Table 4.71
2	Identify Testing, Commissioning, and Handover Requirements	1. Supervision consultant 2. Contractor	1. Contract requirements	Please refer Section 4.3.7.2
3	Identify Sustainability Requirements	1. Supervision consultant 2. Contractor	1. Contract requirements	Please refer Sections 4.3.7.3
4	Develop Testing, Commissioning Scope	1. Supervision consultant 2. Contractor	1. Contract requirements 2. Owner's requirements/ suitable for usage	Please refer Section 4.3.7.4 and Table 4.72
5	Develop Inspection and Testing Plan	1. Supervision consultant 2. Contractor	1. Contract requirements	Please refer Section 4.3.7.5 and Figure 4.67
6	Execute Commissioning of Works/System	1. Supervision consultant 2. Contractor	1. Contract requirements	Please refer Section 4.3.7.6
7	Manage Testing, Commissioning Quality	1. Supervision consultant 2. Contractor	1. Contract requirements	Please refer Section 4.3.7.7
8	Develop Documents	1. Supervision consultant 2. Contractor 3. Subcontractor 4. Vendors	1. Contract requirements	Please refer Section 4.3.7.8 and Table 4.73
9	Train Owner's/End user's personnel	1. Supervision consultant 2. Contractor	1. Contract requirements	Please refer Section 4.3.7.9

(continued)

TABLE 4.76 (Continued)

Total Quality Management Concept for Testing and Commissioning Phase Activities

Serial Number	Phase Activity	Stakeholders/ Team Members	Customer Specifications/ Requirements	System for Managing/Tool
10	Handover Project	1. Owner 2. End user 3. Supervision consultant 4. Contractor	1. Contract requirements	Please refer Section 4.3.7.10 and Figure 4.68, Figure 4.69, and Table 4.74
11	Issue Substantial Completion Certificate	1. Owner 2. Supervision consultant 3. Contractor	1. As per contract documents	Please refer Section 4.3.7.11
12	Lesson Learned	1. Owner 2. Supervision consultant 3. Contractor	As per the organization's requirements for continual improvement	Please refer Section 4.3.7.12
13	Settle Finance	1. Owner 2. Supervision consultant 3. Contractor	1. Settlement of approved payments	Please refer Section 4.3.7.13
14	Settle Claims	1. Owner 2. Supervision consultant 3. Contractor	1. Settle claims	Please refer Section 4.3.7.14
15	Close Contract	1. Owner 2. Supervision consultant 3. Contractor	1. Project is closed 2. Demobilization of resources	Please refer Section 4.3.7.15 and Table 4.75

Appendix: Content of Contractor's Quality Control Plan

The contractor's quality control plan is prepared based on project-specific requirements as specified in the contract documents.

The plan outlines the procedures to be followed during the construction period to attain the specified quality objectives of the project fully complying with the contractual and regulatory requirements.

Following is an outline of such a plan, based on contract documents.

A.1 INTRODUCTION

The contractor's quality control plan (QCP) is developed to meet the contractor's quality control (QC) requirements of (project name) as specified under clause (----) and section (----) of contract documents.

The plan provides the mechanism to achieve the specified quality by identifying the procedures, controls, instructions, and tests required during the construction process to meet the owner's objectives. This QCP does not endeavor to repeat or summarize contract requirements. It describes the process M/S ABC (contractor name) will use to assure compliance with those requirements.

A.2 DESCRIPTION OF PROJECT

(Owner name) has contracted M/S ABC to construct (name of facility) located at (site plan). The facility consists of approximately (---) m² gross building area and approx (---) m² of two basements for car parking and other services.

The facility is to be used as office premises to accommodate (---) personnel. The architecture of the building is of very high quality with a spacious atrium between two towers interconnected with a bridge inside the building, and a well-designed internal landscape area with plants and sky garden to provide a pleasant view inside the building. The building has glazed walls and curtain walling for the atrium area.

Vertical transportation is through panoramic elevators and a designated elevator for VIPs. Additionally, there is a goods elevator and firefighters' elevator to meet an emergency situation.

The structure of the building is of reinforced concrete, and the atrium is of structured steel. External cladding includes a spandrel curtain wall for the block structure and special curtain walling for the atrium. Interior finishes are painted plastered walls and marble and stone finish. Internal partitions have flexibility to adjust the office area. The entire building has a raised floor system, and space under the raised floor is used for distribution of electromechanical services.

The HVAC system is comprised of water-cooled chillers, and electric duct heaters are provided for heating during winter. BMS takes full control of HVAC. Building is provided with all the safety measures against fire. A fire protection system, smoke exhaust fans, and fire alarm system with voice evacuation system are provided to meet an emergency situation. The fire protection system also includes automatic sprinklers, hose reels, extinguishers, and a foam system.

The building is equipped with all the amenities required for a public health system. The plumbing system is consistent with the requirement of the building and includes cold and hot water, drainage, rainwater collection, and an irrigation system. Water fountains are provided at various locations for drinking purposes.

Electrical systems consist of energy-saving lighting with centralized controls. Electrical distribution has all the required safety features. A diesel generator system and uninterrupted power supply system are provided to meet the emergency due to power failure from the main electricity provider. The building to be constructed shall be equipped with the latest technological systems. It shall have an IP-based communication system, and all the low-voltage systems shall be fully integrated. Authorized persons shall have access from anywhere, either from within the building or outside. The building has fully integrated low-voltage systems.

Apart from training and conference facilities, a fully functional auditorium having capacity for (---) people and conferencing is available in the building. It has a sophisticated conferencing system with a rear projection screen.

The contract documents consist of total (----) contract drawings. These are

Architectural	(---) nos.
Architectural Interior	(---) nos.
Structural	(---) nos.
Firefighting	(---) nos.
Public Health	(---) nos.
HVAC	(---) nos.
Electrical	(---) nos.
Smart Building	(---) nos.
Traffic Signage	(---) nos.
Landscape	(---) nos.

A.3 QUALITY CONTROL ORGANIZATION

The QC organization is independent of those persons actually performing the work. They shall be responsible for implementing the quality plan for the entire contract/

project-related activities by scheduling the inspection, testing, sampling, and preparation of mockup and to assure that works are performed per approved shop drawings and contract documents.

An organization chart showing the line of authority and functional relationship is shown in Figure A.1.

QC incharge shall be responsible to ensure the implementation of project quality. He will be supported by QC engineers as follows:

1. QC Engineer (Civil Works)
2. QC Engineer (Concrete Works)
3. QC Engineer (Mechanical Works)
4. QC Engineer (Electrical Works)
5. Foreman (Concrete Works)

These engineers will be responsible for implementing three phases of the quality system at site. The QC incharge will coordinate with the company's head office for all the support and necessary actions. Respective QC engineers shall be responsible to implement the quality program in their respective fields.

A.4 QUALIFICATION OF QC STAFF

All the QC personnel will have adequate experience in their respective fields. Following is the qualification of each of the QC staff. Additional staff shall be provided if required.

QC Incharge: He shall be a qualified civil engineer having minimum ten years of experience as a QC engineer in major construction projects.

QC Engineer (Civil Works): Graduate civil engineer with a minimum three years' experience as a quality engineer on similar projects.

QC Engineer (Concrete Works): Graduate civil engineer with a minimum three years' experience as a quality engineer on similar projects.

QC Engineer (Mechanical Works): Graduate mechanical engineer with a minimum three years' experience as a quality engineer on similar projects.

QC Engineer (Electrical Works): Graduate electrical engineer with a minimum three years' experience as a quality engineer on similar projects.

Foreman (Concrete Works): Diploma in civil engineering with a minimum five years' experience as a quality supervisor on similar projects.

A.5 RESPONSIBILITIES OF QC PERSONNEL

Project Manager: Overall project responsibilities.

QC Incharge: He/she shall be responsible for the following:

- Preparation of QC plan
- Responsible for overall QC/QA responsibilities
- Responsible for monitoring and evaluation of CQP

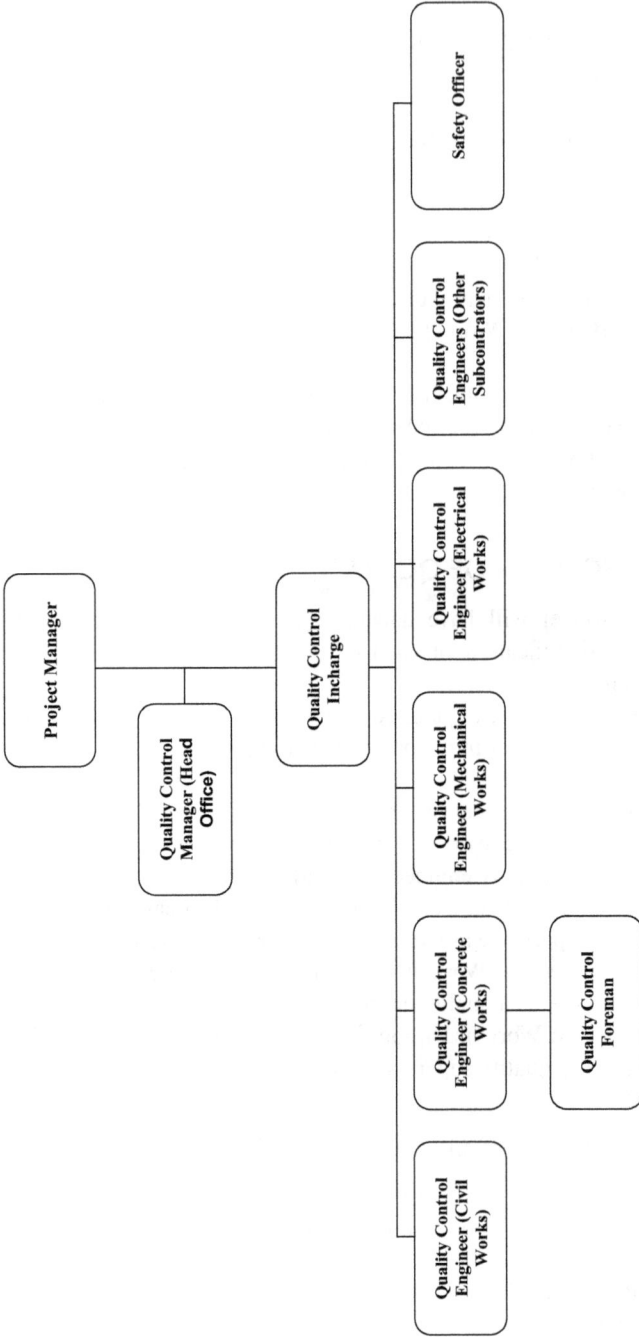

FIGURE A.1 Site quality control organization.

- Implementation of QC plan
- Responsible for implementing QC procedure
- Maintain QC records and documents
- Inspection of works
- Responsible for off-site inspection
- Inspection of incoming material
- Responsible for subcontractor's QC plan
- Coordinating with safety officer to implement safety plan
- Calibration of measuring instruments
- Monitoring equipment certification

QC Engineers: Each trade shall have a responsible QC engineer to ensure that works are carried out in accordance with the contract documents. Following QC engineers shall be available at site. A foreman will assist the QC engineer to maintain the quality of concrete works.

1. QC Engineer (Civil Works)
2. QC Engineer (Concrete Works)
3. QC Engineer (Mechanical Works)
4. QC Engineer (Electrical Works)
5. QC Engineer (Other Subcontractors)
6. Foreman (Concrete Works)

The aforementioned personnel will be responsible for the following:

- Overall quality of the respective trade
- Preparation of method statement
- Preparation of mockup
- Monitor and inspect quality of site works
- Inspection of incoming material
- Coordinate with QC incharge for preparation of CQP
- QC records and documents

A.6 PROCEDURE FOR SUBMITTALS

M/S ABC shall be responsible for the timely submission of subcontractors, shop drawings, and materials to be used in the project.

A.6.1 Submittals of Subcontractor(s)

Prior to submission of the subcontractor's name to the owner/consultant, M/S ABC shall review the capabilities of the subcontractor to perform the contracted works properly per the specified quality and within the time allowed. M/S ABC shall also consider the contractor's performance in the previously executed project. Companies implementing a quality management system plan and having ISO certification shall be given preference while selecting the subcontractor(s) to work on this project.

A.6.2   SUBMITTALS OF SHOP DRAWINGS

Before the start of any activity, M/S ABC shall submit shop drawings based on contract drawings and other related documents. The number of shop drawings shall depend on the requisite details to be prepared against each contract drawings. The shop drawings shall be prepared in the agreed format and shall have coordination certification from all the trades. All the shop drawings shall be submitted through site transmittal for shop drawings. Shop drawing approval transmittals shall be numbered serially for each section of the particular specification. M/S ABC shall be responsible for shop drawings related to subcontracted works.

A.6.3   SUBMITTALS OF MATERIALS

M/S ABC shall submit materials, products, and systems for owner's/consultant's approval from the recommended/nominated manufacturers specified in the contract documents. If material is proposed from a manufacturer who is not nominated, then all the documents to justify that the proposed product is approved and equal to the specified shall be submitted along with the site transmitted for materials approval.

A.6.4   MODIFICATION REQUEST

If during construction, the contractor observes that some work as mentioned in the contract needs modification, then such a modification request shall be submitted to the owner/consultant for their review and approval. Financial obligations shall be clearly specified in the request form.

A.6.5   CONSTRUCTION PROGRAM

M/S ABC shall prepare and submit a construction program based on the contracted time schedule and submit the same to the consultant for approval. Progress of works shall be monitored on a regular basis, and the construction program shall be updated as and when required to overcome the delay (if any). The following reports shall be submitted to the consultant per the following schedule:

1. Daily Report on a daily basis
2. Weekly Report on a weekly basis
3. Monthly Report on a monthly basis
4. Progress Photographs on a monthly basis

A.7   QUALITY CONTROL PROCEDURE

A.7.1   PROCUREMENT

Prior to placing a purchase order or supply contract with the supplier/vendor, ABC will ensure that the material has been approved and complies with the requirements of the contract. A procurement log shall be maintained for all the items for which action has been taken. The procurement log shall be submitted to the owner/consultant every two weeks for review and information.

A.7.2 INSPECTION OF SITE ACTIVITIES (CHECKLISTS)

All the works at site shall be performed per approved shop drawings by using approved material. At least three phases of control shall be conducted by the QC incharge for each activity prior to submission of the checklist to the consultant.
These will be

- Preparatory Phase
- Start-up Phase
- Execution Phase

Preparatory Phase: The concerned engineer shall review the applicable specifications, references, and standards, approved shop drawings and materials, and other submittals (method of statements, etc.) to ensure compliance with the contract.

Start-up Phase: The concerned engineer shall discuss with the foreman responsible to perform the work and shall establish the standards of workmanship.

Execution Phase: The work shall be executed by continuous inspection during this phase.

All the execution/installation works shall be carried out per approved shop drawings utilizing approved material. All the works will be checked for quality before submitting the checklist to the consultant to inspect the works.

The following main categories of site works shall be performed by conducting the QC per above-mentioned phases.

A.7.2.1 Definable Feature of Work

Procedure for scheduling, reviewing, certifying, and managing QC for a definable feature of work shall be agreed upon during the coordination meetings. The definable features of work for concrete shall include formwork, reinforcement, embedded items, design mix, placement of concrete, and concrete finishes.

Likewise, the detailed list for other trades shall be prepared during the coordination meetings. This will consist of embedded conducting works, embedded sleeves, ducts, underground piping, and so on.

A.7.2.2 Earthworks and Site Works

This category shall consist of the following subgroups:

1. Excavation and backfilling
2. Compaction
3. Dewatering

A checklist shall be submitted to the consultant at every stage of work, that is, after backfilling and compaction. Samples shall be taken to make a compaction test. If the compaction test fails the required specifications, then remedial action shall be taken to obtain specified results.

A.7.2.3 Concrete

M/s (approved subcontractor to supply ready-mix concrete) will provide ready-mix concrete for entire concrete structured works of the project. M/s (ready-mix subcontractor) will have their own QC system to maintain the mix design.

Reinforcement steel shall be inspected upon receipt of material at site for proper size and type, and the factory test certificates received with the supply. Small pieces of sample shall be taken from each lot and size for laboratory testing. The testing shall be performed by an approved testing laboratory.

Formwork and the reinforcement steel process shall be performed under a qualified civil engineer(s) and foreman (foremen). They will be continuously inspecting the work for each operation and checklist shall be submitted to the consultant for their approval.

Placement of concrete shall be supervised by the civil engineer along with the foreman and other team members. The respective consultant will witness concrete casting and take concrete samples during casting for laboratory testing and crushing tests. Slump tests shall be performed on the concrete to verify the strength. Curing shall be supervised by the respective foreman.

In case the test results do not comply with the specification limit, the results shall be discussed with the consulting engineer. Failing tests will be followed by appropriate corrective (reworking) efforts and retesting and remedial measures.

M/S ABC shall maintain all the records, laboratory tests, and results regarding the concrete works and submit the same to the consultants for their information.

A.7.2.4 Masonry

The concrete block bricks used on the project shall be procured from the approved source. Material for mortar aggregate and other accessories such as inserts, reinforcement material, and wire mesh shall be submitted to the consultant for approval.

Prior to the start of masonry work, all the related and coordinated drawings shall be reviewed. Marking shall be made for masonry layout. Necessary mockup shall be submitted for each type of unit masonry work. A checklist shall be submitted for visual and other type of inspection.

A.7.2.5 Metal Works

Metal fabrication and installation work shall be carried out by certified welders, per manufacturer's recommendations, and per approved shop drawings. Samples of fittings, brackets, fasteners, anchors, and different types of metal members including plates, bars, pipes, tubes, and any other type of material to be used in the project shall be submitted to the consultants prior to start of fabrication works at site. Finishing material such as primer and paint shall be submitted for approval to the consultant/ owner.

In case the fabrication is carried out at the subcontractor(s)'s workshop, then ABC shall take full responsibility to control the quality.

A checklist shall be submitted at different stages of work to ensure compliance with the specifications and to avoid any rework at the end of completion of fabrication and installation.

A.7.2.6   Wood, Plastics. and Composite

Prior to the start of any woodwork, the material shall be physically and visually inspected for the timber quality to confirm its compliance with the type of the approved material and also to ensure that the wood is free from decay and insects and that necessary treatment has already been done on the wood to be used for the project. A product certificate signed by the woodwork manufacturer certifying that the products comply with specified requirements shall be submitted along with material submittal. All types of fasteners and other hardware to be used shall be submitted for approval per applicable standards, and checklists shall be submitted at various stages of work prior to applying the paint or any other finishes.

A.7.2.7   Doors and Windows

This category shall consist of the following subcategories:

1. Wooden doors and windows
2. Steel doors
3. Aluminum windows
4. Glazing

All doors and windows shall be fabricated from a specified type and material and per approved shop drawings.

Fire-rated types of doors shall comply with relevant standards and local regulations.

Samples of materials used in fabrication of doors and windows shall be submitted to the consultant/owner for approval.

Wooden doors and windows shall be fabricated at an approved subcontractor for carpentry works and aluminum doors and windows shall be fabricated at an approved subcontractor for aluminum works. Finishes shall be as approved by the owner/consultant. Samples of ironmongery coating material and finishing material shall be submitted for approval. Special care shall be taken to maintain the acoustic nature of doors and their sound transmission properties.

Glass and glazing material shall be submitted for approval and shall take care of heat transfer properties to maintain the inside temperature of the building. The glass used in the project shall comply with specified strength, safety, and impact performance requirements.

Inspection at different stages shall be arranged by field inspection at the factory.

The entire fabrication shall be carried out under supervision of ABC to control the quality of finished products.

Doors and windows fabrication shall be coordinated for security requirements. Mockup shall be prepared before the start of installation work.

A.7.2.8   Finishes

This category shall consist of the following subgroups:

1. Acoustic ceiling
2. Specialty ceiling
3. Masonry flooring
4. Tiling
5. Carpet
6. Wall cladding
7. Wall partitioning
8. Paints and coating

All the finishes shall be as specified and approved. Execution of finishing work shall be coordinated with all other trades. Samples and mockup shall be submitted for approval before the start of any finishing work. All the finishing work shall be performed by skilled workmen. Specialist subcontractor(s) shall be submitted for approval to carry out finishing works. QC shall be under direct supervision of ABC.

A.7.2.9   Furnishing

All the furnishing shall be from the specified recommended manufacturer. Special care shall be taken to protect the furnishings. Furnishing material sample(s) shall be submitted for approval. Fabrication of furniture shall be at the approved subcontractor's factory. ABC shall be responsible for controlling the quality.

A.7.2.10   Equipment

This category shall consist of the following subgroups:

1. Maintenance equipment such as window and facade cleaning
2. Stage equipment
3. Parking control equipment
4. Kitchen equipment

All the equipment shall be from specified manufacturers. If the items specified are commercial items—that is, the materials are manufactured and sold to the public as against the materials made to specifications—necessary precaution shall be taken to verify that materials shall be installed per manufacturer's recommended procedure under the supervision of a qualified engineer.

Items to be manufactured per contract documents shall be procured from the manufacturer producing products complying with recognized quality standards. The product and manufacturer's data shall be submitted for approval with all the technical details. Installation of the equipment shall be carried out by skilled workmen under supervision of the manufacturer's authorized representative and under the control of ABC.

A.7.2.11 Conveying System

This category consists of the following subgroups:

1. Passenger elevators
2. Goods elevators
3. Kitchen elevators
4. Escalators

Prior to submission of shop drawings, full technical details along with the catalogs and compliance statement shall be submitted for approval. The elevators shall be from one of the recommended manufacturers specified in the contract documents.

Fabrication of conveying systems shall comply with relevant codes and standards as applicable. Regulatory approval shall be obtained from local authorities for installation of the conveying systems to assure their compliance with local codes and regulations.

Finishing material for the cab, landing door, car control station, and car position indicator shall be submitted for approval.

Coordinated factory shop drawings shall be submitted for approval. Size of shaft and door openings shall be coordinated during structural work. Necessary tests shall be performed by the manufacturer's authorized personnel.

Factory fabrication works shall be performed per the manufacturer's QCP, and installation at site shall be carried out by personnel authorized by the manufacturer.

Third-party inspection shall be arranged per specification requirements.

A.7.2.12 Mechanical Works

This category shall consist of the following subgroups:

1. Fire suppression
2. Water supply, plumbing, and public health
3. HVAC

Mechanical works shall be executed by approved mechanical subcontractor. All the works shall be carried out per approved shop drawings using approved material.

A mechanical engineer from the subcontractor shall be responsible for controlling the quality of work. A QC engineer (mechanical works) shall coordinate all quality-related activities on behalf of ABC. Fire suppression or firefighting equipment shall be carried out per approved drawings by the relevant authority. Materials shall be from the manufacturers approved by the relevant authority having jurisdiction over such materials. Fire pumps and firefighting materials shall comply with regulatory requirements.

Piping for water supply, plumbing, and public health shall be installed with approved material. Leakage and pressure tests shall be performed as specified.

Checklists shall be submitted after installation of piping, accessories, and fixing of equipment. Drainage systems shall be executed per approved shop drawings.

HVAC works shall be supervised by a qualified mechanical engineer. Ductwork and duct accessories shall be fabricated at the subcontractor's workshop and shall be installed at site per approved shop drawings. A mechanical engineer from subcontractor shall be responsible to control the quality at the workshop. A QC engineer (mechanical works) shall coordinate all quality-related activities on behalf of ABC. Duct material shall be submitted for approval prior to start of fabrication works. Chilled/hot water piping shall be from the approved manufacturer. Installation of piping shall be carried out by certified pipe fitters.

Selection of HVAC pumps shall be done with the help of performance curves and technical data from the pump manufacturer. Chillers shall be from the specified manufacturer.

Installation of chillers shall be carried out per manufacturer's recommendation.

A.7.2.13 Automation System

Automation work shall be carried out under supervision of a specialist engineer. The system shall be from the specified manufacturer.

Shop drawings and schematic diagrams shall be submitted for approval configuring approved components/items/equipment.

A.7.2.14 Electrical Works

This category shall consist of the following subgroups:

1. Electrical lighting and power
2. Fire alarm system
3. Communication system (telephone)
4. Public address system
5. Access control and security system
6. Audio-visual system
7. MATV system
8. Emergency generating system
9. Alternate energy system

All types of electrical works shall be executed by an approved electrical subcontractor. All the materials/products to be used shall be from the specified manufacturer and shall be installed by skilled workmen under the supervision of a qualified engineer.

An electrical engineer from the subcontractor shall be responsible to control the quality at the workshop. A QC engineer (electrical works) shall coordinate all quality-related activities on behalf of ABC. Work shall be carried out per approved shop drawings. Quality of the works shall be controlled at different stages such as after installation of conducting/raceways, pulling of wires/cables, and installation of accessories/equipment/panels. Works shall comply with the specifications and recognized codes and standards.

Specified tests shall be performed at every stage. Coordination will be done with other trades to assure that required power supply for their equipment will be available at designated locations.

A.7.2.15 Landscape

Landscape work shall be executed by a specialist-approved subcontractor per the approved shop drawings under direct supervision of ABC. Samples of each and every type of plant, tree, and so on shall be submitted for approval. Soil preparation shall be done per specified standards.

Checklists shall be submitted after completion of activities at each stage.

Work not accepted or rejected shall be repaired/reworked or replaced and checklist shall be resubmitted.

Guarantees and warranties for installed products/items shall be submitted per contract requirements during handover of the project.

A.7.3 INSPECTION AND TESTING PROCEDURE FOR SYSTEMS

The QC procedure for systems (low voltage/low current) and their equipment/components shall be performed in the following stages.

The proposed manufacturer's complete data giving full details of the capability of the proposed manufacturer to substantiate compliance with the specifications and to prove that the proposed manufacturer is following quality management system and producing the product to recognized quality standards as specified shall be submitted. Components/equipment data shall be submitted for all the specified items of the system.

A schematic diagram configuring all the items for the system shall be submitted along with a detailed technical specification and confirmation from the manufacturer that it complies with the contract documents and is equal to or better than the specified products.

Shop drawing(s) fully coordinating with other trades, showing the system location, and the raceways used for installation of the system shall be submitted.

All the work at site shall be submitted for inspection at different stages of execution of work.

Cable testing shall be performed prior to installation of equipment. Checklists shall be submitted to the consultant to witness these tests.

Installation of equipment shall be done per manufacturer's recommendations and per approved installation methods.

Checklists along with testing procedures shall be submitted prior to inspect final inspection and testing.

A.7.4 OFF-SITE MANUFACTURING, INSPECTION, AND TESTING

QC procedures for off-site manufacturing/assembling shall be per contract documents. Factory control inspection for items fabricated or assembled shall be carried out to comply with all the specified requirements and standards under the responsibility of QC personnel at the fabrication/assembly plant. Specified tests shall be carried out per the manufacturer's quality procedures and per specification requirements. Test reports for items fabricated/assembled off-site shall be submitted along with the delivery of material at site. Witnessing/observation of inspection and tests by the consultant/client, if it is required per specifications, shall be arranged. Procedure for such tests and inspection shall be submitted in advance of factory/plant visit.

QC for off-the-shelf items shall be standard QC practices followed by the manufacturer.

A.7.5 PROCEDURE FOR LABORATORY TESTING OF MATERIALS

M/S ABC shall submit name(s) of testing laboratories for approval for performing tests specified in the contract documents.

A.7.6 INSPECTION OF MATERIAL RECEIVED AT SITE

All the material received at site shall be submitted for inspection by the consultant. A material on-site inspection request (MIR) shall be submitted to the consultant. Prior to dispatch of material from the manufacturer's premises, all the tests shall be carried out, and test certificates shall be submitted along with the MIR. Apart from this, all the relevant documents such as the packing list, delivery note, certificate of compliance to recognized standards, country of origin, and any other documents shall be submitted along with the MIR.

Material received at the site shall be properly stored at designated locations and shall maintain original packing until its installation/use.

Inventory records shall be maintained for all the material stored at the site. In case of materials to be directly installed at the site upon its receipt, the inspection shall be carried out before its use or installation on the project.

In case of concrete ready mix, samples will be tested per contract documents before pouring the concrete for castings.

Specified tests shall be performed after delivery of material at site.

A.7.7 PROTECTION OF WORKS

All the works shall be protected at site per specified methods in the contract documents.

A.7.8 MATERIAL STORAGE AND HANDLING

M/S ABC shall arrange for a storage facility on the project site to keep the material in safe condition. Material at site shall be fully protected from damages and properly stored. Necessary care shall be taken to store paints and liquid types of materials by maintaining appropriate temperature to prevent damage and deterioration of such products. Materials, products, and equipment shall be properly packed and protected to prevent damage during transportation and handling.

A.8 METHOD STATEMENT FOR VARIOUS INSTALLATION ACTIVITIES

Method statement shall be submitted to the consultant for their approval per the contract documents. The method statement shall describe the steps involved in execution/installation of work by ensuring safety at each stage. It shall have the following information:

1. Scope of Work: Brief description of work/activity
2. Documentation: Relevant technical documents to undertake this work/activity
3. Personnel Involved
4. Safety Arrangement
5. Equipment and Plant Required
6. Personal Protective Equipment
7. Permits/Authorities' Approval to Work
8. Possible Hazards
9. Description of the Work/Activity: Detail method of sequence of each operation/key steps to complete the work/activity

Figures A.2–A.9 are illustrative examples of the method of sequence for major activities, for different trades, performed during the construction process.

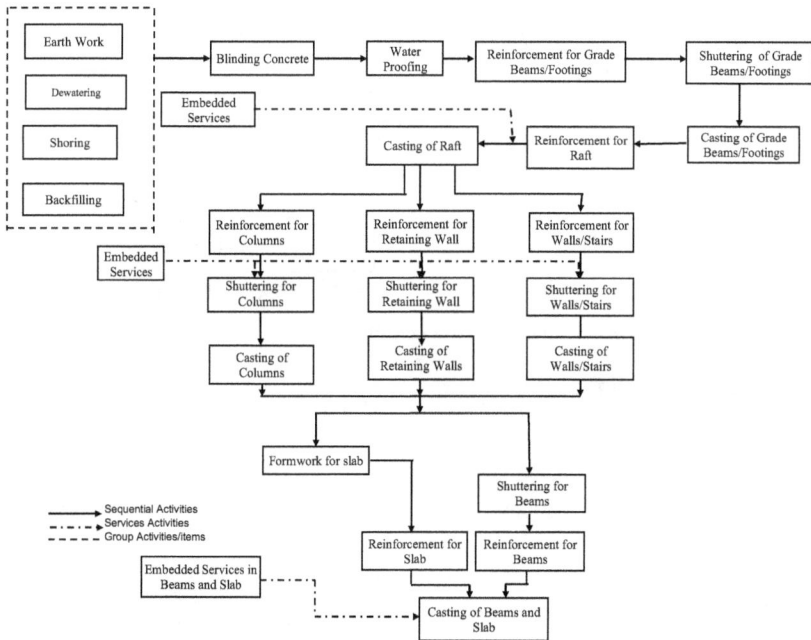

FIGURE A.2   Method of sequence for concrete structure work.

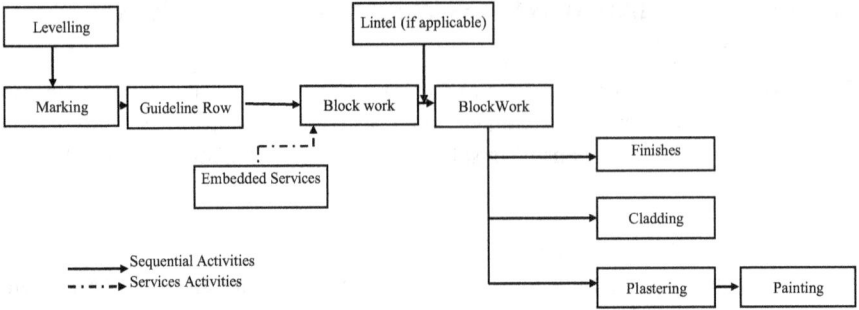

FIGURE A.3 Method of sequence for block masonry work.

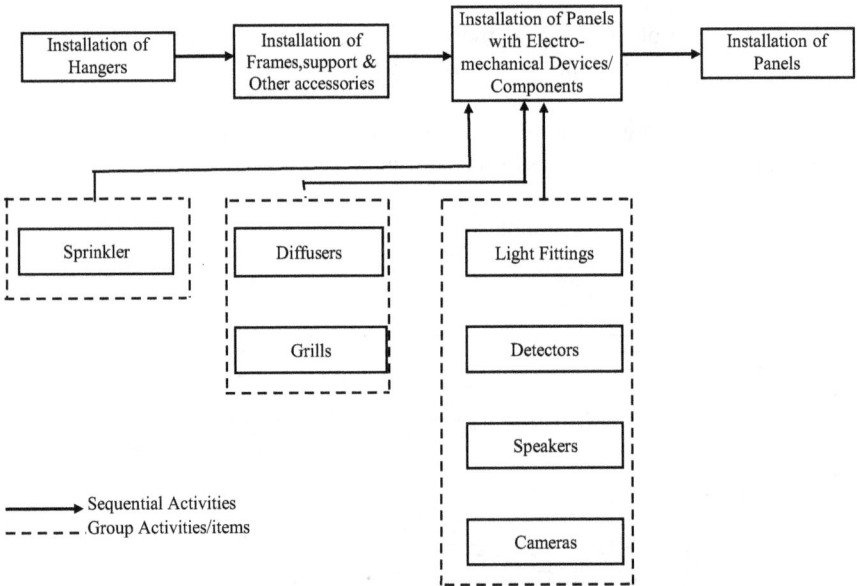

FIGURE A.4 Method of sequence for false ceiling work.

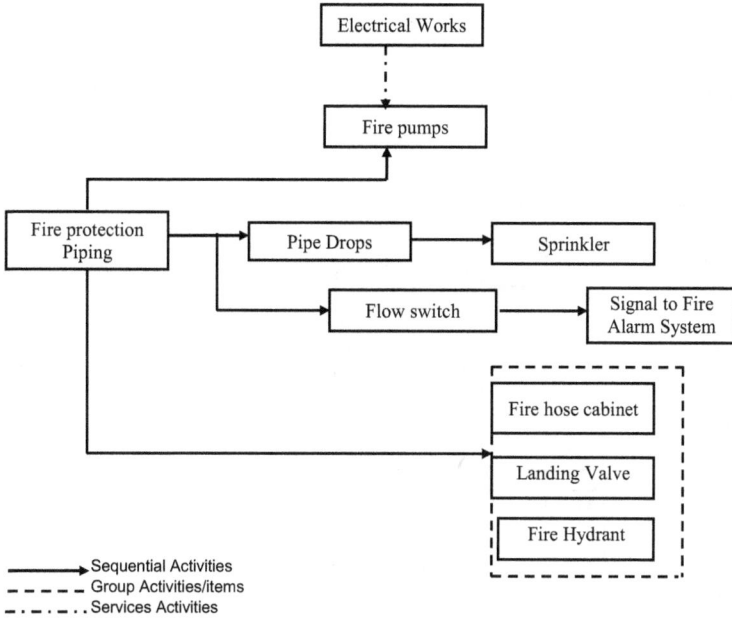

FIGURE A.5 Method of sequence for HVAC work.

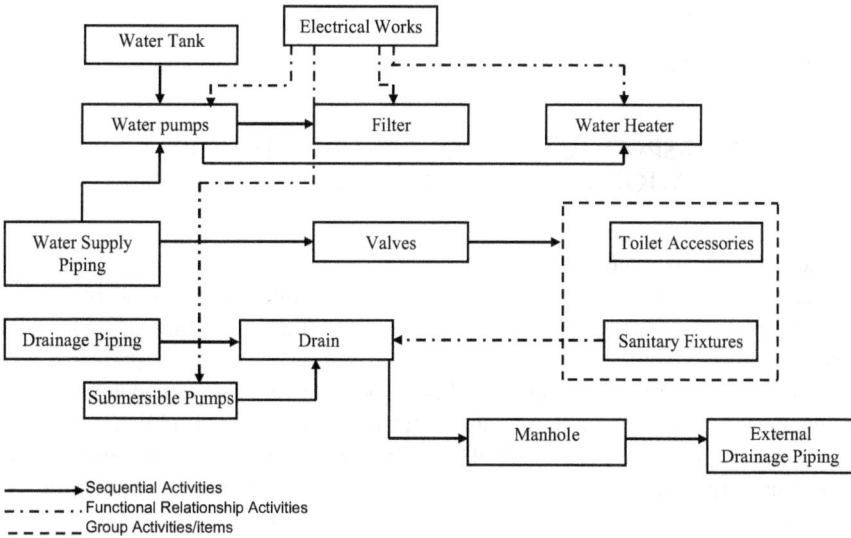

FIGURE A.6 Method of sequence for mechanical work (public health).

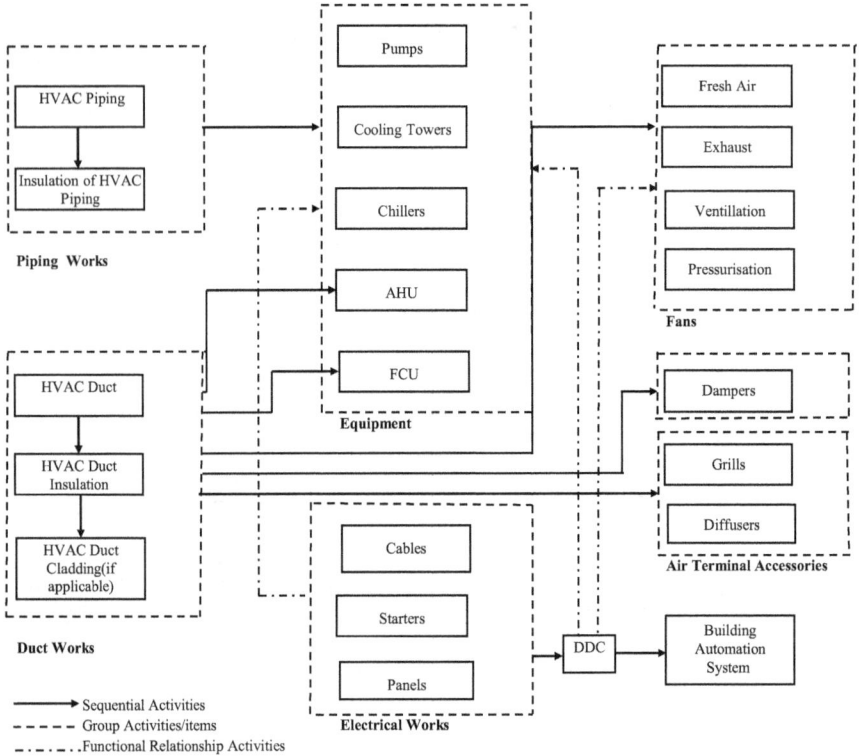

FIGURE A.7   Method of sequence for mechanical work (fire suppression).

A.9   PROJECT-SPECIFIC PROCEDURES FOR SITE WORK INSTALLATIONS REMEDIAL NOTICE

QC for such installations shall be as specified in the contract documents or as mentioned in the instructions.

Remedial notices received shall be actioned immediately and replied to inspect the works having corrected/performed per contract documents. Similarly, nonconformance report shall be actioned and replied.

All repaired and reworked items shall be inspected by the concerned engineer and QC engineer before submitting a "Request for Inspection" or "Checklist."

Appropriate preventive actions shall be taken to avoid repetition of nonconformance work. The following steps shall be taken to deal with any problem requiring preventive action:

- Detect
- Identify the potential causes
- Analyze the potential causes
- Eliminate the potential causes
- Ensure the effectiveness of preventive action

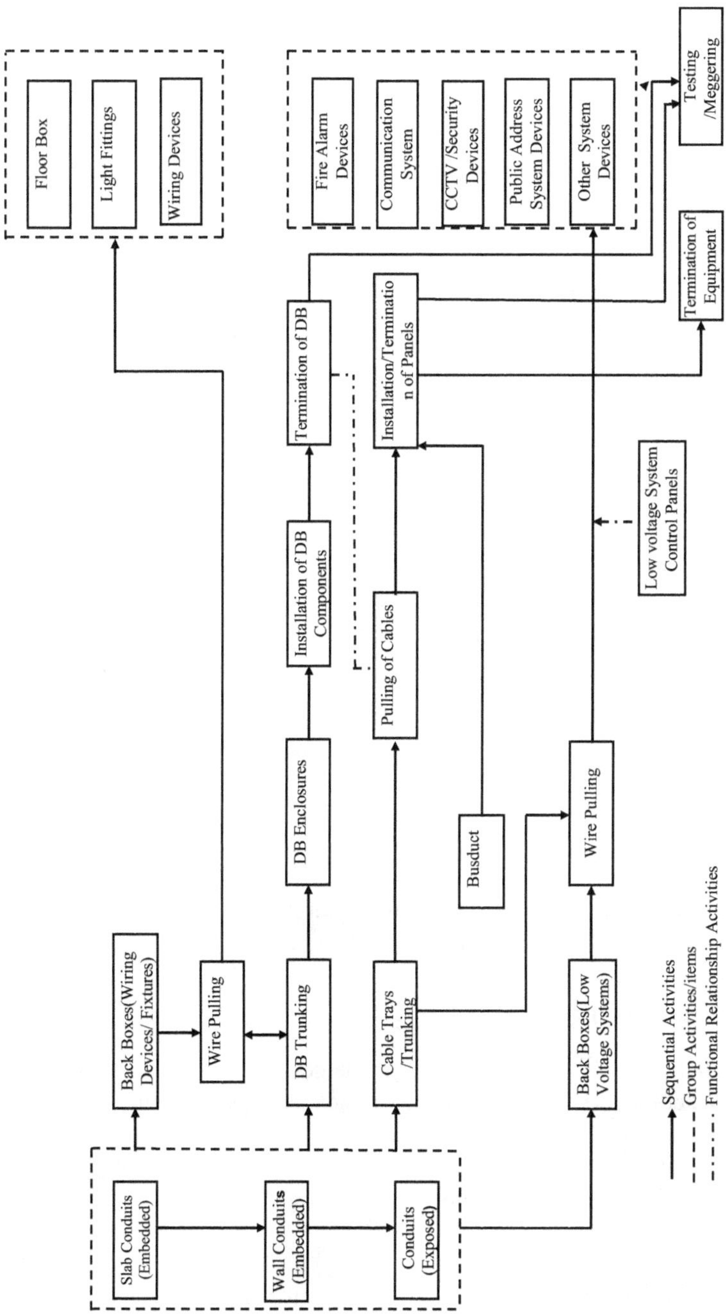

FIGURE A.8 Method of sequence for electrical work.

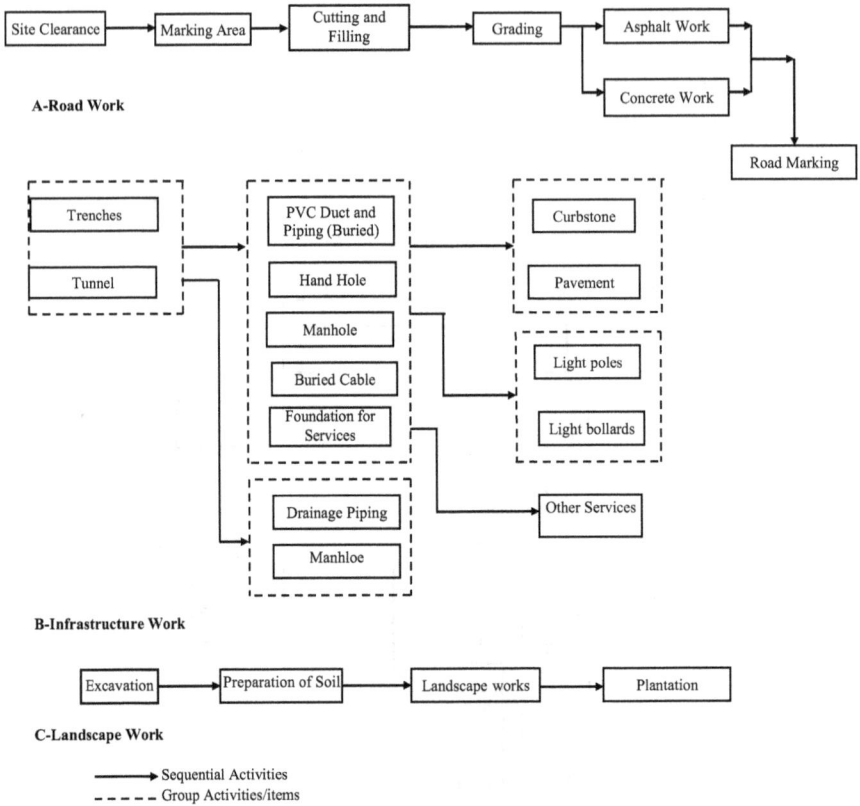

FIGURE A.9   Method of sequence for external works.

A.10   RISK MANAGEMENT

The contractor shall maintain a risk register throughout the contract period and shall resolve/mitigate the risk with appropriate response(s)

A.11   QUALITY CONTROL RECORDS

QC records shall be maintained by the QC incharge and shall be accessible to authorized personnel for review and information. The records shall include

- Contract drawings and revisions
- Contract specifications and revisions
- Approved shop drawings
- Record drawings
- Material approval record
- Checklists
- Test reports
- Material inspection reports

- Minutes of QC meetings
- Site works instructions
- Remedial notes
- Other records specified in the contract documents or as requested by the owner/consultant

A.12 COMPANY'S QUALITY MANUAL AND PROCEDURES

The following documents from the company's manual shall be part of contractor's QCP.

1. Documents for specific projects
2. Subcontractor evaluation
3. Supplier evaluation
4. Client supplied items
5. Receiving inspection and testing
6. Site inspection and testing
7. Final inspection, testing, and commissioning
8. Material storage and handling
9. Risk management system
10. Control of quality records
11. Internal audit

A.13 PERIODICAL TESTING OF CONSTRUCTION EQUIPMENT

Hoists, cranes, and other lifting equipment used at site shall be tested periodically by third-party inspection authority and test certificates shall be submitted to the consultant. Measuring instruments shall have calibration tests performed by the approved testing agencies/government agencies at regular intervals, and a calibration certificate shall be submitted to consultants. Other equipment shall be tested/calibrated per contract requirements and per the schedule for such tests.

A.14 QUALITY UPDATING PROGRAM

The QC program shall be reviewed every six months and necessary updates shall be done, if required. All such information shall be given to the consultants and a record shall be made for such revisions.

A.15 QUALITY AUDITING PROGRAM

Internal auditing shall be performed by the company's internal auditor to ensure that specified quality procedures are followed by the site quality personnel.

A.16 TESTING, COMMISSIONING, AND HANDOVER

The test shall be carried out per contract documents. Necessary test procedures shall be submitted prior to the start of testing of installed systems/equipment.

Engineers from the respective trades will be responsible for arranging orderly performance of testing and commissioning of the systems/equipment.

A.17   HEALTH, SAFETY, AND THE ENVIRONMENT

The health, safety, and the environment officer shall be responsible for implementing health and safety measures per OHSAS 18000 and the project's environmental requirements per local regulatory authority. ABC shall prepare and submit a health and safety management and accident prevention program. The safety program shall embody the prevention of accidents, injuries, occupational illnesses, and property damage. All personnel at the site shall be provided with safety gear. Regular meetings and awareness training shall be conducted at the site on a regular basis. ABC shall ensure that first aid and emergency medical facilities are available at the site all the time. ABC shall comply with all safety measures specified in the contract documents. ABC shall take all necessary measures to comply with regulatory requirements and environmental considerations.

Bibliography

AACE International Recommended Practice No. 18R-97 (2010).

AACE International Recommended Practice No. 27R-03 (2010).

AACE International Recommended Practice No. 37R-06 (2010).

Construction Industry Institute (June 1992), *Guidance for Implementing Total Quality Management in Engineering and Construction Industry*, CII Source Document 74, University of Texas, Austin, Texas.

Abdul Razzak Rumane (2013), *Quality Tools for Managing Construction Projects*, CRC Press, Florida (Taylor & Francis).

Abdul Razzak Rumane (2016), *Handbook of Construction Management: Scope, Schedule, and Cost Control*, CRC Press, Florida (Taylor & Francis).

Abdul Razzak Rumane (2017), *Quality Management in Construction Projects*, Second Edition, CRC Press, Florida (Taylor & Francis).

Abdul Razzak Rumane (2019), *Quality Auditing in Construction Projects*, Routledge, UK (Taylor & Francis).

Abdul Razzak Rumane (2021), *Quality Management in Oil and Gas Projects*, CRC Press, Florida (Taylor & Francis).

Index

www.ingramcontent.com/pod-product-compliance
Lightning Source LLC
Chambersburg PA
CBHW060424220326
41598CB00021BA/2280